C000040956

BIOCONTROL POTENTIAL AND ITS EXPLOITATION IN SUSTAINABLE AGRICULTURE

VOLUME 1: CROP DISEASES, WEEDS, AND NEMATODES

BIOCONTROL POTENTIAL AND ITS EXPLOITATION IN SUSTAINABLE AGRICULTURE

SERIES EDITOR

K. G. Mukerji, *University of Delhi, Delhi, India*

BIOCONTROL POTENTIAL AND ITS EXPLOITATION IN SUSTAINABLE AGRICULTURE

VOLUME 1: CROP DISEASES, WEEDS, AND NEMATODES

EDITED BY

R. K. UPADHYAY
Directorate of Plant Protection, Quarantine and Storage
Faridabad, India

K. G. MUKERJI

AND

B. P. CHAMOLA
University of Delhi
Delhi, India

KLUWER ACADEMIC / PLENUM PUBLISHERS
New York, Boston, Dordrecht, London, Moscow

Library of Congress Cataloging-in-Publication Data

Biocontrol potential and its exploitation in sustainable agriculture/edited by R.K.
Upadhyay, K.G. Mukerji and B.P. Chamola.
    p. cm.
   Includes bibliographical references and index (p.).
   ISBN 0-306-46460-8 (v. 1)
    1. Biological pest control agents. 2. Agricultural pests—Biological control. I.
Upadhyay, R. K., 1953– II. Mukerji, K. G. III. Chamola, B. P.

SB975 .B5 2000
632′.96—dc21

00-044399

ISBN 0-306-46460-8

©2000 Kluwer Academic / Plenum Publishers, New York
233 Spring Street, New York, N.Y. 10013

http://www.wkap.nl

10 9 8 7 6 5 4 3 2 1

A C.I.P. record for this book is available from the Library of Congress

Printed in the United States of America

PREFACE

In India about 300 million people are undernourished, so is the case in other under developed countries. The successful effects of the green revolution which helped avoid death and famine in India are beginning to fade with agricultural output decreasing rapidly due to poor growth of plants and diseases and pests. With Indias population expected to reach approximately 1.5 billion people by 2050, agricultural production needs to increase 100%, shortage of irrigation water, increasing salinity levels and water logging of soil further add to the problem of food security. According to a recent report of World Watch Institute, global food supply will be reduced by an additional 10% on account of spreading water shortages. However major portion of food loss is due to pests and pathogens. Therefore, if these problems of future food scarcity are left unaddressed it will lead to widespread hunger and civil unrest.

Plant based biotechnology has come to represent a means of mitigating the problems of global food security in the 21st century. Products and processes in the agriculture are increasingly becoming linked to science and cutting edge technology. Although every technically feasible application may not be economically attractive, biologically safe, eco-friendly or ethically acceptable the underlying approach is virtually limitless in the array of products that can emerge. For centuries, the selection of and improvement of economically important plant species has been attained (attempted) by breeding with desired cultivars. This process is however lengthy and sometimes results in some suceptive varieties.

Scientific revolution that occured in molecular biology over the past two decades has enabled the engineering of what are in effect designer plants which express novel and desired characteristics. By using the tools of modern molecular biology scientists are now able to introduce genes or chemical recieps for making proteins into plants for the specific traits they want (better growth, yield and nutrition value). Since the biotech revolution is currently being experimented in other parts of the world it becomes imperative for India to venture on this with caution.

In recent years agriculturists, in the U.S., U.K., Canada, Australia, France, Mexico, South Africa and even China have adopted new plant varieties developed through modern technologies. These plant varieties are referred to as transgenic or genetically engineered (modified) resistant plants generated by transforming them with a natural bacterial enzyme and insect resistant-plants created by the incorporation of genes for one of the proteins produced by *Bacillus thuringiensis* a naturally occuring soil bacterium that infects and kills most insect pests.

Several major crop plants have been engineered with genes that make them resistant to insect pests or to herbicides so that farmers can apply the weed killer without fear of wiping out their standing crop. The benifits derived from these varieties include the reduced use of insecticides and herbicides thus reducing soil and ground water pollution and reduced tillage that results in topsoil loss. In 1988, at least 30 million hectares worldwide were planted with these genetically engineered plants. As a result, more than one half of the world's soybean harvest and about one third, of the maize (corn) harvest now comes from plants engineered with genes for hervicides or disease resistance.

Advances in agricultural biotechnology have created opportunities for efficient crop improvement. However, this process necessitates caution to prevent unforeseen problems associated with the adoption of new crop varieties. Biosafety issues need to be dealt with carefully. Comprehensive testing followed by open discussion among the different users (stakeholders) would help create public confidence.

Although the first generation transgenics based on simple changes that add a single gene are successful, they are the Model. As of agricultural biotechnology in the future genetically engineered plants promise entirely new hopes. It is, therefore necessary to adopt this technology now to reap the benefits in the future.

One of the most successful, non-chemical approaches to pest management and disease control is biological control using biocontrol agents in which the active principle is living organisms for regulating the incidence of pests/pathogens. Biological control envisages use of parasites, predators, antagonists, fast growing microoogranisms and sex attractants. Numerous organisms exist in nature that feed upon or infect insect pests, pathogens and weeds. Collectively these organisms provide a significant level of 'natural control', in many instances preventing insect species from ever reaching the status of pests. Since ancient times, man has practiced biological control of plant pathogens through cultural practices which aim at biological destruction and/or suppression of pathogens/pests. Biological control of plant pathogens seeks a solution in terms of restoring and maintaining the biological balance within the ecosystem and must be considered part of modern agriculture.

The present book has been compiled in two volumes :
1.   Crop Diseases, Weeds and Nematodes
2.   Insect Pests

First volume contains twenty chapters and Second volume twenty three chapters. These chapters have been written by authorities in the field. We would like to express our deep appreciation to each contributor for his/her work, patience and attention in detail during the entire production process. We trust that these volumes will be an asset for all users who have interest in sustainable agriculture and crop productivity. We particularly hope that this work will serve as a useful focal point for further studies and research.

It has been a pleasure to edit these volumes, primarily due to the stimulating cooperation of the contributors. We would like to thank staff of Kluwer Academic/Plenum Press for their help and active cooperation during the preparation of this work. We are greatful to serveral fellow workers for their helpful comments on the manuscripts.

We are confident that the joint efforts of authors and editors will contribute to a better understanding of advances in biocontrol of plant diseases in relation to sustainable agriculture.

A word of appreciation is also due to M/s. Neelam Graphics for active cooperation in preparing the neat electronic copies of the text of the chapters.

Delhi
31st March, 2000

R.K. Upadhyay
K.G. Mukerji
B.P. Chamola

LIST OF CONTRIBUTORS

1. Adholeya, A.
    Microbial Biotechnology
    Tata Energy Research Institute
    Darbari Seth Block
    Habitat Place, Lodhi Road
    New Delhi-110 003, INDIA
    Tel. : 91-11-4622246
    Fax : 91-11-4632609
    E-mail : aloka@teri.res.

2. Aneja, K.R.
    Department of Microbiology
    Kurukshetra University
    Kurukshetra-136 119
    Haryana, INDIA
    Tel. : 91-1744-20196/592
    Fax : 91-1744-20277

3. Baby, U.I.
    Centre for Advanced Studies in Botany
    University of Madras, Guindy Campus
    Chennai-600 025
    Tamil Nadu, INDIA
    Tel. : 91-44-235041
    Fax : 91-44-566693

4. Balasubramanian, P.
    Department of Plant Pathology
    Tamil Nadu Agricultural University
    Coimbatore-641 003
    Tamil Nadu, INDIA
    Tel. : 91-422-431222
    Fax : 91-422-431672

5. Baranwal, V.K.
    National Centre for Integrated
      Pest Management (ICAR)
    Pusa Campus
    New Delhi-110 012, INDIA
    Tel. : 91-11-5781474
    Fax : 91-11-5766420
    E-mail : baranwal@pic.iari.ren.nic.in

6. Chaudhary, B.L.
    Department of Microbiology
    School of Life Sciences
    North Maharashtra University
    P.B. No. 80
    Jalgaon-425 001
    Maharashtra, INDIA
    Tel. : 91-257-222187/411
    Fax : 91-257-222183

7. Chincholkar, S.B.
    Department of Microbiology
    School of Life Sciences
    North Maharashtra University
    P.B. No. 80
    Jalgaon-425 001
    Maharashtra, INDIA
    Tel. : 91-257-222187/411
    Fax : 91-257-222183

8. Chowdhury, A. K.
    Department of Plant Pathology
    Bidhan Chandra Krishi Vishwavidyalaya
    Mohanpur-741 252
    West Bengal, INDIA
    Tel. : 91-3162-98407
    Fax : 91-3162-98407

9. Gnanamanickam, S.S.
Centre for Advanced Studies in Botany
Univesity of Madras, Guindy Campus
Chennai-600 025
Tamil Nadu, INDIA
Tel. : 91-44-2350401
Fax : 91-44-4997005
E-mail : samg@giasmdol.net.in

10. Gupta, Rajni
Department of Botany
Univesity of Delhi
Delhi-110 007, INDIA
Tel. : 91-11-7257573
Fax : 91-11-7257830

11. Jayanth, K.P.
Biocontrol Reseach Laboratories
A Division of Pest Control (India) Ltd.
Post Box No. 3228, 479, 5th Cross
HMT Layout, R.T. Nagar
Bangalore-560 032
Karnataka, INDIA
Tel. : 91-80-3330168
Fax : 91-80-3334316

12. Jeyarajan, R.
Department of Plant Pathology
Tamil Nadu, Agricultural University
Coimbatore-641 003
Tamil Nadu, INDIA
Tel. : 91-44-423041
Fax : 91-44-440107

13. Joseph, P.J.
College of Agriculture
Kerala Agricultural University
Vellayani, Thiruvananthapuram-695 522
Kerala, INDIA
Tel. : 91-4868-37207
Fax : 91-4868-37285

14. Kaushal, S.
Department of Microbiology
Kurukshetra University
Kurukshetra-136 119
Haryana, INDIA
Tel. : 91-1744-21827
Fax : 91-1744-20410

15. Khan, S.A.
Department of Microbiology
Kurukshetra University
Kurukshetra-136 119
Haryana, INDIA
Tel. : 91-1744-20592
Fax : 91-1744-20410

16. Kothari, R.M.
Department of Microbiology
School of Life Sciences
North Maharashtra University
P.B. No. 80, Jalgaon-425 001
Maharashtra, INDIA
Tel. : 91-257-222187/411
Fax : 91-257-222183

17. Krishnamurthy, K.
Centre for Advance Studies in Botany
University of Madras
Chennai-600 025
Tamil Nadu, INDIA
Tel. : 91-44-423041
Fax : 91-44-499705
E-mail : samg@giasmd01.vsnl.net.in

18. Kumar, Vasanthi U.
Downy Mildew Research Laboratory
Department of Applied Botany
   and Biotechnology
Univesity of Mysore, Mysore-570 006
Karnataka, INDIA
Tel. : 91-821-515126
Fax : 91-821-411467
E-mail : appbot@blr.vsnl.net.in

19. Mahadevan, A.
Centre for Advanced Studies in Botany
University of Madras, Guindy Campus
Chennai-600 025
Tamil Nadu, INDIA
Tel. : 91-44-423041
Fax : 91-44-499705
E-mail : samg@giasmd01.vsnl.net.in

20. Manibhushanrao, K.
Centre for Advanced Studies in Botany
University of Madras Guindy Campus
Chennai-600 025
Tamil Nadu, INDIA
Tel. : 91-44-235041
Fax : 91-44-566692

21. Mukerji, K.G.
    Applied Mycology Laboratory
    Department of Botany
    University of Delhi
    Delhi-110 007, INDIA
    Tel. : 91-11-7654874
    Fax : 91-11-7256708

22. Nakkeeran, S.
    Department of Plant Pathology
    Tamil Nadu Agricultural Univesity
    Coimbatore-641 003
    Tamil Nadu, INDIA
    Tel. : 91-422-431222
    Fax : 91-422-431672

23. Nautiyal, C. S.
    Microbiology Group
    National Botanical Research Institute
    Rana Pratap Marg, P.O. Box 436
    Lucknow-220 061
    Uttar Pradesh, INDIA
    Tel. : 91-522-272031
    Fax : 91-522-282849

24. Pal, K.K.
    National Research Centre for Groundnut
    Timbawadi P.O.
    District-Junagadh-362 015
    Gujrat, INDIA

25. Pramila Devi, T.
    Depatment of Plant Pathology
    National Centre for Integrated
      Pest Management (ICAR)
    Pusa Campus,
    New Delhi-110 012, INDIA
    Tel. : 91-11-5765935
    Fax : 91-11-5765472

26. Reddy, M.V.
    Regional Agricultural Research Station
    Lam, Guntur-522 034
    Andhra Pradesh, INDIA
    Tel. : 91-863-230517
    Fax : 91-863-354556

27. Saxena, A.K.
    Division of Microbiology
    Indian Agricultural Research Institute
    New Delhi-110 012, INDIA
    Tel. : 91-11-5787649
    Fax : 91-11-5751719

28. Sen, C.
    Department of Plant Pathology
    Bidhan Chandra Krishi Vishwavidyalaya
    Kalyani-741 235
    West Bengal, INDIA
    Tel. : 91-33-5828297
    Fax : 91-33-5828297

29. Sharma, M. P.
    Microbial Biotechnology
    Tata Energy Research Institute
    Darbari Seth Block
    Habitat Place, Lodhi Road
    New Delhi-110 003, INDIA
    Tel. : 91-11-4622246
    Fax : 91-11-4632609

30. Sharma, S. B.
    International Crops Research Institute for
      the Semi-Arid Tropics (ICRISAT)
    Patancheru-502 324
    Andhra Pradesh, INDIA
    Tel. : 91-40-596161
    Fax : 91-40-241239

31. Shetty, H.S.
    Downy Mildew Research Laboratory
    Department of Applied Botany
      and Biotechnology
    University of Mysore
    Mysore-570 006
    Karnataka, INDIA
    Tel. : 91-821-515126
    Fax : 91-821-411467
    E-mail : appbot@blr.vsnl.net.in

32. Sivaprasad, P.
    College of Agriculture
    Kerala Agricultural University
    Vellayani- 695 522
    Thiruvanathapuram, Kerala, INDIA
    Tel. : 91-4868-37207
    Fax : 91-4868-37285

33. Srinivasulu, B.
    Regional Agricultural Research Station
    Lam, Guntur-522 034
    Andhra Pradesh, INDIA
    Tel. : 91-8574-50666
    Fax : 91-8574-27499

34. Sumeet
    Applied Mycology Laboratory
    Department of Botany
    University of Delhi
    Delhi-110 007, INDIA
    Tel. : 91-11-575573
    Fax : 91-11-7257830

35. Talegaonkar, S.K.
    Department of Microbiology
    School of Life Sciences
    North Maharastra University
    P.B. No. 80, Jalgaon-425 001
    Maharashtra, INDIA
    Tel. : 91-257-222187/411
    Fax : 91-257-222183

36. Thomas, J.,
    Indian Cardamom Research Institute
      (Spices Board), Myladumpara,
    P.O. Kailasanadu-685 553,
    Kerala, INDIA.
    Tel. : 91-4868-37207
    Fax : 91-4868-37285

37. Tilak, K.V.B.R.
    Division of Microbiology
    Indian Agricutural Research Institute
    New Delhi-110 012, INDIA
    Tel. : 91-11-5787649
    Fax : 91-11-5751719

38. Vats, R.
    Department of Nematology
    CCS Haryana Agricultural University
    Hisar-125 004, Haryana, INDIA
    Tel. : 91-1662-73721
    Fax : 91-1662-43952

39. Velazhahan, R.
    Department of Plant Pathology
    Tamil Nadu Agriculture University
    Coimbatore - 641 003
    Tamil Nadu, INDIA
    Tel. : 91-422-431222
    Fax : 91-422-431672

40. Verma, H.N.,
    Department of Botany
    Lucknow University
    Lucknow-226 007,
    Uttar Pradesh, INDIA.
    Tel. : 91-522-324794
    Fax : 91-522-206307

41. Vidhyasekaran, P.
    Department of Plant Pathology
    Tamil Nadu Agricultural University
    Coimbatore-641 003
    Tamil Nadu, INDIA
    Tel. : 91-422-431222
    Fax : 91-422-431672

42. Walia, R.K.
    Department of Nematology
    CCS Haryana Agricultural University
    Hisar-125 004
    Haryana, INDIA
    Tel. : 91-1662-73721
    Fax : 91-1662-34952

CONTENTS

BIOLOGICAL CONTROL OF CROP DISEASES EXPLOITING GENES INVOLVED IN SYSTEMIC INDUCED RESISTANCE

P. Vidhyasekaran, R. Velazhahan and P. Balasubramanian

Department of Plant Pathology
Tamil Nadu Agricultural University
Coimbatore - 641 003, Tamil Nadu, INDIA.

1. INTRODUCTION

Plants are endowed with various defense genes and disease resistance genes to combat the onslaught of fungal, bacterial, viral, viroid and phytoplasma pathogens (Vidhyasekaran, 1997). Defense genes are functional genes and they are involved in production of various antimicrobial compounds like phenolics, phytoalexins, pathogenesis – related (PR) proteins, and active oxygen species or involved in reinforcement of cell wall by accumulating callose, lignin, wall–bound phenolics and hydroxyproline – rich glycoproteins (Vidhyasekaran, 1993; Vidhyasekaran and Balasubramanian, 1995; Vidhyasekaran and Velazhahan, 1996; Vidhyasekaran et al., 1996). Contrastingly, resistance genes are not functional genes and they do not encode for synthesis of any toxic chemicals. But they are regulatory genes which regulate the functions of defense genes (Baker et al., 1997). Disease resistance genes are involved in signal transduction system (Song et al., 1995). Defense genes are 'sleeping' genes and signals are needed to activate them. These signals have to pass through a signal transduction system before reaching the nucleus of host cells for de novo synthesis of mRNAs of defense genes. Signal molecules of pathogen and host origin have been reported. By manipulating these signals through a signal transduction system, the defense genes can be activated to induce resistance against pathogens. Resistance genes play only a role in signal transduction in genetically resistant culvitars; but the signal transduction system can be activated even in the absence of resistance genes. Several biocontrol agents have been identified to induce systemic resistance (ISR) by activating the signal transduction system in genetically susceptible cultivars. This paper critically reviews the possibility of exploiting defense genes in control of crop diseases by using biocontrol agents.

1

2. SIGNAL MOLECULES FROM BIOCONTROL AGENTS

Elicitors, the signal molecules, elicit defense mechanisms in plants by activating defense genes which are quiescent in healthy plants (Vidhyasekaran, 1988a, b). Elicitors have been isolated from pathogenic and saprophytic fungi and bacteria (Dean and Anderson,1990; Duijff et al., 1997; He et al., 1993; Schaffrath et al., 1995). Signal molecules have been detected in plants also and they have been identified as oligogalacturonides. They are released from the hosts by pectinases produced by both parasitic and saprophytic fungi and bacteria (Movahedi and Heale, 1990). Several biocontrol agents are known to produce elicitors (Dean and Anderson, 1990; Duijff et al., 1997).

3. SIGNAL TRANSDUCTION SYSTEM

Second messenger systems to transmit the elicitation signals released by the microbes into host cells exist (Dixon, 1986). Calcium ions act as second messengers by transducting extracellular primary stimuli into intracellular events (Pitt and Kaile, 1990). The plant cells possess a system of proteins that interact with calcium ions and govern the transmission of the intercellular message. The important calcium- dependent enzymes are protein kinases, phospholipases and hydrogen ion transport ATP-ases (Breviario et al., 1995). Phosphorylation of proteins catalyzed by protein kinases is an important component in the integration of internal and external stimuli (Yu et al., 1993). The activated protein kinases enter the nucleus to induce gene expression (Martin et al., 1994). Products of most of the resistance genes which have been so far cloned are invovled in protein phosphorylation (Baker et al., 1997; Bent et al., 1994; Ori et al., 1997; Salmeron et al., 1996; Song et al., 1995; Wang et al., 1996).

Homologs of these resistance genes have been detected in susceptible varieties also (Jones et al., 1994; Mindrinos et al., 1994). Analogs of resistance genes which have been so far cloned have been detected in several crops like rice, barley, wheat, cotton, lettuce, soybean, pepper, chickpea, flax, tomato, tobacco and Brassica napus (Graham et al., 1988; Lyon et al., 1998; Vallad et al., 1998). It suggests that all resistance genes are actively involved in protein phosphorylation process.

The active oxygen species, H_2O_2, superoxide anion, singlet oxygen and hydroxyl radical are involved in signal transduction system. Among them H_2O_2 is the important active oxygen species (Conrath et al., 1995). Salicylic acid is another important signal molecule which acts locally in intracellular signal transduction. It is a phenolic compound commonly present in plant kingdom (Shah et al., 1997). Salicylic acid binding protein has been detected in plants and it has been identified as catalase. (Shirasu et al., 1997). Salicylic acid may enhance release of H_2O_2 and H_2O_2 - derived active oxygen species and induce activities of defense genes. Salicylic acid may suppress the H_2O_2 degrading activities of catalase and ascorbate peroxidase (Durner and Klessig, 1995). Elevated levels of H_2O_2 resulting from the inhibition of catalase and ascorbate peroxidase may be involved in the activation of defense response (Du and Klessig, 1997).

Thus the intracellular signal transduction system may involve pathogen signals, calcium ion flux, protein phosphorylation, active oxygen species and salicylic acid. These signals when they reach the nucleus of the cells, locally induce synthesis of messenger RNAs encoding various defense related proteins. Several systemic signals are released and the entire plant receives the message and defense genes are activated in all parts of the plant. Salicylic acid has been implicated as one of the key compound in the systemic signal transduction pathway leading to plant resistance to various pathogens. (Ryals et al., 1996). When plants are inoculated with fluorescent pseudomonads, salicylic acid accumulates systemically in all leaves (Rasmussen et al., 1991).

Some systemic signals may act in the upstream of salicylic acid signal transduction (Rasmussen et al., 1991). Application of high concentrations of H_2O_2 was found to stimulate salicylic acid accumulation. H_2O_2 appears to function upstream of salicylic acid systemic signal transduction pathway (Sharma et al., 1996). In the downstream of salicylic acid signal transduction system methyl jasmonate and jasmonic acid may act in many plants (Schweizer et al., 1997). Jasmonic acid and its methyl ester have been shown to originate from linoleic acid. Lipoxygenase converts linoleic acid into jasmonic acid (Sembdner and Partier, 1993). Ethylene has been also found to be a systemic signal molecule in many plants (Xu et al., 1994).

Signal molecules may differ in inducing different defense genes. Some times a combination of signal molecules may be needed to induce a single defense gene (Xu et al., 1994). Multiple signal transduction system may induce more defense genes (Schweizer et al., 1997).

4. BIOCONTROL AGENTS PRODUCING SIGNAL MOLECULES

Pseudomonas fluorescens, Pseudomonas putida, Bacillus spp. and Trichoderma spp. are the major biocontrol agents which have been reported to control several diseases. Some of the strains of these microbes are known to produce signal molecules. Lipopolysaccharides (LPS) are constituents of outer membrane of many bacteria. Some of the LPS have been shown to elicit plant defense genes. LPS of P. fluorescens strain WCS417r acts as elicitor and induces resistance against different diseases (Duijff et al., 1997). Fluorescent pseudomonad strains are known to produce salicylic acid which acts as local and systemic signal molecules (Dowling and O' Gara, 1994). H_2O_2 production due to P. fluorescens in plants has also been reported (Jakobek and Lindgren, 1993). Thus it is possible to select strains of P. fluorescens, P. putida and Bacillus spp., which can induce systemic resistance by activating the signal transduction system.

5. INDUCTION OF SYSTEMIC RESISTANCE BY BIOCONTROL AGENTS

Induced resistance to Fusarium wilt of carnation by P. fluorescens strain WCS417r has been reported (Van Peer et al., 1991). Fusarium wilt of radish is controlled by Pseudomonas fluorescens strain WCS 374 by inducing systemic resistance (Hoffland et al., 1995; Leeman et al., 1995a, b; Raaijmakers et al., 1995). Wei et al.(1991) reported that cucumber seeds treated with Pseudomonas putida resulted in a significant reduction in anthracnose disease caused by Colletotrichum orbiculare. Induced resistance in cucumber against Pythium aphanidermatum by Pseudomonas sp. has been reported (Zhou and Paulitz, 1994). Treatment of cucumber seeds with P. fluorescens reduced the symptoms of mosaic disease caused by cucumber mosaic virus in cucumber (Liu et al., 1992). Pseudomonas putida induced resistance against Fusarium wilt of cotton (Chen et al., 1995). Treatment of rice seeds with P. fluorescens has been shown to reduce the incidence of blast disease caused by Pyricularia oryzae (Vidhyasekaran et al., 1997), sheath blight caused by Rhizoctonia solani (Rabindran and Vidhyasekaran, 1996) and bacterial blight caused by Xanthomonas oryzae pv. oryzae. P. fluorescens strain CHAO when applied to root zone of tobacco reduced tobacco necrosis virus infection in tobacco leaves (Maurhofer et al., 1994). Fluorescent pseudomonads induced systemic resistance in

3

rice against bacterial blight pathogen (Ohno *et al.*, 1992). Several other studies have also indicated that the fluorescent pseudomonads induce systemic resistance against many pathogens and control diseases caused by them. (Alstrom, 1991; Benhamou *et al.*, 1996a, b, c; Kempe and Sequeira, 1983; Kloepper *et al.*, 1993; Liu *et al.*, 1995; M'Piga *et al.*, 1997; Raaijmakers *et al.*, 1995; Van Peer and Schippers, 1992; Zdor and Anderson 1992).

Among the fungal antagonists *Penicillium oxalicum* induced resistance against tomato wilt caused by *Fusarium oxysporum* f. sp. *lycopersici* (Cal *et al.*, 1997).

6. MODE OF ACTION OF INDUCED RESISTANCE

The various fluorescent pseudomonad strains which induce resistance have been reported to be endophytic (Benhamou *et al.*, 1996 b; Duijff *et al.*, 1997; Vidhyasekaran *et al.*, 1997a). *P. fluorescens* strain 63 - 28 penetrated and colonized some host epidermal cells and the outer cortex of tomato plants (M'Piga *et al.*, 1997). Penetration of the epidermis by *P. fluorescens* correlated with a number of structural changes, mainly characterized by the frequent occlusion of epidermal cells with a densely stained amorphous material and the formation of polymorphic wall thickenings along the host epidermal walls (M'Piga *et al.*, 1997). Pea root bacterization with *P. fluorescens* or *Bacillus pumilus* triggered a set of plant defense reactions (Benhamou *et al.*, 1996c). Marked host metabolic changes culminating in a number of structural (accumulation of callose and lignin) and biochemical responses (synthesis of chitinases) occurred at the onset of bacteria antagonists- mediated induced resistance in several plants (Kloepper *et al.*, 1993; M'Piga *et al.*, 1997). Maurhofer *et al.*(1994) reported that resistance induced by root colonizing *Pseudomonas fluorescens* strain CHAO in tobacco against the spread of tobacco necrosis virus (TNV) in leaves is correlated with induction of PR-1 proteins, ß-1, 3 - glucanase and endochitinases. Treatment with *P. fluorescens* caused increase in activities of peroxidase, lysozyme and phenylalanine ammonia - lyase (PAL) in tobacco (Schneider and Ullrich, 1994). A massive accumulation of phytoalexins could be detected in bacterized carnation roots after pathogen challenge (Van Peer *et al.*, 1991). In tomato plants treated with *P. fluorescens* phenolic compounds accumulated after challenge inoculation with *Fusarium oxysporum radicis - lycopersici* (M'Piga *et al.*, 1997). An increase in the activity of peroxidase and the level of mRNAs encoding PAL and chalcone synthase could be seen in bean roots colonized by the bacterial antagonists (Zdor and Anderson, 1992).

7. METHOD OF EXPLOITING BIOCONTROL AGENTS INDUCING DEFENSE GENES

The antagonists which control diseases by inducing disease resistance can be exploited for effective disease management by applying them before the disease out break. They cannot be applied once the disease out break is noticed. *P. fluorescens* applied even two days after inoculation with *X. oryzae* pv. *oryzae* was ineffective in control of bacterial blight of rice. *P. fluorescens* applied 4-5 days before inoculation with the pathogen, effectively controlled the disease. Fluorescent pseudomonad strains should be sprayed at least 3 to 4 days before inoculation with the rice blast pathogen to control the disease effectively (Vidhyasekaran *et al.*, 1997). Seed treatment with the antagonists gives protection against pathogens by inducing defense mechanisms before the pathogen attack (Benhamou *et al.*, 1996b, M'Piga *et al.*, 1997). However, efficacy of seed treatment is

4

lost in 14 days in bean (Zdor and Anderson, 1992) and 45 days in rice (Vidhyasekaran *et al.*, 1997a). Foliar application of the antagonists also does not provide long term control. Efficacy of *P. fluorescens* was observed only up to 15 days after spraying (Vidhyasekaran *et al.*, 1997a). Soil application of fluorescent pseudomonads also induce resistance and control diseases effectively (Hagedorn *et al.*, 1993). However, effective doses of the inoculum should be applied. The antagonist inoculum dose determines the efficacy of antagonist in control of diseases (Johnson, 1994). A threshold population density of fluorescent pseudomonad strain of approximately 10^5 colony forming unit per g of root is required for significant suppression of *Fusarium* wilt of radish (Raaijmakers *et al.*, 1995). The efficacy of biocontrol agents is also affected drastically by increasing the disease pressure (Williams and Asher, 1996). The antagonist may be highly useful in moderately resistant varieties rather than in highly susceptible varieties (Leeman *et al.*, 1995a).

8. CONCLUSIONS

Systemic induced resistance is an interesting phenomenon existing in plants. Induction of systemic resistance has been described in over 100 host pathogen systems (Lusso and Kuc, 1995; Sequeira, 1983). More recent studies have clearly indicated that some saprophytic bacteria and fungi can also induce resistance. These studies have opened a new era in management of crop diseases by exploiting systemic induced resistance. For effective control of diseases, suitable strains of bacteria and fungi have to be selected. Fluorescent pseudomonads appear to be the most potential candidates for this purpose. Besides inducing resistance, these bacteria are known to promote plant growth and increase crop yield. The major advantage of induced systemic resistance (ISR) is that the induced resistance is a general resistance mechanism conferring resistance against all types of pathogens, fungi, bacteria and viruses. However, since these microbes act by inducing resistance, they cannot be used as curative treatment. They should be applied before the pathogens occur. ISR is not a permanent one; it is transient. The duration of the protection ranges from 15 to 40 days. Hence, repeated application of these agents are required. Further ISR may not be effective in high disease pressure area.

ISR may be highly useful in moderately resistant varieties. The alternate strategy to exploit ISR under high disease pressure may be using fungicides or other chemicals along with the antagonists to reduce disease pressure. Another challenging task is in the development of suitable formulations for these antagonists with long storage life for easy handling and transport. However, these ISR- inducing agents induce resistance against several diseases in various crops and hence they can be developed as a potential commerical product. Some commercial formulations of these products are already available. ISR- inducing agents may be the most potential biocontrol agents and intensive studies on these agents may be highly useful in the development of biocontrol strategy in near future.

REFERENCES

Alstrom, S. 1991, Induction of disease resistance in common bean susceptible to halo blight bacterial pathogen after seed bacterization with rhizosphere pseudomonads, *J. Gen. Appl. Microbiol.* 37: 495-501.

Baker, B., Zambryski, P., Staskawicz, B. and Dinesh Kumar, S.P. 1997, Signaling in plant- microbe interactions, *Science* 276 : 726 – 733.

Benhamou, N., Belanger, R.R. and Paulitz, T. 1996a, Induction of differential host responses by *Pseudomonas fluorescens* in Ri T-DNA transformed pea roots upon challenge with *Fusarium oxysporum* f.sp. *pisi* and *Pythium ultimum* Trow, *Phytopath.* 86: 1174 – 1185.

Benhamou, N., Belanger, R.R. and Paulitz, T. 1996b, Ultrastructural and cytochemical aspects of the interaction between *Pseudomonas fluorescens* and Ri T-DNA transformed pea roots : host response to colonization by *Pythium ultimum* Trow, *Planta* 199: 105 – 117.

Benhamou, N., Kloepper, J.W., Quadt–Hallmann, A. and Tuzun, S. 1996c, Induction of defense – related ultrastructural modifications in pea root tissues inoculated with endophytic bacteria, *Plant Physiol.* 112: 919 – 929.

Bent, A., Kunkel, B.N., Dahlbeck, D., Brown, K., Schmidt, R., Giraudat, J., Leung, J. and Staskawicz, B.J. 1994, *RPS2* of *Arabidopsis thaliana* : A leucine – rich repeat class of plant disease resistance genes, *Science* 265 : 1856 – 1860.

Breviario, D., Morello, L. and Ginanai, S. 1995, Molecular cloning of two novel rice cDNA sequences encoding putative calcium dependent protein kinases, *Plant Mol. Biol.* 27: 953 – 967.

Cal, A. De., Pascual, S. and Melgarejo, P. 1997, Involvement of resistance induction by *Penicillium oxalicum* in the biocontrol of tomato wilt, *Pl. Pathol.* 46 : 72 – 79.

Chen,C., Bauske, E.M., Musson, G., Rodriquez – Kabana, R. and Kloepper, J.W. 1995, Biological control of *Fusarium* wilt in cotton by use of endophytic bacteria, *Biol. Cont.* 5: 83 – 91.

Conrath, U., Chen, Z., Ricigliano, J.R. and Klessig, D. 1995, Two inducers of plant defense responses, 2-6 dichloroisonicotinic acid and salicylic acid, inhibit catalase activity in tobacco, *Proc. Natl. Acad. Sci. USA* 92 : 7143 – 7147.

Dean, J. F. D. and Anderson, J. D. 1990, Ethylene biosynthesis- inducing xylanase, II, Purification and physical characterization of the enzyme produced by *Trichoderma viride, Plant Physiol.* 94: 1849-1854.

Dixon, R.A. 1986, The phytoalexin response: elicitation, signaling and control of host gene expression, *Biol. Rev.* 61: 239 – 291.

Dowling, D.N. and O' Gara, F. 1994, Metabolites of *Pseudomonas* involved in the biocontrol of plant disease, *TIBTECH* 12 : 133 –141.

Du, H. and Klessig, D.F. 1997, Role for salicylic acid in the activation of defense responses in catalase – deficient transgenic tobacco, *Mol. Plant – Microbe Interact.* 10 : 922 – 925.

Duijff, B.J., Gianinazzi– Pearson, V. and Lemanceau, P. 1997, Involvement of the outer membrane lipopolysaccharides in the endophytic colonization of tomato roots by biocontrol *Pseudomonas fluorescens* strain WCS417r, *New Phytol.* 135 : 325 – 334.

Durner, J. and Klessig, D.F. 1995, Inhibition of ascorbate peroxidase by salicylic acid and 2,6-dichloroisonicotinic acid, two inducers of plant defense responses, *Proc. Natl. Acad. Sci. USA* 92 : 11312 – 11316.

Graham, M.A., Marek, L.F. and Shoemaker, R.C. 1998, Analysis of resistance gene analog cDNA clones from soybean, In : *Proc. Sixth Intl. Conf. Plant and Animal Genome Research*, San Diego, California, p. 83.

Hagedorn, C., Gould, W. D. and Bardinelli, T. R. 1993, Field evaluation of bacterial inoculants to control seedling disease pathogens on cotton, *Pl. Dis.* 77: 278-282.

He, S.Y., Huang, H.C. and Collmer, A. 1993, *Pseudomonas syringae* pv. *syringae* Harpin $_{Pss.}$ A protein that is secreted via the Hrp pathway and elicits the hypersensitive response in plants, *Cell* 73 : 1255 – 1266.

Hoffland, E., Peterse, C.M.J., Bik, L. and Van Pelt, J.A. 1995, Induced systemic resistance is not associated with accumulation of pathogenesis – related proteins, *Physiol. Mol. Plant Pathol.* 46 : 309 – 320.

Jakobek, J. L. and Lindgren, P. B. 1993, Generalized induction of defense responses in bean is not correlated with the induction of the hypersensitive response, *Plant Cell* 5: 49-56.

Johnson, K. B. 1994, Dose- response relationships and inundative biological control, *Phytopath.* 84: 780-784.

Jones, D.A., Thomas, C.A., Hammond – Kosack, K.E. and Balint – Kurti, P.J. and Jones, J.D.G. 1994, Isolation of tomato Cf-9 gene for resistance to *Cladosporium fulvum* by transposon tagging, *Science* 266 : 789.

Kempe, J. and Sequeira, L. 1983, Biological control of bacterial wilt of potatoes : attempts to induce resistance by treating tubers with bacteria, *Plant Dis.* 67 : 499 – 503.

Kloepper, J.W., Tuzun, S., Liu,L. and Wei,G. 1993, Plant- growth promoting rhizobacteria as inducers of systemic resistance.In: *Pest Management: Biologically Based Technologies*, eds. R.D. Lumsden.and J.L. Waughn, ASC Cong. Proc. Series, American Chemical Society Press, pp. 156 – 165.

Leeman, M., Van Pelt, J.A., Den Ouden, F.M., Heinsbroek, M., Bakker, P.A.H.M. and Schippers, B. 1995a, Induction of systemic resistance by *Pseudomonas fluorescens* in radish cultivars differing in susceptibility to *Fusarium* wilt, using a novel bioassay, *Eur. J. Plant Pathol.* 101 : 655 – 664.

6

Leeman, M., Van Pelt, J.A., Den Ouden, F. M., Heinsbroek, M., Bakker, P.A.H.M. and Schippers, B. 1995b, Induction of systemic resistance against *Fusarium* wilt of radish by lipopolysaccharides of *Pseudomonas fluorescens, Phytopath.* 85 : 1021 – 1027.

Leeman, M., Van Pelt, J.A., Hendrickx, M.J., Scheffer, R.J., Bakker, P.A.H.M. and Schippers, B. 1995c, Biocontrol of *Fusarium* wilt of radish in commercial greenhouse trials by seed treatment with *Pseudomonas fluorescens* WCS 374, *Phytopath.* 85: 1301 – 1305.

Liu, L., Kloepper, J.W. and Tuzun, S. 1992, Induction of systemic resistance against cucumber mosaic virus by seed inoculation with selected rhizobacterial strains, *Phytopath.* 82: 1109.

Liu, L., Kloepper, J. W. and Tuzun, S. 1995, Induction of systemic resistance in cucumber against *Fusarium* wilt by plant growth- promoting rhizobacteria, *Phytopath.* 85: 695 – 698.

Lusso, M. and Kuc, J. 1995, Increased activities of ribonuclease and protease after challenge in tobacco plants with induced systemic resistance, *Physiol. Mol. Plant Pathol.* 47 : 419 – 428.

Lyon, B.R., Hill, M.K., Kota, R. and Lyon, K.J. 1998, Isolation and characterization of genes associated with enhanced tolerance to phytopathogenic fungi in cotton, In : *Proc. Sixth Int. Conf. Plant and Animal Genome Research,* San Diego, California, p. 86.

M'Piga, P., Belanger, R.R., Paulitz, T.C. and Benhamou, N. 1997, Increased resistance to *Fusarium oxysporum* f.sp. *radicis – lycopersici* in tomato plants treated with the endophytic bacterium *Pseudomonas fluorescens* strain 63 – 28, *Physiol. Mol. Plant. Pathol.* 50: 301 – 320.

Martin, G.B., Frary, A., Wu, T., Brommonschenkel, S., Chunwongse, J., Earle, E.D. and Tanksley, S.D. 1994, A member of the tomato *Pto* gene family confers sensitivity to fenthion resulting in rapid cell death, *Plant Cell* 6 : 1043.

Maurhofer, M., Hase, C., Meuwly, J.P. and Defago, G. 1994, Induction of systemic resistance of tobacco to tobacco necrosis virus by the root – colonizing *Pseudomonas fluorescens* strain CHAO : influence of the *gacA* gene and of pyoverdine production, *Phytopath.* 84 : 139 – 146.

Mindrinos, M., Katagiri, F., Yu, G.L. and Ausubel, F.M. 1994, The *A. thaliana* disease resistance gene *RPS2* encodes a protein containing a nucleotide- binding site and leucine – rich repeats, *Cell* 78 : 1089 – 1099.

Movahedi, S. and Heale, J.B. 1990, The roles of aspartic proteinase and endo-pectin lyase enzymes in the primary stages of infection and pathogenesis of various host tissues by different isolates of *Botrytis cinerea* Pers, *Physiol. Mol. Plant Pathol.* 36 : 303 – 324.

Ohno, Y., Okuda, S., Natsuaki, T. and Teranaka, M. 1992, Control of bacterial seedling blight of rice by fluorescent *Pseudomonas* spp., *Proc. Kanto – Tosan Plant Prot. Soc.* 39: 9-11.

Ori, N., Eshed, Y., Paran, Y., Presting, G., Aviv, D., Tanksley,S., Zamir,D. and Fluhr, R. 1997, The *I2C* family from the wilt disease resistance locus *I2* belongs to the nucleotide binding, leucine – rich repeat superfamily of plant resistance genes, *Plant Cell* 9 : 521 – 532.

Pitt, D. and Kaile, A. 1990, Transduction of the calcium signal with special reference to Ca^{2+} - induced conidiation in *Penicillium notatum,* In: *Biochemistry of Cell Walls and Membranes in Fungi,* eds. J. Kuhn, A.P.J. Trinci, M.J. Jung, M.W. Goosey and L.G. Copping , Springer – Verlag, Berlin, pp. 283 – 298.

Raaijmakers, J.M., Leeman, M., Van Oorschot, M.M.P., Van der Sluis, I., Schippers, B. and Bakker, P.A.H.M. 1995, Dose – response relationship in biological control of *Fusarium* wilt of radish by *Pseudomonas* sp., *Phytopath.* 85 : 1075 – 1081.

Rabindran, R. and Vidhyasekaran, P. 1996, Development of a formulation of *Pseudomonas fluorescens* PfALR 2 for management of rice sheath blight, *Crop Protec.* 15: 715 – 721.

Rasmussen, J.B., Hammerschmidt, R. and Zook, M.N. 1991, Systemic induction of salicylic acid accumulation in cucumber after inoculation with *Pseudomonas syringae* pv. *syringae, Plant Physiol.* 97 : 1342 – 1347.

Ryals, Neuenschwander, U.H., Willits, M.G., Molina, A., Steiner, H.Y. and Hunt, M.D. 1996, Systemic acquired resistance, *Plant Cell* 8 : 1809 – 1819.

Salmeron, J.M., Oldroyd, G.E., Rommens, C.M., Scofield, S.R., Kim, H.S., Lavelle, D.T., Dahlbeck, D. and Staskawicz, B.J. 1996, Tomato *Prf* is a member of the leucine - rich repeat class of plant disease resistance genes and lies embedded within the *Pto kinase* gene cluster, *Cell* 86: 123 - 133.

Schaffrath, U., Scheinpflug, H. and Reisner, H. 1995, An elicitor from *Pyricularia oryzae* induces resistance responses in rice : isolation, characterization and physiological properties, *Physiol. Mol. Plant Pathol.* 46 : 293 – 307.

Schneider, S. and Ullrich, W. R. 1994, Differential induction of resistance and enhanced enzyme activities in cucumber and tobacco caused by treatment with various abiotic and biotic inducers, *Physiol. Mol. Plant Pathol.* 45: 291-304.

Schweizer, P., Buchala, A. and Metraux, J.P. 1997, Gene– expression patterns and levels of jasmonic acid in rice treated with the resistance inducer 2,6 – dichloroisonicotinic acid, *Plant Physiol.* 115 : 61 – 70.

7

Sembdner, G. and Partier, B. 1993, The biochemistry and the physiological and molecular actions of jasmonates, *Annu. Rev. Plant Physiol. Plant Mol. Biol.* 44 : 569 – 589.

Sequeira, L. 1983, Mechanisms of induced resistance in plants, *Annu. Rev. Microbiol.* 37: 51-79.

Shah, J., Tsui, F. and Klessig, D.F. 1997, Characterization of a salicylic acid - insensitive mutant (*sai 1*) of *Arabidopsis thaliana*, identified in a selective screen utilizing the SA - inducible expression of *tms 2* gene, *Mol. Plant - Microbe Interact.* 10 : 69 - 78.

Sharma, Y.K., Leon, J., Raskin, I. and Davis, K.R. 1996, Ozone - induced responses in *Arabidopsis thaliana* : the role of salicylic acid in the accumulation of defense - related transcripts and induced resistance, *Proc. Natl. Acad. Sci. USA* 93 : 5099 - 5104.

Shirasu, K., Nakajima, H., Rajasekhar, V.K., Dixon, R.A. and Lamb, C. 1997, Salicylic acid potentiates an agonist - dependent gain control that amplifies pathogen signals in the activation of defense mechanisms, *Plant Cell* 9 : 261 - 270.

Song, W.Y., Wang, G.L., Chen L-L., Kim, H.S., Pi, L.Y., Holsten, T., Gardner, J., Wang, B., Zhai, W. X., Zhu, L. H., Fauquet, C. and Ronald, P. 1995, A receptor kinase- like protein encoded by the rice disease resistance gene Xa21, *Science* 274 : 1804 – 1806.

Vallad, G., Rivkin, M.I., Vallajos, E. and Mcclean, P. 1998, Cloning of *Pto* - like sequences in common bean. In : *Proc. Sixth Int. Conf. Plant* and *Animal Genome Research*, San Diego, California, p. 137.

Van Peer, R. and Schippers, B.1992, Lipopolysaccharides of plant growth – promoting *Pseudomonas* sp. strain WCS 417r induce resistance in carnation to *Fusarium* wilt, *Neth. J. Plant Path.* 98 : 129 – 39.

Van Peer, R., Niemann, G.J. and Schippers, B. 1991, Induced resistance and phytoalexin accumulation in biological control of *Fusarium* wilt of carnation by *Pseudomonas* sp. strain WCS417r, *Phytopath.* 81 : 728 – 734.

Vidhyasekaran, P. 1988a, *Physiology of Disease Resistance in Plants*, Vol. I., CRC Press, Florida, pp. 149.

Vidhyasekaran, P. 1988b, *Physiology of Disease Resistance in Plants*, Vol.II, CRC Press, Florida, pp. 127.

Vidhyasekaran, P. 1993, Defense genes for crop disease management, In: *Genetic Engineering, Molecular Biology and Tissue Culture for Crop Pest and Disease Management*, ed. P. Vidhyasekaran, Daya Publishing House, Delhi, pp. 17 – 30.

Vidhyasekaran, P. 1997, *Fungal Pathogenesis in Plants and Crops : Molecular Biology and Host Defense Mechanisms*, Marcel Dekker Inc. New York, USA, pp. 553.

Vidhyasekaran, P. and Balasubramanian, P. 1995, Disease management by exploiting pathogenesis- related proteins with lytic action, In: *Detection of Plant Pathogens and their Management*, eds. J. P. Verma, Anupam Varma and Dinesh Kumar, Angkor Publishers, New Delhi, pp. 259-272.

Vidhyasekaran, P. and Velazhahan, R. 1996, Elicitor induced defense responses in suspension cultured rice cells, In: *Agricultural Biotechnology*, eds. V.C. Chopra, R.P. Sharma, and M.S. Swaminathan, Oxford and IBH Publishing Co., New Delhi, pp. 249 – 255.

Vidhyasekaran, P., Rabindran, R., Muthamilan, M., Nayar, K., Rajappan, K., Subramanian, N.˙and Vasumathi, K. 1997, Development of a powder formulation of *Pseudomonas fluorescens* for control of rice blast, *Pl. Pathol.* 46: 291-297.

Vidhyasekaran, P., Velazhahan, R., Samiyappan, R., Ruby Ponmalar, T. and Muthukrishnan, S. 1996, Isolation of a 35-kDa chitinase from suspension – cultured rice cells and its potential in the development of sheath blight-resistant transgenic rice plant, In: *Rice Genetics-III*, ed. G. Khush, International Rice Research Institute, Philippines, pp. 868 – 872.

Wang, G. L., Song, W. Y., Ruan, D. L., Sidetis, S. and Ronald, P. C. 1996, The cloned genes, *Xa21* confers resistance to multiple *Xanthomonas oryzae* pv. *oryzae* isolates in transgenic plants, *Mol. Plant-Microbe Interact.* 9: 850-855.

Wei, G., Kloepper, J.W. and Tuzun,S. 1991, Induction of systemic resistance of cucumber to *Colletotrichum orbiculare* by select strains of plant growth – promoting rhizobacteria, *Phytopath.* 81 : 1508 – 1512.

Williams, G.E. and Asher, M.J.C. 1996, Selection of rhizobacteria for the control of *Pythium ultimum* and *Aphanomyces cochlioides* on sugar beet seedlings, *Crop Protec.* 15 : 479 – 486.

Xu,Y., Chang, P.L., Liu,D., Narasimhan, M.L., Raghotthama, K.G., Hasegawa, P.M. and Bressan, R.A. 1994, Plant defence genes are synergistically induced by ethylene and methyl jasmonate, *Plant Cell* 6: 1077 – 1085.

Yu, G. L., Katagini, F. and Ausubel, F. M. 1993, *Arabidopsis* mutations at the *RPS2* locus result in loss of resistance to *Pseudomonas syringae* strain expressing the avirulence gene *avrRpt2*, *Mol. Plant-Microbe Interact.* 6: 434 – 443.

Zdor, R.E. and Anderson, A.J. 1992, Influence of root colonizing bacteria on the defense responses of bean, *Plant Soil* 140 : 99 – 107.

Zhou, T. and Paulitz, T.C. 1994, Induced resistance in the biocontrol of *Pythium aphanidermatum* by *Pseudomonas* spp. on cucumber, *J. Phytopathol.* 142 : 51 – 63.

8

BIOCONTROL OF PLANT DISEASES FOR AGRICULTURAL SUSTAINABILITY

C.S. Nautiyal

Microbiology Group
National Botanical Research Institute
Rana Pratap Marg, P.B. No. 436
Lucknow - 226 001, U.P., INDIA

1. INTRODUCTION

High input agriculture is increasingly recognised as environment and health degrading and not profitable due to its dependence on chemical inputs. Green revolution in India during 1970s no doubt brought about self sufficiency in food and millions escaped starvation. There is serious concern for food security of developing countries including that of India for the following reasons as described by Khanna-Chopra and Sinha (1998) : (i) Increasing food demand for the rapidly burgeoning population which will be further enhanced due to improved economic growth, (ii) Stagnating or declining productivity in high productivity regions, often described as "Green Revolution" fatigue, and (iii) Increasing vulnerability to agriculture as a result of potential climate change. Moreover, major advances in development in general and agricultural production in particular have also brought in its wake serious environmental degradation in term of salinity, water logging, soil erosion, air and water pollution and poor soil health. Therefore, there is a conscious effort to improve production through use of environment friendly products such as bioinoculants, instead of chemicals. This may ensure that the nature is not exploited in the production process but is instead harmonised so that the entropy of the environment decreases and sustainability in agricultural production is promoted (Narain, 1998; Purohit, 1995; Sinha, 1996, 1997).

One of the most frequent justifications given for the need to expand industrial agriculture and introduce genetically engineered crops is that sustainable agriculture is less productive. US Secretary for Agriculture Mr. Earl Butz recently responded to calls for support for organic farming in the United States by saying, "Show me the first 10,000 Americans who are prepared to starve to death and then I will do something" (de Selincourt, 1996). Hardly anyone argues that modern agriculture is sustainable and a better use of land will be possible only through better family planning (Swaminathan, 1992).

However, it would also be fraudulent to claim that through the use of organic farming or, genetically engineered crops alone, it would be possible to feed expanding human population. Crops can be raised purely on either resources, if they are not compared against yields, economics, nutritional quality and environmental damages - all integrated on the index of sustainability. There is irrefutable evidence on all counts of the superiority-significant edge of organics (de Selincourt, 1996).

The heightened scientific interest in biological control of plant pathogens is a response, in part, to growing public concerns over chemical fertilisers, fungicides and pesticides. Clearly, a crop growing in a soil fertilised by a rich mix of organically derived compounds (manure and compost) will contain a wider range of nutrients and microflora, than one repeatedly fed with only nitrogen, phosphorus and potassium from artificial fertilisers and protected from diseases by plant protection chemicals. Paroda (1997) has stated that excessive use of chemical fertilisers and plant protection chemicals for maximising crops yields and change in traditional cultivation practices, resulted in deterioration of chemical, physical and biological health of the cultivated land. This situation eliminated the desirable types of microbes from soil and allowed enrichment of undesirable pathogens and pests which forced the farmers to use innumerable types of plant protection chemicals. The situation has become alarming and it is now imperative to understand the role of soil microorganisms in sustainable crop productivity without further deterioration in soil health, as the biological control offers the chance of improving crop production within existing resources, avoiding pathogen resistance to chemicals, and relatively pollution and risk-free control. The use of such microorganisms as inoculants helps as non-chemical means of disease management and may help in economising crop production without affecting adversely the soil health (Mukerji and Garg, 1988a,b). Therefore there is no doubt about the exciting potential of this approach, borne out by the recent explosion in research worldwide on hundreds of individual species, subspecies, and strains of antagonists shown experimentally to provide biological control to protect pruning wounds or germinating seeds and seedlings, control wilts and root diseases, control nematodes, control foliar diseases, and protect harvested fruits and vegetables.

We have to increase the productivity and production of wide range of agricultural commodities under conditions of shrinking land resources and diminution in both the biological potential of the soil and in the biological wealth. Obviously, there is no simple or single solution to the complex ecological, socioeconomic and technological problems facing those engaged in promoting sustainable advances in agricultural biotechnology (Swaminathan, 1992). Swaminathan (1996) has proposed several requirements which any strategy for agricultural research should consider and stated that the protection of long-term productivity of land, water, flora and fauna, and atmosphere must be a primary goal of all R & D efforts. The principal basis of sustainable land use is the long-term maintenance of the productive capacity of soils (Thomas and Kevan, 1993). This involves employing practices that enhance the long-term productivity of the soil or soil health. Biocontrol strategies are highly compatible with the sustainable agricultural practices that are required for conoserving natural resources of agriculture, as biocontrol of plant pathogens possess advantages other than non-hazardous nature. For example, it offers a non-chemical means of disease management besides, most chemical products have only a temporary effect and usually requires repeated applications during the plant growth season. Biological control agents have the ability to reproduce, to establish themselves in the soil ecosystem, and to colonise seeds, spermosphere, rhizosphere, rhizoplane, and foliage.

Biological control may be achieved by a number of strategies, both direct and indirect. Indirect strategies include the use of crop or cultivar rotation, various tilling strategies, addition of organic amendments such as compost, mulch or manures, flooding before planting, soil aeration or solarization, and so forth. The direct approach involves
10

deliberate use of specific antagonistic organisms for prevention or management of specific diseases, increasing or augmenting the density of local native microbes that enhance the activity of indigenous microbial antagonists against pathogens, competitively displacing the antagonistic microbes from the site of infestation, manipulating the environment or other introduced microbes to favour native microbes and conferring resistance on the host plant (Cook, 1993; Mukerji *et al.*, 1992; Upadhyay *et al.*, 1996).

Research into controlling plant pathogens by biological means has mostly been directed to manipulation of the environment to favour the introduced antagonist in the appropriate ecological niche in order to be active against the pathogen. Among the several types of microorganisms i.e., bacteria, fungi and viruses, bacteria have been used most extensively against plant pathogens. Bacterial antagonists are popular for several reasons, including the relative ease of isolation, growing, identification, tracking, genetic manipulation, laboratory and commercial applications and are the focus of this review article. There are numerous reviews dealing with various aspects of biological control that they are becoming loquacious. Because of space constraints, it has not been possible to cite all papers relevant to the subject. An attempt has been made to cite at least one paper or review article on the subject covered by this review, and those citations will hopefully provide access to the complete literature on the subject.

This review, suggests ways and means by which populations of introduced biocontrol bacteria can be manipulated ecologically and genetically and most importantly constraints of biological control both fundamental and applied, that must be met if we are to make successful commercial use of bacteria as reliable biocontrol agents, contributing significantly to modern agriculture in the form of effective products or practices, as a consequence, soil health.

2. RECENT STATUS

Biological control of soilborne pathogens by introduced microorganisms has been studied for over 7 decades, but during most of that time it has not been considered commercially feasible. Schroth and Hancock (1981) stated that the biological control of plant pathogens is a fascinating, challenging, but elusive and frustrating area of study for plant scientists and microbiologists. Their statement holds true even as of to date. Although biological control occurs naturally and is the principle reason diseases are not naturally catastrophic, sufficient knowledge in most cases is not available to explain how biological control operates naturally or how the many abiotic and biotic factors can be manipulated to effect economic control of a pathogen. However, the general lack of specific examples of success has not dampened interest in developing biological controls and research in this area have increased steadily, as reflected by the number of reviews (Cook 1993; Johri *et al.*, 1997; Lam and Gaffney 1993; Mukerji *et al.*, 1999; O'Sullivan and O'Gara 1992; Raaijmakers *et al.*, 1999; Upadhyay *et al.*, 1996) about it that have appeared. This renewed interest in biocontrol is in part a response to public concern about hazards associated with chemical fungicides and pesticides and the inordinately high cost of developing chemicals to control pathogens and the lack of resistance of plants to many diseases. Concurrently, there has been a shift to the opinion that biological control can have an important role in agriculture in the future, and it is encouraging that several major biotechnology companies world over like Novartis and Monsanto, now have programs to develop bacterial biocontrol agents as commercial products.

Biocontrol of plant disease involves the use of an organism or organisms to reduce disease. Biocontrol solutions to disease problems are receiving increasing attention and are gaining popularity owing to the concern over the use of pesticides in agriculture and because these solutions may be applicable to disease that can not be controlled by genetic resistance or chemical pesticides. However, despite a few notable successes, the field of biocontrol has contributed little to modern agriculture in the form of effective products or practices. Biocontrol research has been plagued by variable and unpredictable success, which has often resulted in the abandonment of potential biocontrol agents. If biocontrol is ever to be a alternative to chemical control of plant disease, we must begin to understand how and why biocontrol works; most important, why it fails? This in turn will require an understanding of the interactions between the biocontrol agent and the complex microbial community in which the agent must function. From this understanding we may be able to develop models that will predict the conditions under which a biocontrol strategy will succeed or fail (Mukerji and Garg, 1988a,b).

Developing predictive models for biocontrol is necessarily a complex endeavour because of the multicomponent nature of the system. Biocontrol often requires the introduction and establishment of an organism or organisms in the microbial community on or near the plant surface. The mechanism of disease control must be appropriate to the plant, the pathogen, the site of activity, and the developmental stage of the target organism. We must develop a thorough understanding of the environmental and host factors that affect the microbial community and the efficacy of a particular mechanism of disease suppression in the plant environment. A massive research effort on population biology and mechanisms of interactions of biocontrol agents is required to accumulate and integrate the information that will supply the basis for predictive models. Although population biology of biocontrol agents has been studied in both the laboratory and field, research directed toward elucidating mechanisms of interactions between biocontrol agents and the target pathogens has been primarily laboratory-based. One of the most difficult challenges in developing realistic predictive models will be to make accurate determinations of how biocontrol agents and pathogens interact under field conditions. In addition to its importance for developing models, an understanding of the mechanisms underlying these interactions may allow us to manipulate the soil environment to create conditions conducive to successful biocontrol. In addition, an understanding of these mechanisms may provide a logical basis for selection and construction of more effective biocontrol agents.

It is encouraging that there are now so many examples of biological control with bacteria in the field. Unfortunately, one characteristic that is common to most biocontrol systems with introduced bacteria is the inconsistency of disease control. However it is easier said than done. A multitude of factors could account for inconsistent results, given the complex interactions among host, pathogen, antagonist, and the environment. For example loss of ecological competence of the bacteria to compete and survive in nature. Many bacterial traits (most of them unknown) contribute to ecological competence in the rhizosphere and loss of any one can reduce the ability of the bacteria to become established or function on or near the root. Inconsistent performance of the biocontrol agent could also be due to the fact that target pathogen may be absent or nontarget pathogen interference may be occurring. Because bacterial biocontrol agents improve plant growth by reducing damage from pathogens, a positive response to their introduction does not occur when the target pathogen(s) is absent, or when environmental conditions are unsuitable for disease development. The effect of pathogens other than pathogen is another concern. If a bacterium suppresses only one pathogen, but another becomes predominant, the treatment will appear ineffective. Thus, an understanding of the pathogens in the agroecosystem and the conditions that favour each is essential. Therefore the real potential for microbial biocontrol of plant diseases may well be in the use of many different locally adapted strains for each disease, possibly on each crop, and possibly for each soil. Biological control of take-all of wheat with fluorescent *Pseudomonas* species, will almost certainly require different strains for different soils. It has been shown that isogenic mutants of strain 2-79, some positive and others negative for phenazine (Phz) production, that biocontrol activity of Phz+ strains in 10 different soils was negatively correlated with % clay, % silt, exchangeable acidity, the iron and manganese contents, % organic matter, total carbon, and total nitrogen, and positively correlated with % sand, soil pH, sulfate-sulfur, ammonium nitrogen, sodium and zinc contents of the soils. *A. radiobacter* K-84 with its close relationship with wounds and applied directly by bare-root dip to the infection courts might be expected to work in almost as many kinds of soils as can be used to grow the host plant, whereas *P. fluorescens* 2-79, by comparison, which amust establish in the rhizosphere from a seed inoculation and grow in soil around as well as on the root, expectedly would be more sensitive in its ecology to soil conditions (Cook, 1992). The discovery of a genetically controlled mechanism that would bring about a closer relationship between root and biocontrol agent, and that could be expressed in either the root and biocontrol agent, or both, would open the way for many advances in biological control of root pathogens, including, possibly, a means to lessen the influence of soil factors.

One controversial topic is the tendency to treat introduced microbes in a chemical rather than a biological paradigm is an unrealistic expectation (Cook, 1991). This epitome probably traces to product based on *Bacillus thuringiensis* (*Bt*). *Bt* is a marvellous success story for insect control, but probably is a typical of microbial biocontrol of the future. *Bt* works dead or alive. But it would be interesting to know whether this agent would be successful if, like other introduced biocontrol agents, in order to work it had to establish and function temporally and spatially within an ecosystem. Similarly *Agrobacterium radiobacter* K-84 used for biological control of crown gall is adapted to use in many environments, possibly because of the root-dip method of application to root wounds (Moore, 1979). This perception that microbes are chemicals is further reinforced because to apply them repeatedly usually involves formulations, adjuvants, and application equipment similar if not identical to that used for chemicals. Still another factor may be the emphasis on antibiotic production as a mechanism of biological control. Many examples of biological controls, including many examples of host plant resistance, can be explained in terms of chemical mechanisms, yet only microbes have come to be called "biopesticides".

Till to date no detailed study has been undertaken to improve the consistency of the performance of bacterial biocontrol agent(s), so that it can reliably be used for commercial use. Accomplishing this will require research in many diverse areas, because biological control is the culmination of complex interactions among the host plant, pathogen(s), antagonist and environment. Research to identify bacterial traits that function in plant colonisation and pathogen antagonism is critically important, and molecular genetics offers the best approach to such studies. Identifying important traits would allow more efficient selection of new strains. Further, such traits can be altered to make a strain more effective. There is obviously considerable potential to improve and construct biocontrol agents by the techniques of molecular genetics. Perhaps a word of caution is necessary. At present, there are strict limitations to what can be achieved by these methods. The synthesis of most antibiotics is under the control of many genes not always in juxtaposition. It is still enormously difficult to combine genes from several locations, and to transfer them to and have them expressed in a new organism, even a bacterium. However, molecular genetics can still be used to analyse and elucidate such system (Mukerji *et al.*, 1999).

3. POTENTIAL BOTTLENECKS

Although the concept of using biocontrol agents to increase plant yields goes back over 100 years (Hellriegel, 1886), relatively very few reliable bioinoculant products are currently available in the market. This is mainly because of the lack of confidence among the end users about the performance of the products being marketed. The experience about the commercialisation of bioinoculant world over is same. Therefore the most pertinent question at this juncture is why, in spite of the enormous scientific literature on the biocontrol, are so few bacteria are actually in commercial use? and more fundamental question at this critical juncture in the work is why, it fails?

There are several reasons which could result into the failure of the optimal performance and wider use of the biocontrol agents. The majority of the commonly used fungicides, insecticides, nematicides and herbicides are toxic to soil microflora. High fertility level or adverse physical condition of the soil due to extreme pH etc. or presence of acidic chemical fertilisers also cause the failure or diminish the performance of the inoculum.

If the inoculum is not rhizosphere competent then in that case also it will fail to establish itself into the introduced environment. This is one of the major causes of the failure of the inoculum as even as of today no systematic study has been done on the ecology of nitrogen fixing microbes. Customarily biocontrol agents are applied without any consideration for the site of application and the place of origin.

The overwhelming problem in the area of biocontrol agents is to get repeatable results (especially in field, rather than in the laboratory or greenhouse) which are consistent from year to year and over different climatic and soil types. The variability has many causes including the sensitivity of many potential control agents to these environmental factors, especially when control depends upon the growth and spread of the antagonist, as it usually does. The soil, though very variable, is also remarkably stable: the organisms present are often assumed to be well adapted community in which there may be changes in individual species or populations but which overall remain quite constant. It is, therefore, difficult to introduce 'foreign' organisms into the environment where it does not already exist. Thus the expectation to promote and recommend a universal biocontrol agent for all the 30 crops grown across the 16 agroclimatic zones of the country covering a legion of soil types is unscientific and absurd.

Ideally biocontrol organism would be so well adapted to the environment, the host plant and the pathogen that it would essentially survive and flourish for a very long time i.e., one application gives control for ever. However, from the point of view of environment safety, one can make an argument that it would be easier to use an antagonist that will stop being active after reasonably short time (the survival time on the crop is critical, but varies with the disease): there could be serious concern about introducing an organism that appeared to survive for ever in its new environment. Also for the development of a commercial biocontrol agent it is desirable to apply, and therefore to sell, the product at least once every crop, so that research can be justified by continued sales and production.

There is also a very serious problem of low quality biocontrol inoculum. This problem arises because of the incongruous people handling the inoculum development work at the government & non-government level and because of some of the small time regional biotechnology companies who jumped into biotechnology bandwagon to make fast bucks. This unfortunately has resulted into the supply of inferior quality products, impairing the product reliability and subsequently adverse name to the inoculum industry. Despite all the encouragements form the government, poor quality of biocontrol agents manufactured is one of the most important factors resulting in the failure of popularising

14

biocontrol agents to the desired extent. Status of quality of randomly collected samples of biocontrol agents produced by number of major National and State level manufactures both from government and non government sectors have shown 10-60 per cent samples below standard. However, If the inoculant is prepared, marketed and applied correctly then significant increase in the yields can be obtained.

There is ample literature available on potentially useful biocontrol agents but in reality a minuscule amount of these strains actually are put in use beyond few small scale field trials conducted by the researchers. Even fewer are actually used for industrial scale production. It is very difficult to find a bioinoculant with ideally required 15 or 20 site years (number of sites x number of years) of performance data before considering the strain for scale up (Cook, 1993). There is need for the development of industrial scale production of biocontrol agents. It is due to lack of interest among both researchers and bioinoculant manufacturers to coordinate efforts to look into the need for technology transfer from basic research to product development.

Irrespective of their advantages, the use of biocontrol agents in India has been low. This is partially due to lack of dissemination of proper knowledge in the farming community and lack of extension support. Farmers should be educated about the utility of biocontrol agents by conducting field demonstrations. This will increase awareness about bioinoculant use and will help in a better adaptation of the technology. Biological products are based on living organisms. This is a fundamental concept which must be fully appreciated because it affects every aspect of commercial development. Products based on biologicals can be successful in any market as long as they are as good as or better than existing technology in terms of cost, efficacy or reliability. The biggest lacuna for the low key success of commercialised biologicals inclusive of bioinoculant is because the performance of products based on biologicals falls below that of chemicals. There is simply due to the fact that by virtue of being a product based on microorganisms, the performance of the product is based on how the living organisms reacts to the alien environment of its application, unlike that of chemicals. Uniform performance of biocontrol agents under different geographical conditions can not be guaranteed by any expert familiar with this arena. This answer can only be elucidated by conducting field trials in the desired locations on the crop(s) of interest, because biological products although congenial to the environment unlike chemical treatments are not uniform/ predictable in its performance. Farmers should be educated that besides environment damages caused by the usage of chemical fungicides/pesticides/fertilisers, biocontrol of plant pathogens have other advantages. For example, most fungicides have only a temporary effect and usually require repeated applications during the growth season unlike biocontrol agents.

Above all the take home message world over is same, and that is, "Only quality of the product is the key to success". Thus, the government is under compulsion to have a serious second look and redraft the policy based on the experience of the past and the reasonable expectations of the future and accordingly the use of bioinoculants by our farmers is being promoted by the government. Fortunately, several companies, mainly private manufacturers have entered into the market with a strong quality base.

4. MANIPULATION OF RHIZOSPHERE BIOCONTROL AGENTS

Bacteria have been used as a biocontrol agents to increase yields and reducing the use of chemicals is an attractive development and the potential benefits may be considerable. The heightened interest in biological control with microorganisms is also a response to the availability of the new tools of biotechnology, which are particularly applicable to microorganisms. Other than disease-resistant plants, the biocontrol agents available for use against plant diseases and most nematodes are microorganisms. Moreover, plants and microorganisms can now be manipulated to "deliver" the same mechanisms of biological control, as done with the *Bt* gene for production of the delta endotoxin transferred from *Bacillus thuringiensis* to plants to control sensitive insects. We can now think of microorganisms with inhibitory activity against phytopathogens as potential source of genes for disease resistance. The establishment of beneficial bacteria on root systems of plants via seed bacterisation has long been of a major interest to agricultural researchers (Paroda, 1997). This is due to the secretions of nutrients into the surrounding environment. This environment, or the volume of soil that is influenced biologically and biochemically by the living root, is known as the rhizosphere (Savka and Farrand, 1997). Root exudates and secretions create a "rhizosphere effect" which manifests itself in the intense microbial activity that is associated with the immediate vicinity of the root. Root-associated microorganisms must compete with other soil microbes for sources of carbon, nitrogen and/or energy. The ability to utilise these nutritional resources present in the rhizosphere contributes to the survival and competitiveness (rhizosphere competence) of root-associated bacteria (rhizobacteria). Processes in the rhizosphere influence plant disease and plant nutrition by affecting the dynamics of microbial populations and communities. The action of both plant-beneficial and plant-deleterious soilborne microbes depends on the ability to establish in the rhizosphere. Therefore the knowledge of bacterial growth conditions in the rhizosphere is important for understanding rhizosphere colonisation. In general, a proper characterisation of target soils and rhizosphere as habitats for introduced microbes, as well as adequate strategies is the key to success, to enhance the plant-microbial interactions. As climatic conditions, soils, plants and microbes are all variable and/or diverse, there is no general rule for how introductions into soils can be optimised. However, it is clear that since soil generally represents a hostile environment to microbial introduction of an alien microbe and since the microbes in the soil are subjected to a range of adverse abiotic and biotic conditions, the rhizosphere competence of the introduced microbe depends to a large extent on how favourable to its survival and functioning the target environment is or can be made, in terms of either natural or induced ecological selectivity (through genetic engineering) or available protective niches. Therefore, it is anticipated that knowledge of key ecological and molecular characteristics that determine bacterial survival and adaptation in soil may also lead to methods for genetic manipulations directed to improve the survival of bacteria introduced into soil. This in turn is critical for the application of beneficial microorganisms as inoculants to support plant growth. Although the inability of commercial inocula to compete in the rhizosphere has been well documented, the factors involved in the colonisation remain elusive. Hardly anything is known about the ecological and molecular basis of rhizosphere colonisation, despite its importance for plant-microbe interactions. Therefore an understanding of the interactions between the biocontrol agent and complex rhizosphere microbial community in which the agent must functions, is essential. From this understanding, we may be able to develop models that will predict the conditions under which a biocontrol strategy will succeed or fail. However, developing predictive models for bioncontrol is a complex endeavour because of the multicomponent nature of the system. Biocontrol often requires the introduction and establishment of biocontrol agent in the microbial community on or near the plant surface. Once introduced to the rhizosphere, various mechanisms account for the successful application, or failure of biocontrol agent to control phytopathogenic fungi, by complex interactions between the plant rhizosphere microflora, phytopathogenic fungi, the biocontrol agent, and various biotic and abiotic factors encountered in soil, including plant released compounds, induction of systemic resistance, niche exclusion, and environment stress, e.g., salt, pH, temperature, mositure, chemical fertilisers, pesticides, fungicides, etc.

16

(Fig.1). It is envisaged that an understanding of the mechanisms underling these interactions may allow us to provide a logical basis for selection and manipulation of more effective biocontrol agents.

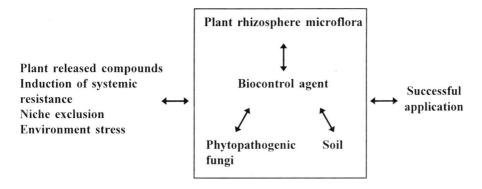

Figure 1. The effect of interactions of plant rhizosphere microflora, phytopathogenic fungi, biocontrol agent and soil on the success of biocontrol agent.

Microorganisms that can grow in the rhizosphere are ideal for use as biocontrol agents, since the rhizosphere provides the front-line defence for roots against attack by pathogens. Pathogens encounter antagonism from rhizosphere microorganisms before and during primary infection and also during secondary spread on the root. In some soils described as microbiologically suppressive to pathogens microbial antagonism of the pathogen is especially great, leading to substantial disease control (Schroth and Hancock, 1982). Although pathogen-suppressive soils are rare, those identified are excellent examples of the full potential of biological control of soil pathogens. Therefore, it is generally assumed that colonisation by introduced bacteria is essential for biocontrol of root pathogens and that increasing the population of an introduced bacterium on the root should enhance disease control. Bacterial root colonisation comprises a series of steps: migration towards plant roots, attachment, distribution along the root and finally growth and establishment of the population. Research efforts have focused on the determination of the roles, if any, of known bacterial structures and behavioural traits in these steps. The colonising ability of a strain can at present be evaluated only *in vivo*. Sufficient information on the factors involved in this complex process is not yet available to enable an assessment to be made by investigating the biochemical or genetic makeup of a strain. However, through the use of genetic means, some factors which play a role in colonisation of root surfaces are being recognised.

The isolation and development of plant beneficial bacterial strains applicable to a variety of crops, soils and locations will depend upon the development of improved detection and screening procedures that more rapidly screen and identify beneficial strains (Nautiyal, 1997a). Few methods have been developed for selecting rhizosphere competent bacterial strains. Recently we have developed a greenhouse assay to evaluate the root colonising capability of the native chickpea rhizospheric bacterial population (Nautiyal, 1997a). In this assay system screening time was reduced on two counts. First, spontaneous chromosomal Rifr strains were directly inoculated to seeds without any check for the stability of the mutation and secondly, no attempts were made to taxonomically identify all the strains being screened for chickpea rhizosphere competence. These findings indicated that the isogenic or equally rhizospheric competitive second non-isogenic Rifr strains should be compared for their survival and competition with that of

17

the isogenic parent and with each other for specific ecological niche, before using mixture of isolates, for stable and consistent biological seed treatment to control soilborne pathogens or pests or to promote plant growth (Nautiyal, 1997a). Using this method we have isolated a chickpea rhizosphere-competent bacteria *Pseudomonas fluorescens* NBRI1303 for suppression of the chickpea pathogenic fungi *Fusarium oxysporum* f. sp. *ciceri, Rhizoctonia bataticola* and *Pythium* sp. This is the first report of a single biocontrol bacterium active against three most devastating pathogenic fungi of chickpea (Nautiyal, 1997b). In greenhouse test chickpea seed bacterization with *P. fluorescens* NBRI1303 increased the germination of seedlings by 25%, reduced the number of diseased plants by 45%, compared to non-bacterized controls. Increases in seedling dry weight, shoot length and root length ranged from 16 to 18%. Significant growth increases in shoot length, dry weight and grain yield, averaging 11.59%, 17.58% and 22.61% respectively above untreated controls were attained in field trials in Agra and Jhansi. A rifampicin-resistant mutant *P. fluorescens* NBRI1303R of the *P. fluorescens* NBRI1303, used to monitor chickpea root colonisation confirmed the rapid and aggressive colonisation by the bacterium, making it a potential biocontrol agent against chickpea phytopathogenic fungi (Nautiyal, 1997b). The results demonstrate an increase in the efficiency of screening and detection of plant beneficial strains that will greatly benefit future studies. In addition to the properties noted above, the novel strain *P. fluorescens* NBRI1303 also have the following qualities which are important for use as a biological control agent for chickpea by :

(i) being a naturally occurring isolate that does not require any genetic manipulations to be effective

(ii) being easily cultured and developed for commercial purposes

(iii) exhibiting capability to be rhizosphere competent for the full growing season

(iv) being suppressive to one or more phytopathogenic fungi

(v) exhibiting capability to solubilise insoluble phosphorus

(vi) enhancing the yield of host plant

(vii) being environment friendly unlike chemical fungicides and fertilisers.

It will be difficult to identify the biocontrol mechanism(s) for the ability of our novel strain *P. fluorescens* NBRI1303 to control the chickpea pathogenic fungi *Fusarium oxysporum* f. sp. *ciceri, Rhizoctonia bataticola* and *Pythium* sp., as various mechanisms, including competition for iron and other nutrients, niche exclusion, induction of systemic resistance, and production of antimicrobial metabolites, could be operative (Pieterse *et al.,* 1996; Press *et al.,* 1997; Raaijmakers *et al.,* 1999). Commercial use of *P. fluorescens* NBRI1303 as inoculants or use of its purified antifungal metabolite(s), due to its novel propeties stated as above, may help in maintaining sustainable crop productivity without further deterioration in soil health (Nautiyal, 1997b,c).

Engineering of the rhizosphere is another emerging field in which little information is available, both at ecological and molecular level. There is interest in engineering the rhizosphere for several reasons. Such plants might resist soilborne pathogens more effectively, be better hosts to beneficial microbes, remediate toxic waste, or attract communities of soil microbes that enhance plant health (O'Connell *et al.,* 1996).The ecology of rhizosphere is a relatively new research area. Consequently, there is little understanding of how environmental factors will affect bacterial colonisation effects and persistence on roots and the resulting effect on plant. However, it is known that the bioinoculants which are introduced to the rhizosphere are involved in a complex of biological interactions with the host plant and with the surrounding rhizosphere microbes. The introduced bacteria are nourished by the root exudates and are thus dependent on the host plant. At the same time, the introduced bacteria may affect the host by inducing physiological changes in the plant. Interactions with indigenous rhizosphere microbes may

18

be neutral, antagonistic (e.g., competition for nutrients, production of antibiotic compounds, parasitism, or predation), or synergistic (i.e., the promotion of *Rhizobium* induced nodulation of legumes). These microbial interactions are greatly influenced by many environmental parameters, including, temperature, moisture, fertiliser regimen, and soil type. At an ecological level, Bashan (1986) and Li and Alexander (1990) showed that ecological selectivity worked in the rhizosphere by using the concept of antibiosis. They temporarily suppressed the competing or antagonistic indigenous microbes by using streptomycin, in conjugation with an antibiotic resistant inoculant strain. Similarly, a Tn5 carrying *P. fluorescens* strain was shown to be selected by streptomycin in the soil and rhizosphere (Britto *et al.*, 1995), based on the streptomycin resistance provided by Tn5. Ecological selectivity can also be based on the use of a specific substrate which is unavailable to a majority of other soil microbes. Devliegher *et al.* (1995) showed the effectiveness of this principle, in the use of a combined addition of specific detergents as carbon sources and detergent degradative *Pseudomonas* spp. added to soil. Treatment of soil with certain detergents resulted in a 100 to 1,000 fold increase in the detergent-adapted inoculant populations and significantly enhanced the colonisation of maize roots. van Elsas and van Overbeek (1993) showed the existence of genes that could be activated by specific root exudates. In particular, a strong promoter was identified in *P. fluorescens*, which was specifically induced by proline present in the rhizosphere of gramineous plants (van Elsas and van Overbeek, 1995). This induction was found to be prevalent in the rhizosphere of wheat both in soil microcosms and in the field (van Overbeek *et al.*, 1997) and further in that of maize and grass.

Central among the strategies to engineer the rhizosphere is the effort to create a "biased rhizosphere", which involves engineering plants to secrete nutrients that specifically enhance the growth of desirable microbes (O'Connell *et al.*, 1996). Perhaps the most elegant example of the hypothesis that novel substrate utilization is a component of a microbe-host association confers the relationship between *Agrobacterium tumefaciens* and the plants on which it induces crown gall tumors. These neoplasias result from expression by the plant genes transferred to them by the infecting bacteria. The tumors are characterized, in part, by the production of novel, low molecular-weight compounds generically called opines. This trait is also coded by the DNA transferred from the bacterium to the plant during tumor initiation. In turn, the bacteria can catabolize those opine classes produced by the tumors that they induce. The observation that the families of opines produced by a given tumor are dependent upon the inducing bacterium, and that the bacterium can use only those opines that are produced by the tumor it induces led to the opine concept. In its general form, the opine concept states that a specific interaction between a microbe and a host may be driven in part by the capacity of the microbe to use a novel resource produced by the host.

We were the first one to report the capability to catabolise opines by the pseudomonads, as carbon and nitrogen sources (Nautiyal and Dion, 1990). Later on the impact of a novel substrate, mannopine (MOP) by creating the nutritional biasing on rhizobacterial colonisation was examined using mannopine utilisation (Mut) as the model system (Fig.2). The relative competitiveness of two *Pseudomonas* strains that differed only in Mut (one strain was a Tn5 generated Mut⁻ mutant of its parent wild type Mut+) was determined. Near isogenic tobacco lines differing in mannopine production (MOP+) were obtained by transformation with *Agrobcterium rhizogenes*. Tobacco seeds were inoculated with a mixture of the two bacterial strains and the ratio between the two was determined (input ratio). Four weeks after planting, bacteria were recovered from seedling roots and ratio between the Mut+ and Mut⁻ were again determined (output ratio). On MOP⁻ plants (no mannopine in the environment), the input and output ratios were essentially the same, indicating that the two bacterial strains were equally rhizosphere competent in the absence of MOP. On MOP+ plants (mannopine in the environment), however, Mut+ bacteria increased relative to the Mut⁻ bacteria, indicating that, in the presence of MOP, the ability to utilise MOP conferred a competitive advantage (Lam *et al.*, 1997). Recently, modification of rhizobacterial populations by engineering bacterium utilisation of opines has been achieved based on this concept (Oger *et al.*, 1997, Savka and Farrand 1997). It is anticipated that the discovery of a genetically controlled mechanism that would bring about a closer relationship between root and biocontrol agent, and that could be expressed in either the root, biocontrol agent, or both, would open the way for many advances in biological control of root pathogens, including, possibly, a means to lessen the influence of soil factors.

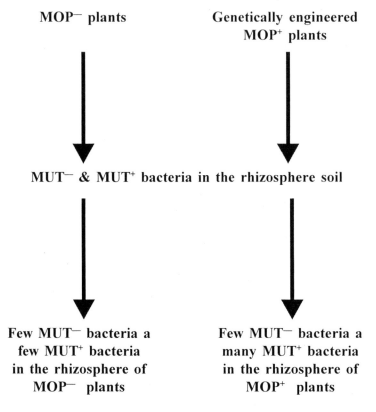

Figure 2. Genetically engineeered plants, which code for the biosynthesis of mannopine (MOP+), release the mannopine into the rhizosphere thus creating a biaesed rhizosphere where only the bacteria containing mannopine catabolic (MUT) genes, use it as a unique nitrogen and carbon source. This provides a selective growth advantage for bacteria that can catabolize the mannopine (MUT+), in contrast to mannopine non-utilizers (MUT⁻) in the rhizosphese of MOP+ plants, compeared with MOP⁻ plants.

5. CONCLUSIONS

Confirmed market demand is the precursor to all commercial research and development. For a commercial venture it is important to distinguish between societal need for biological products and actual customer demand. Biofertilzers decreases the social costs of agriculture by enhancing soil health and decreasing water and soil pollution. These factors are generally overlooked by farmers making-short term decisions based solely on their costs and benefits. It will be unrealistic to expect from a commercial concern to be driven to bankruptcy by pursuing the development of unprofitable products. However, social issues are having increasing effects on farm management decisions as demonstrated by recent trend towards sustainable agriculture, organic farming, integrated pest management and intensive crop management. Thus, societal needs for biological products may increasingly affect farmer demand and the farmers' perception of the value of biological products.

Economic, toxicological and environmental advantages are continuing to expand the market opportunities for biocontrol agents. Bioinoculant industry enjoys an edge over the traditional chemicals because smaller market niches are more suitable for biological products because of lower discovery and regulatory costs and reduced lead time to get a new product to market. Critical success factors for the selection and commercialisation of biocontrol agents are adequate customer demand and market size to ensure return on investment in a reasonable time; cost effective manufacturing; effective, easy-to-use formulations; compatibility with distribution and agricultural practices; highly efficacious and reliable field performance and adequate revenues to support sales and to continue product improvements. The areas of production, formulation, application, efficacy, regulatory and reliability occur more simultaneously, and there must be continuous communication and feedback between each area. In each area, particular attention and resources should be allocated to research that is critical to producing a product which meets farmers' expectations.

The successful development of biocontrol agents that meet farmers' expectation must deliver function and value. This requires development of new technologies on a constant basis and assimilate the same in their production set up. To succeed, those manufacturing the biocontrol agents must deliver the knowledge. Indian farmer is conscious of the fact that chemicals adversely affect soil health. It may probably be more important for us to convince our farmers about the ecofriendly and soil rejuvenating nature of biocontrol agents. Addressing the limitations of biocontrol agents up front is critical to setting realistic expectations for product performance. For example, farmers must understand that biocontrol agents application is prone to environment like any other living object, unlike chemicals. Once these limitations are understood, the benefits of biocontrol agents such as efficacious season long control, flexible timing of application, crop, soil health and environment safety may be established. Limitations and benefits may then be combined to arrive at consistently attainable expectations of value of the biocontrol agents. Finally, the success of the wider use of biocontrol agents will partly depend on the total commitment of companies for maintaining the quality of the biocontrol agents and partly on our government regulators to implement regulations to ensure that companies are responsible for maintaining consistently high standards, as the bitterness of poor quality remains long after the sweetness of low price is forgotten. Biocontrol agents products that meet and exceed farmers' expectations will drive productivity gains of benefit to all.

The complex studies involved in the process of bioinoculant manufacturing are beyond the scope of this article. It is important to realise, however, that the time has come when farmer expect a product which has consistently high quality. It is also important to realise that now the production of biocontrol agents is a very refined technology, unlike old perception of the process being simple, which attracted the novices to begin with to give a bad name to this otherwise very profitable industry. It is one thing to identify an organism and show it can do something useful, but it is quite another to put it into a user-friendly formulation which is stable, efficacious and can withstand the rigours of commercial distribution and handling.

There is a lot of awareness even among developed nations who can easily afford the application of chemical products to put in million of dollars in developing and marketing biological products. Two of the major fortune 500 companies of the United States of America, Novartis and Monsanto are committed towards this goal.

REFERENCES

Bashan, Y. 1986, Enhancement of wheat root colonisation and plant development by *Azospirillum brasiliense* Cd. following temporary depression of rhizosphere microflora, *Appl. Environ. Microbiol.* 51: 1067-1071.

Britto de Oliveira, R. G., Wolters, A. C. and van Elsas, J. D. 1995, Effects of antibiotics in soil on the population dynamics of transposon Tn5 carrying *Pseudomonas fluorescens*, *Plant Soil* 175: 323-333.

Cook, R. J. 1991, Success with biological control requires new thinking by industry, Counterpoint, *Impact AgBioBusiness,*. Oxon, UK (CAB Int. June) pp. 3-4.

Cook, R.J. 1993, Making greater use of introduced microorganisms for biological control of plant pathogens, *Ann. Rev. Microbiol.* 31: 53-80.

de Selincourt, K. 1996, Intensifying agriculture-the organic way, *The Ecologist* 26: 271-272.

Devliegher, W., Arif, M. A. S. and Verstaete, W. 1995, Survival and plant growth promotion of detergent-adapted *Pseudomonas fluorescens* ANP15 and *Pseudomonas fluorescens* 7NSK2, *Appl. Environ. Microbiol.* 61: 3865-3871.

Hellriegel, H. 1886, Welche stickstoffquellen stehen der Pflanze zu Gebote? Tageblatt der 59 Versammlung Detscher Naturforcher und Aerzte in Berlin, 18-24, Sept., p. 290.

Johri, B. N., Rao, C. V. S. and Goel, R. 1997, Fluorescent pseudomonads in plant disease management, In : *Biotechnological Approaches in Soil Microorganisms for Sustainable Crop Production*, ed. K. R. Dadarwal, Scientific Publishers, Jodhpur, India.

Khanna-Chopra, R. and Sinha, S. K. 1998, Prospects of success of biotechnological approaches for improving tolerance to drought stress in crop plants, *Curr. Sci.* 74: 25-34.

Lam, S. T. and Gaffney, T. D. 1993, Biological activities of bacteria used in plant pathogen control, In : *Biotechnology in Plant Disease Control*, ed. I. Chet, Wiley-Liss Press, New York, pp. 291-320.

Lam, S.T., Torkewitz, N.R., Nautiyal, C.S. and Dion, P. 1997, Microorganisms with mannopine catabolizing ability, United States of America Patent Number 5,610, 044.

Li, D.M. and Alexander, M. 1990, Factors affecting co-inoculation with antibiotic-producing bacteria to enhance rhizobial colonisation and nodulation, *Plant Soil* 129: 195-201.

Moore, L. W. 1979, Practical use and success of *Agrobacterium radiobacter* strain 84 for crown gall control, In : *Soil-borne Plant Pathogens*, eds, B. Schippers and W. Gams, Academic Press, London, pp. 553-568.

Mukerji, K.G. and Garg, K.L. (eds.), 1988a, *Biocontrol of Plant Disease*, Vol. I, CRC Press Inc., Florida, USA.

Mukerji, K.G. and Garg, K.L. (eds.) 1988b, *Biocontrol of Plant Disease*, Vol. II, CRC Press Inc., Florida, USA.

Mukerji, K.G., Chamola, B.P. and Upadhyay, R.K. (eds.) 1999, *Biotechnological Approaches in Biocontrol of Plant Pathogens*, Kluwer Academic/Plenum Press, New York, USA.

Mukerji, K.G., Tewari, J.P., Arora, D.K. and Saxena, G. (eds.) 1992, *Recent Development in Biocontrol of Plant Diseases*, Aditya Books Pvt. Ltd. New Delhi, India.

Narain, P. 1998, A dialectical perspective of agricultural research for sustainable development, *Curr. Sci.* 74: 663-665.

Nautiyal, C.S. and Dion, P. 1990, Characterisation of opine-utilizing microflora associated with samples of soil and plants, *Appl. Environ. Microbiol.* 56: 2576-2579.

Nautiyal, C. S. 1997a, A method for selection and characterisation of rhizosphere competent bacteria of chickpea, *Curr. Microbiol.* 34: 12-17.

Nautiyal, C. S. 1997b, Selection of Chickpea-Rhizosphere Competent *Pseudomonas fluorescens* NBRI1303, Antagonistic to *Fusarium oxysporum* f. sp. *ciceri*, *Rhizoctonia bataticola* and *Pythium* sp., *Curr. Microbiol.* 35: 52-58.

Nautiyal, C. S. 1997c, Rhizosphere competence of *Pseudomonas* sp. NBRI9926 and *Rhizobium* sp. NBRI9513 involved in the suppression of chickpea (*Cicer arietinum* L.) pathogenic fungi, *FEMS Microbiol. Ecol.* 23: 145-158.

O'Connell, K. P., Goodman, R. M. and Handelsman, J. 1996, Engineering the rhizosphere expressing a bias, *Trends Biotechnol.* 14: 83-88.

Oger, P., Petit, A. and Dessaux, Y. 1997, Genetically engineered plants producing opines alter their biological environment, *Mol. Plant-Microbe Interact.* 15: 369-372.

O'Sullivan, D. J. and O'Gara, F. 1992, Traits of fluorescent *Pseudomonas* spp. involved in suppression of plant root pathogens, *Microbiol. Rev.* 56: 662-676.

Paroda, R.S. 1997, Foreword, In : *Biotechnological Approaches in Soil Microorganisms for Sustainable Crop Production*, ed., K. R. Dadarwal, Scientific Publishers, Jodhpur,India.

Purohit, A. N. 1995, *The Murmuring Man*, Bisen Singh Mahendra Pal Singh Publishers, Dehradun, U.P. India.

Savka, M. A. and Farrand, S. K. 1997, Modification of rhizobacterial populations by engineering bacterium utilisation of a novel plant-produced resource, *Mol. Plant-Microbe Interact.* 15: 363-368.

Schroth, M. N. and Hancock, J. G. 1982, Diseases suppresive soil and root colonising bacteria, *Science* 216: 1376-1381.

Sinha, S. K. 1996, Food security in 2020 and beyond-Will developing countries be dependent on the developed world? *Curr. Sci.* 71: 732-734.

Sinha, S. K. 1997, Global change scenario: Current and future with reference to land cover changes and sustainable agriculture-South and South-East asian context, *Curr. Sci.* 72: 846-854.

Swaminathan, M.S. 1992, Biodiversity and Biotechnology, In: *Biodiversity and Global Food Security*, eds. M.S. Swaminathan and S. Jana, MacMillan India Ltd., Chennai, India.

Swaminathan, M.S. 1996, Biotechnology - biovillages and a better biofuture, In: *Trends in Molecular Biology and Biotechnology*, eds. S. Srivastava, P.S. Srivastava and B.N. Tiwari, CBS Publications and Distributors, New Delhi, India.

Thomas, V.G. and Kevan, P.G. 1993, Basic principales of agroecology and sustainable agriculture, *J. Agric. Environ. Ethics* 6: 1-19.

Upadhyay, R.K., Mukerji, K.G. and Rajak, R.L.(eds.) 1996, *IPM System in Agriculture,* Vol. II, *Biocontrol in Emerging Biotechnology*, Aditya Books Pvt. Ltd., New Delhi, India.

van Elsas, J. D. and van Overbeek, L. S. 1993, Bacterial responses to soil stimuli, In : *Starvation in Bacteria*, ed. S. Kjellebrg, Plenum Press, New York, pp. 55-79.

van Elsas, J. D. and van Overbeek, L. S. 1995, Root exudate induced promoter activity in *Pseudomonas fluorescens* mutants in the wheat rhizosphere, *Appl. Environ. Microbiol.* 61: 890-898.

van Overbeek, L. S., van Veen J. A. and van Elsas, J. D. 1997, Induced reporter gene activity, enhanced stress resistance, and competitive ability of a genetically modified *Pseudomonas fluorescens* strain released into a field plot planted with wheat, *Appl. Environ. Microbiol.* 63: 1965-1973.

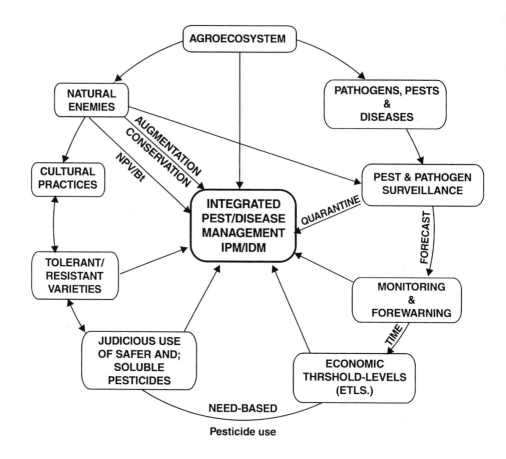

SCHEMATIC MODEL FOR IPM

(Original by K.G. Mukerji)

Bt   = *Bacillus thuringiensis*
NPV = Nuclear polyhearosis virus - *Autographa californica*

BACTERIAL BIOCONTROL AGENTS AND THEIR ROLE IN PLANT DISEASE MANAGEMENT

A.K. Saxena[1], K.K. Pal[2] and K.V.B.R. Tilak[1]

[1]Division of Microbiology
Indian Agricultural Research Institute
New Delhi-110012, INDIA

[2]National Research Centre for Groundnut
Timbawadi P.O. District Junagadh - 362 015, Gujrat, INDIA

1. INTRODUCTION

The newer innovations in the field of agriculture has led to enhanced destructive potential of diseases. These Innovations include practice of continuous monoculture and use of high doses of nitrogenous fertilizers that enhance disease susceptibility. To combat high incidence of fungal, bacterial and viral diseases, modern agriculture is now largely dependent on use of different formulations of agrochemicals. Increased environmental pollution by agrochemicals has led to considerable changes in attitude towards the use of pesticides in agriculture. It has led to more stringent regulations on their use and, in some cases, removal from the market. Consequently, interest has been focused on alternatives to chemicals for controlling pests and diseases. The use of biological control measure is one of the best strategies available to combat the pests and diseases in an ecofriendly manner and much experimental work is being carried out all over world to assess its commercial acceptability and applicability.

The world agrochemical market is about $12 billion out of which only 1 per cent is contributed by biological control agents (BCA) (Powell, 1993). *Bacillus thuringiensis* formulations contribute 92 per cent of the BCA.

Boswell (1965) described biological control as "a landmark in a great renaissance of interest and research in microecological balance in relation to soil-borne plant disease, and in the development of more enduringly profitable and wiser farming practices". To overcome the public concerns over chemical pesticide there is a greater need for biological control of pathogens that go uncontrolled or only partially controlled. There is no replacement for *Agrobacterium radiobacter* strain K-84 (Kerr, 1980) for biological control of crown-gall and ice-minus bacteria for control of ice-nucleation active bacteria

(Lindow, 1983) responsible for frost damage to potatoes. Biological control should be justified on its own merits without compromising its perceived importance to chemical controls.

In the era of rhizosphere manipulations, organisms that can grow in the rhizosphere are ideal for use as biocontrol agents, since rhizosphere provides the front-line defense for root against attack by pathogens (Weller, 1988). Pathogens face a complicated phenomena of antagonisms from roots. They also compete with each other for site of colonisation on the root surface.

Biocontrol methods and strategies involve timely manipulation of antagonist populations to suppress pathogens in various inoculum sources or on host plants. The ideal biocontrol introduces or promote the antagonists only when and where they are needed or are most effective and minimizes wasteful application of inoculum to non-targets.

Bacterial biocontrol agents improve plant growth by suppressing either major or minor pathogens and / or producing plant growth promoting substances like auxins, gibberellins, etc. Major pathogens cause well known rot of vascular diseases with obvious symptoms whereas minor pathogens are parasites or saprophytes that damage mainly juvenile tissue such as root hairs and tips and cortical cells (Salt, 1979).

Many genera belonging to fungi, bacteria, actinomycetes and viruses are used as a biocontrol agent to combat several important plant diseases. The present review focus only on the bacterial biocontrol agents active against soil - borne fungal pathogens.

2. RHIZOBACTERIA AS BIOCONTROL AGENTS

During the last two decades, several examples of rhizobacteria capable of providing substantial disease control in the field have been reported. The term "rhizobacteria" was coined for bacteria with ability to colonise roots aggressively (Kloepper and Schroth, 1978).

Many bacterial genera have shown their potential for biocontrol both under *in vitro* and *in vivo* conditions. *Agrobacterium* (Kerr, 1972, 1980; Thomson, 1987), *Arthrobacter* (Mitchell and Hurwitz, 1965), *Alcaligenes* (Martinetti and Loper, 1992), *Azotobacter* (Meshram and Jager, 1983), *Bacillus* (Asaka and Shoda, 1996; Fiddaman and Rossall, 1995; Hwang, 1994; Michereff *et al.*, 1994a; Sharga, 1997; Silo-Suh *et al.*, 1994), *Escherichia coli* (Roberts *et al.*, 1997a, b), *Enterobacter* (Costa and Loper, 1994; Roberts *et al.*, 1997 b),*Pseudomonas* (Dowling and O'Gara, 1994; Gomes *et al.*, 1996; Hallman *et al.*, 1996; Hebbar *et al.*, 1992; Kloepper *et al.*, 1980; Moulin *et al.*, 1994b, 1996; Pierson and Weller, 1994; Thara and Gnanamanickam, 1994), *Burkholderia* (King and Parke, 1993; Roberts *et al.*, 1997a,b), *Rhizobium* and *Bradyrhizobium* (Chakraborty and Purkayastha, 1984; Tu, 1980), *Serratia* (Berg and Behl, 1997) and *Stenotrophomonas* (Berg and Behl, 1997) were found to be potent for suppression of soil-borne fungal pathogens. Many of these biocontrol agents exhibited their effectiveness under field condition also.

Agrobacterium radiobacter strain 84 was the first bacterium used commercially for biocontrol of crown gall caused by *A. tumefaciens* and got worldwide success (Kerr, 1980).

Martnetti and Loper (1992) reported that *Alcaligenes* sp. strain MFAI inhibited microconidial germination and germ tube elongation of *Fusarium oxysporum* f. sp. *dianthi* and also reduced the severity of disease presumably as a result of siderophore production.

The usefulness of *Bacillus* as a source of antagonist for many plant pathogens (Campbell, 1989) is well known. Several potent strains from different species of *Bacillus* have been tested on a wide variety of plant species for their ability to control several diseases. *Bacillus* has ecological advantages because it produces endospores that are tolerant to extreme environmental conditions such as heat and desiccation. The greatest interest is *Bacillus subtilis* A 13, which was isolated from lysed mycelium of *Sclerotium rolfsii* by Broadbent *et al.* (1971). It was found to be antagonistic to several pathogens and as seed inoculant, increased the yield of carrots by 48%, oats by 33% (Merriman *et al.*,1974) and peanuts by 37 % (Turner, 1987). Since, 1983 it has been sold as a bioinoculant under the trade name QUANTUM - 4000 (Turner, 1987). *Bacillus* spp. and actinomycetes share several features that make them attractive for biological control agents, including their abundance in soil and the production of various biologically potential metabolites active against a range of fungi (Silo-Suh *et al.*, 1994). In this respect *Bacillus cereus* UM85 has shown a wide range of biological effects on plants. It was found to protect alfalfa seedlings from damping-off caused by *Phytophthora medicaginis* (Handelsman *et al.*, 1990), tobacco seedlings from *Phytophthora nicotianae* (Handelsman *et al.*, 1991), cucumber fruits from rot caused by *Pythium aphanidermatum* (Smith *et al.*, 1993), and peanut from *Sclerotinia minor* (Phipps, 1992). At least, suppression of alfalfa damping-off was due to the production of two fungistatic antibiotics (Silo-Suh *et al.*, 1994). Likewise, *Bacillus subtilis* RB14, produces antibiotics iturin and surfactin and could suppress *Rhizoctonia solani* damping-off of tomato (Asaka and Shoda, 1996). *Bacillus* BS 153 was able to prevent chocolate spot symptoms (causal agents *Botrytis cinerea* and *B. fabae)* on faba beans both in green house and field studies (Sharga,1997). *Bacillus subtilis* strain could reduce root rot of lentil caused by *Fusarium avenaceum* (Fr.) Sacc (Hwang, 1994), northern leaf blight of corn (causal organism *Exserohilum turcicum*) (Reis *et al.*, 1994) rust infection of safflower caused by *Puccinia carthami* (Tosi and Zazzerini, 1994) *Fusarium* wilt of cotton caused by *Fusarium oxysporum* (Zhang JinXu *et al.*, 1995) Yam leaf spot disease (causal organism *Curvularia eragrostridis*) (Michereff *et al.*, 1994b) and anthracnose fruit rot of chilli caused by *Colletotrichum capsici* and *C. gloeosporioides* (Sariah - Meon, 1995). *Bacillus subtilis* and other *Bacillus* spp. were also reported to control post-harvest diseases of apple caused by *Penicillium expansum* and *Botrytis cinerea* (Sholberg *et al.*, 1995).

Enterobacter cloacae, a potent root coloniser, was reported as biocontrol agent against *Pythium ultimum*, a prevalent phytopathogen that causes damping-off of many crop plants (Costa and Loper, 1994; Howell *et al.*, 1988; Nelson, 1988). Nelson (1988) reported the potentiality of *Enterobacter cloacae* strain ECct-501 to prevent pre-emergence damping-off of cotton. Howell *et al.*(1988) applied *E. cloacae* strain ECH-1, which produced ammonia, to control the damping-off of cotton caused by *P. ultimum*. Costa and Loper (1994) evaluated and characterised that siderophore of *E. cloacae* strain EcCT-501 was responsible for disease suppression. *E. cloacae* strain 501 R3, a rifampicin resistant mutant of EcCT 501 was demonstrated to control damping off cucumber by *Pythium ultimum* both alone and in combination with *Burkholderia cepacia* (Roberts *et al.*, 1997b).

Fluorescent pseudomonads have revolutionized the field of biological control of soil-borne plant pathogenic fungi. During the last 25 years, they have emerged as the largest potentially most promising group of plant growth promoting rhizobacteria involved in the biocontrol of plant diseases (Barbosa *et al.*,1995; Burr *et al.*,1978; Fravel, 1988; Gomes *et al.*, 1996; Pierson and Weller, 1994; Wei *et al.*, 1996). Fluorescent pseudomonads have received the most attention for several compelling reasons. First, fluorescent pseudomonads readily colonize roots in nature, where they are frequently the most common of microorganisms (Weller, 1988). The simple nutritional requirement and the ability to use many carbon sources that exude from roots and to compete with indigenous microflora, may explain their ability to colonize the rhizosphere (Mazzola and Cook, 1991). Additionally, pseudomonads are amenable to genetic manipulation. These characteristics make them useful vehicle for the delivery of antimicrobial and insecticidal compounds and plant hormones to the rhizosphere. O'Sullivan and O'Gara (1992) reviewed the traits of fluorescent pseudomonads such as production of antibiotics, hydrogen cyanide, siderophore which are involved in suppression of plant root pathogens.

There are numerous examples of biocontrol of several devastating fungal plant pathogens of important crops by fluorescent pseudomonads and has been reviewed from time to time (Cook, 1993; O'Sullivan and O'Gara, 1992; Weller,1988). Fluorescent *Pseudomonas* spp. have been implicated in the control of several wilt diseases caused by *Fusarium* spp. (Chen *et al.*,1995; Hebbar *et al.*,1992; Pal, 1995; Roberts *et al.*, 1997a); damping-off due to *Pseudomons* spp.(Anith, 1997; Barbosa *et al.*, 1995; Gomes *et al.*, 1996; Pal, 1995); root rot of important crops like wheat, cucumber, tulip caused by *Pythium* spp. (Moulin *et al.*, 1994b, 1996; Paulitz *et al.*, 1992; Roberts *et al.*, 1997a,b; Weststeijn, 1990); take all disease of wheat (causal organism *Gaeumannomyces graminis* var. *tritici*) (Bull *et al.*, 1991; Duffy *et al.*, 1996; Pierson and Weller, 1994) and several other fungal diseases like charcoal rot (causal organism *Macrophomina* spp.) (Pal, 1995) and grey mould (causal organism *Botrytis cinerea*) (Walker *et al.*, 1996), sorghum anthracnose (Causal organism *Colletotrichum graminicola*) (Michereff *et al.*, 1994a) and seedling blight of cucumber and poinsettia caused by *Pythium aphanidermatum, P. splendens* and *Rhizoctonia solani* (Wu - Ws, 1995).

In recent years many potent biocontrol *Pseudomonas* strains have been developed and investigated. *Pseudomonas fluorescens* 89B-61 is a root colonist which has been shown to reduce the incidence of *Fusarium* wilt of cotton (Chen *et al.*,1995), cucumber anthracnose (causal agent *Colletotrichum orbiculare)*, and bacterial angular leaf spot (causal agent *Pseudomonas syringe* pv. *lachrymans)* disease (Wei *et al.*, 1996). *Pseudomonas glumae* EM 85 could control spread of damping off of cotton (*Rhizoctonia solani)*; charcoal rot of maize and *Fusarium* wilt (Pal, 1995). The strain was found to produce antifungal antibiotic the gene for which was found to be located on a 23 kb fragment of chromosomal DNA (Anith, 1997). *P. fluorescens* NBRI 1303 , a good chickpea coloniser, could inhibit three fungi *in vitro F. oxysporum* f. sp. *ciceri, Macrophomina phaseolina* and *Pythium* sp. and was found to increase the plant dry weight and grain yield in field trials (Nautiyal, 1997). *P. fluorescens* WCS 374 could control *Fusarium* wilt of radish (Leeman *et al.*, 1995) where as *P. putida* WCS 358 could suppress *Fusarium* wilt of carnation (Duijff *et al.*, 1994). *Sclerotinia* wilt of sunflower could be controlled by *P. fluorescens* GL-92 and *P. putida* T1-5 (Expert and Digat, 1995). Gomes *et al.* (1996) reported specificity between bacterial strain and pathogen race. *P. putida* strain FR-38 was the best biocontrol agent in relation to RS-2 isolate of *R. solani* while CR-26 showed the best performance in relation to isolate RS-4. Fluorescent *Pseudomonas* strain CH31 suppress *Pythium* root rot of cucumber (Moulin *et al.*,1994b) and could do so by decreasing root colonization of *Pythium aphanidermatum* OP4 (Moulin *et al.*, 1996). Barbosa *et al.* (1995) found *P. fluorescens* isolate CR-20 to reduce disease intensity caused by *Rhizoctonia solani* in cowpea under field conditions.

Pseudomonas aureofaciens isolate PA 147-2 produces an antibiotic inhibitory to a number of fungal pathogens *in vitro* (Carruthers *et al.*,1994) and could suppress root rot of *Asparagus officinalis* seedlings caused by *Phytophthora megasperma* var. *sojae* (Carruthers *et al.*, 1995). *P. aureofaciens* strain Q2-87 producing the antibiotic 2,4 - diacetyl - phloroglucinol (DAPG) was inhibitory to *G. graminis* var. *tritici* (Vincent *et al.*, 1991). *Pseudomonas cepacia* J82 rif and J51 rif were also reported to suppress the sunflower wilt caused by *Sclerotinia sclerotiorum* (McLoughlin *et al.*,1992). Hill *et al.*

28

(1994) isolated *P. fluorescens* (B/L 915) producing pyrrolnitrin, which was found to be an effective antagonist of *R. solani*-induced damping-off of cotton. Cartwright *et al.* (1995) reported *P. cepacia* strain 5, 5B producing pyrrolnitrin and phenazine, as a potent biocontrol agent of *R. solani*, causing stem rot of Poinsettia. *P. aureofaciens* PA 1472 has also been used for biological control of *Phytophthora megasperma* rot of *Asparagus* (Carruthers *et al.*, 1995). *P. cepacia* AMMD could control *Aphanomyces* root rot and *Pythium* damping-off of four pea cultivars (King and Parke, 1993).

Attention has now been diverted to other genera of bacteria especially nitrogen-fixers, capable of showing antifungal activities. Therefore, there is an ample scope to explore potent biocontrol agents.

A more ecologically sound approach for biocontrol involves the use of multiple microbial strains. Increasing the genetic diversity of the biological control system through the use of mixtures may result in treatments that persists longer in the rhizosphere and express wider array of biocontrol mechanism under a broader range of environmental conditions (Duffy *et al.*, 1996; Pierson and Weller, 1994; Roberts *et al.*,1997b).Fluorescent *Pseudomonas* strains were tested alone and in combinations for the ability to suppress take-all disease of wheat. Certain combinations of strains performed better than individual strains and suggests the importance of additive and interactive effects among introduced bacteria in biological control (Pierson and Weller, 1994). Duffy *et al.* (1996) also reported better biocontrol of take-all disease of wheat when fluorescent *Pseudomonas* was applied in combination with *Trichoderma koningii*. Seed treatments containing combinations of *Escherichia coli* S17R1 and *Burkholderia cepacia* BC-B provided significantly greater suppression of cucumber seedling pathogenesis in a field soil naturally infested with *Pythium* and *Fusarium* spp. than seeds treated with strain BC-B, S17R1 or *Enterobacter cloacae* 501 R3 (Roberts *et al.*,1997b). Mazzola *et al.* (1995) observed that isolates of *G. graminis* var. *tritici* insensitive to antibiotics phenanzine-1-carboxylic acid (PCA) and 2,4-diacetyl phloroglucinol (Phl) are present in pathogen population and provide additional justification for the use of mixtures of *Pseudomonas* spp. that employ different mechanisms of pathogen suppression to manage this disease.

3. MECHANISMS OF DISEASE SUPPRESSION

The above cited literature clearly indicate the importance of bacterial inoculants in biocontrol of fungal diseases of plants. Inspite of umpteen reports on exploitation of bacterial endophytes as a means for controlling plant diseases, the exact mechanisms by which these microorganisms confer increased plant protection have not been fully investigated.

Numerous hypotheses for protection have been proposed including production of siderophores, accumulation of antifungal metabolites, nutrient competition, and niche exclusion (Chen *et al.*,1995). Traits of fluorescent *Pseudomonas* spp. involved in suppression of plant root pathogens have been critically reviewed by O'Sullivan and O'Gara(1992). Some of the reports on protection mechanisms in different groups of endophytes which appeared recently are included in this review.

Costa and Loper (1994) reported that *Enterobacter cloacae* EcCT 501, which suppressed *Pythium* damping-off of cucumber, produced two kinds of siderophores, aerobactin (hydroxamate) and enterobactin (catechol). Mutants deficient in the biosynthesis of the either siderophores were found to be similar to the potential strains in suppressing the disease. Therefore, aerobactin and enterobactin production by *E. cloacae* did not contribute to its activity in biological control of *Pythium* damping-off of cucumber or

29

cotton. Lim and Kim (1997) suggested the importance of siderophore production by *P. fluorescens* GL 20 for the biocontrol of *Fusarium solani.*

Many bacterial strains are known to suppress fungal growth *in vitro* by the production of one or more antifungal antibiotics (Carruthers *et al.*, 1994; Mazzola *et al.*, 1995; Pal, 1995). Some of these antibiotic producing strains were also shown to suppress fungal plant diseases *in vitro* (Carruthers *et al.*,1995; Leifert *et al.*, 1995; Pal., 1995). Several strains of *Pseudomonas* and *Bacillus* has been shown to produce wide array of antibiotics and includes phenazines, pyoluteorin, pyrrolnintrin, tropolone, pycocyanin, 2-4 diacetyl phloroglucinol, oomycine, iturin and surfactin, agrocin 84 (Asaka and Shoda,1996; Defago, 1993; Fujimoto *et al.*, 1995; Kerr,1980; Maurhofer *et al.*,1995; Mazzola *et al.*, 1995; Michereff *et al.*, 1994a).In most of the studies demonstrating the involvement of secondary metabolites in biocontrol, evidence has been generated through developing Tn5 induced deficient mutants (for respective metabolite) (Carruthers *et al.*,1995; Keel *et al.*,1992; Shanahan *et al.*,1992; Toyota and Ikeda, 1997), UV induced antibiotic negative mutants (Leifert *et al.*,1995) or through selection of antibiotic over producing mutants (Maurhofer *et al.*, 1992, 1995 ; Pal, 1995).

Burkhead *et al.* (1994) reported that *Pseudomonas cepacia* B 37 w produced pyrrolnitrin antibiotic inhibitory to *F. sambucinum*. Michereff *et al. (*1994a) could correlate the *in vitro* inhibition of *Pythium* and *Rhizoctonia* by 2,4 diacetyl pholoroglucinol, an antibiotic produced by *P. fluorescens* PF 5 and *in vivo* control of sorghum anthracnose caused by *Colletotrichum graminicola. P. fluorescens* F113 laç zy produces an antifungal pholoroglucinols which protects sugarbeet against *Pythium-*mediated damping-off (Russo *et al.*,1996). The work of Carruthers *et al.* (1994, 1995) conclusively implicated the role of antibiotic in disease suppression. *Pseudomonas aureofaciens* PA 147-2 produces an antibiotic (AF+) which inhibits the growth of phytopathogens *in vitro* (Carruthers *et al.*, 1994). They developed Tn5 antibiotic deficient mutant PA109 and tested along with the wild type for their ability to suppress root rot of *Asparagus officinalis*. Seedlings coinoculated with the pathogen and wild type strain showed a significantly reduced level of infection and disease severity compared to seedlings inoculated with the pathogen alone. However, 100% of seedlings treated with Af - mutant PA 109 were diseased (Carruthers *et al.*, 1995).

Many *Bacillus* strains primarily belonging to *B. subtilis* are known to suppress fungal growth both *in vitro* and *in vivo* through production of antifungal antibiotics (Brannen l995; Kim *et al.*, 1997; Leifert *et al* .,1995). *B. subtilis* YB-70 produces an antifungal antibiotic which could suppress *Fusarium solani* (Kim *et al.*, 1997). The mechanism involved in the suppression of *Rhizoctonia solani* lesions on cotton by *B. subtilis* GB03 is also suspected to be antibiosis (Brannen, 1995). Antibiotic production was clearly indicated as the mode of action of *in vivo* biocontrol by *Bacillus subtilis* CL27 against damping-off Astilbe caused by *Botrytis cinerea*. CL27 produced three different antibiotics, one with activity against *Alternaria brassicola*, one with activity against *Botrytis cinerea* and one with activity against both fungus (Leifert *et al.*, 1995). Biocontrol of *Rhizoctonia solani* damping-off of tomato by *Bacillus subtilis* RB14 involved two antibiotics, iturin and surfactin (Asaka and Shoda, 1996). Likewise, *Bacillus cereus* UM85 could suppress alfalfa damping-off due to the production of two fungistatic antibiotics (Silo-Suh *et al.*, 1994), one designated as zwittermicin A and second as antibiotic B.

4. LYTIC ENZYMES

Production of lytic enzymes like chitinases and B-1,3-glucanases by certain bacteria forms the basis of control of plant-pathogenic fungi in the rhizosphere (Mauch and

Staehelin, 1989). Recently there are few reports on the exploitation of lytic enzyme production trait for suppression of fungal diseases (Chernin *et al.*, 1995; Frandberg and Shnurer, 1994; Inbar and Chet, 1991; Lim and Kim, 1995). Inbar and Chet (1991) reported the involvement of chitinase enzyme produced by *Aeromonas caviae*, in the biological control of several pathogens. Frandberg and Schnurer (1994) isolated a chitinolytic *Bacillus pabuli* K 1 from heavily mould infested grain and reported the physiology of chitinase production in this isolate. Koby *et al.* (1994) reported the introduction of Tn7 based chiA gene into *P. fluorescens* and the construct could improve the biocontrol activity against *Rhizoctonia solani*. Three *Enterobacter agglomerans* strains produce and excrete proteins with chitinolytic activity and were able to decrease the incidence of disease caused by *Rhizoctonia solani* in cotton by 64 to 86 %. The study provides the most complete evidence for the presence of a complex chitinolytic enzymes in *Enterobacter* strains. Four enzymes were detected : two N-acetyl-beta-D-glucosaminidases of 89 and 67 KDa, and endochitinase with a mass of 59 KDa and a chitobasidase of 50 KDa (Chernin *et al.*, 1995).

Enzyme B 1,3 glucanase produced by *Pseudomonas stutzeri* YPL-1 could suppress *Fusarium solani* (Lim and Kim, 1995) whereas the one produced by *P. cepacia* could control *Rhizoctonia solani, Sclerotium rolfsii* and *Pythium ultimum* by 85, 48 and 71 % respectively (Fridlender *et al.*, 1993).

Production of hydrogen cyanide and ammonia has been reported as a mechanism of disease suppression by few bacteria (Howell *et al.*, 1988; Stutz *et al.*, 1986). *Bacillus* spp. produce a range of other metabolites including biosurfactants (Edwards and Seddon, 1992), volatiles (Fiddaman and Rossall, 1993, 1994) and compounds which elicit plant resistance mechanisms (Kehlenbeck *et al.*,1994).

5. RHIZOSPHERE COMPETENCE AND COLONIZATION

Root colonization is defined as the process whereby introduced bacteria become distributed along roots in non-sterile soil, multiply and survive for several weeks in the presence of indigenous soil micro biota (Weller, 1988). Root colonization involves two phases; phase I is the attachment to roots, and phase II is the multiplication on roots (Howie *et al.*,1987). Rhizosphere competence describes the relative root colonizing ability of a rhizobacteria (Weller and Thomashow, 1994).

Bacterial traits that are linked to poor rhizosphere competence are poorly understood (Bull *et al.*, 1991; Handelsman *et al.*,1990; Liu and Sinclair, 1992), but some that may be important are characterized into three classes. Class I involves cell surface polysaccharides, fimbriae, flagella and chemotaxis toward seed or root exudates. Duffy *et al.* (1997) reported the involvement of the outer membrane lipopolysaccharides in the endophyte colonization of tomato roots by biocontrol *P. fluorescens* strain WC S417. It has been reported that non motile isolates were poorer root colonisers than motile isolates (Misaghi *et al.*, 1992; Toyata and Ikeda,1997). Class II involves growth rate and ability to utilize complex carbohydrates. Class III involves osmotolerance, which is necessarily for survival. The ability of an organism to produce certain antibiotics have also been linked to its superior colonizing ability, Mazzola *et al.* (1992) reported that phenazine antibiotics production contributes to the ecological competence of *P. fluorescens* in the rhizosphere of wheat. However, for *P. fluorescens* MelRC2Rif, antibiotic deficient Tn5 mutant showed no loss of rhizosphreric competence suggesting little contribution of antibiotics production in the rhizospheric competence of strain. Similar results were obtained by Carroll *et al.* (1995) for *P. fluorescens* F113.

Even though the underlying mechanisms for rhizospheric competence are varied and poorly understood, several reports stress the importance of colonization of the rhizosphere for effective biocontrol with introduced bacteria (Pan *et al.,* 1997). Accordingly, the biocontrol agent should grow and persists, or colonize, the surface or endorhizosphere of plant it protects (Benhamous *et al.,* 1996; Forlani *et al.,* 1995; McInvoy and Kloepper, 1995; Quadt-Hallman *et al.,* 1997). Scanning EM observations showed that the cells of good colonizing isolates (B 16 and V 15) of fluorescent *Pseudomonas* were arranged linearly on the growing root axis and existed continuously on the root tip. Whereas the cells of poorly colonizing isolates were scattered at random on the root surface (Kang *et al.,*1997). In a study showing the suppression of damping off of peas by *Burkholderia cepacia,* there was a significant relationship between population size of the biocontrol agent and the degree of disease suppression (Parke,1990). Also suppression of take all of wheat and *Fusarium* wilt of radish was correlated with colonization of roots by *Pseudomonas* strains (Bull *et al.,*1991; Raaijmakers *et al.,* 1991). Pau *et al.* (1997) though ultrastructural studies demonstrated the colonization of banana tissue and the pathogenic fungus, *Fusarium oxysporum* f. sp. *cubense* race 4 by the endophytic bacterium *B. cepacia.* Berger *et al* (1996) conclusively showed strong relationship between rhizosphere colonization by biocontrol agent *Bacillus subtilis* cot 1 and disease suppression of *Phytopthora* and *Pythium* damping-'off of *Photimia* and *Brassica.* However, Roberts *et al.* (1997a) reported that root colonization deficient strains of *E. coli* S17R1 could suppress damping off of cucumber caused by *Pythium ultimum.* They concluded that extensive root colonization by *E. coli* was not required because of the brief period of susceptibility of cucumber to this disease.

6. CONCLUSION

Bacterial biocontrol agents have been used extensively both *in vivo* and *in vitro* to suppress the fungal pathogens and the plant diseases. Although numerous microorganisms capable of suppressing plant diseases have been identified only few biological agents have been commercialised, primarily due to their inconsistent performance (Weller, 1988). This has many causes, the major being variability in root colonization by introduced bacteria which in turn is influenced by survival rate in the environment, effects of environmental and edaphic factors on the live organisms and interactions with other microorganisms. Another factor that can contribute to inconsistent performance is variable production or inactivation *in situ* of bacterial metabolites responsible for disease control. Part of this lack of consistency results from the method(s) of selection of the biocontrol microorganisms, and in their spectrum of activity in relation to the races of the pathogen. Inconsistent performance also may result from the occurrence of diseases caused by other non target pathogens. Another important problem associated with biocontrol is the development of resistance in the target organisms as a result of continuous application of biocontrol agent (Li and Leifert, 1994).

To overcome these inconsistencies in the results, approach towards use and selection of biocontrol agent has to be changed. One approach could be to use multiple microbial strains with different modes of action. This would increase the genetic diversity and will help in controlling different races of the same pathogen, non-target pathogens and better establishment under different edaphic and environmental conditions. Another approach could be to use genetically modified microorganisms. Two or three traits involved in biocontrol like production of antibiotic, siderophore and antifungal volatiles could be combined in an otherwise good colonizer. However, it requires extensive knowledge of the exact mode of action of a biocontrol agent.

Approach of integrated biological and chemical control has yielded good results for certain fungal pathogens (Hwang, 1994; Krebs *et al.*,1993). However, it needs judicious selection of the two control agents as it always did not affect the efficacy of disease control (Mathre *et al.*,1995).

The practice of biological control of diseases has come a long way. Several potent biocontrol agents have been developed and tested both under laboratory and field conditions. However, still the full potential of biocontrol has not been realised and practiced and needs a concerted effort from pathologists, microbiologists, agronomists and extension scientists.

REFERENCES

Anith, K.N. 1997, Molecular basis of antifungal activity of a fluorescent pseudomonad, Ph.D. Thesis, Indian Agricultural Research Institute, New Delhi.

Asaka, O. and Shoda, M. 1996, Biocontrol of *Rhizoctonia solani* damping-off of tomato with *Bacillus subtilis* RB 14, *Appl. Environ. Microbiol.* 62 : 4081-4085.

Barbosa,M.A.G., Michereff,S.J.; Mariano, R.L.R. and Maranhao, E. 1995, Biocontrol of *Rhizoctonia solani* in cowpea by seed treatment with fluorescent *Pseudomonas* spp., *Summa-Phytopath.* 21 : 151-157.

Benhamou, N.,Belanger, R.R. and Paulitz, T. 1996, Ultrastructural and cytochemical aspects of the interactions between host and parasite in time and space, In: *Vascular Diseases of Plants*, eds. E.C. Tjamos and C.H. Beckman, Spinger Verlag, Heidelberg, Berlin, pp. 19-32.

Berg, G. and Bahl, H. 1997, Characterization of beneficial rhizobacteria of oilseed rape for biological control of *Verticillium* wilt, *Gesunde Pflanzen* 49 : 76-82.

Berger, F., Hong, Li, White, D., Frazer, R. and Leifert, C. 1996, Effect of pathogen inoculum, antagonist density, and plant species on biological control of *Phytopthora* and *Pythium* damping-off by *Bacillus subtilis* Cot 1 in high humidity, frogging glasshouses, *Phytopath.* 86 : 428-433.

Boswell, V.R. 1965. A landmark in biology, In : *Ecology of Soil-borne Plant Pathogens*, eds. K.F. Baker and W.C. Snyder, Prelude to Biological Control, Univ. of California Press, Berkeley, California, p.3.

Brannen, P.M. 1995, Potential modes of action for suppression of root diseases and yield enhancement when using *Bacillus subtilis* seed inoculants on cotton, In : *Proc. Beltwide Conf.,* 1995, San Antonio, TX, USA, Vol. I , pp. 205-228.

Broadbent, P., Baker, K.F. and Waterworth, Y. 1971, Bacteria and actinomycetes antagonistic to fungal root pathogens in Australian soils, *Aust. J. Biol. Sci.* 24 : 975.

Bull,C.T., Weller, D.M. and Thomashow, L.S. 1991, Relationship between root colonization and suppression of *Gaeumannomyus graminis* var. *tritici* by *Pseudomonas fluorescens* strain 2-79, *Phytopath.* 81: 954-959.

Burkhead, K.D., Schisler, D.A. and Slininger, P.J. 1994, Pyrrolnitrin production by biological control agent *Pseudomonas cepacia* B 37w in culture and in colonized wounds of potatoes, *Appl. Environ. Microbiol.* 60 : 2031-2039.

Burr, T.J., Schroth, M.N. and Suslow, T. 1978, Increased potato yields by treatment of seed pieces with specific strains of *Pseudomonas* strains, *J. Biol. Chem.* 261 : 791-794.

Campbell, R. 1989, Biological control of microbial plant pathogens, Cambridge University Press, Cambridge, pp. 218.

Carrol, H., Moenne-Laccoz, Y., Dowling, D.N. and O'Gara, F. 1995, Mutational disruption of the biosynthesis genes coding for the antifungal metabolite 2,4-Diacetyl phloroglucinol does not influence the ecological fitness of *Pseudomonas fluorescens* F 113 in the rhizosphere of sugarbeets, *Appl. Environ. Microbiol.* 61 : 3002-3007.

Carruthers, F.L., Conner, A.J. and Mahanty, H.K. 1994, Identification of a genetic locus in *Pseudomonas aureofaciens* involved in fungal inhibition, *Appl. Environ. Microbiol.* 60 :71-77.

Carruthers, F.L., Shum-Thomas, T., Conner, A.J. and Mahanty, H.K. 1995, The significance of antibiotic production by *Pseudomonas aureofaciens* PA 147-2 for biological control of *Phytophthora megasperma* root rot of asparagus, *Plant Soil* 170 : 339-344.

Cartwright, D.K., Chiton, W.S. and Benson, D.M. 1995, Pyrrolnitriin and Phenazine production by *Pseudomonas cepacia,* strain 5.5B, a biocontrol agent of *Rhizoctonia solani, Appl. Microbiol. Biotechnol.* 433: 211-216.

Chakraborty, U. and Purkayastha, R.P. 1984, Role of rhizobiotoxine in protecting soybean roots from *Macrophomina phaseolina* infection, *Can. J. Microbiol.* 30 : 285-289.

Chen, C., Bauske, E.M.,Musson, G., Rodriguez-Kabana, R. and Kloepper, J.W. 1995, Biological control of *Fusarium* wilt on cotton by use of endophytic bacteria, *Biol. Cont.* 5 : 83-91.

Chernin, L, Ismailov, Z., Havan, S. and Chet, I. 1995, Citinolytic *Enterobacter agglomerans* antagonistic to fungal plant pathogens, *Appl. Environ. Microbiol.* 61 : 1720-1726.

Cook, R.J. 1993, Making greater use of introduced microorganisms for biological control of plant pathogens, *Annu. Rev. Phytopath.* 31 : 53-80.

Costa, J.M. and Loper, J.E. 1994, Characterisation of siderophore production by the biological control agent *Enterobacter cloacae*, *Mol. Plant-Microbe Interact.* 7 : 440-448.

Dowling, D.N. and O'Gara, F. 1994, Metabolites of *Pseudomonas* involved in the biocontrol of plant diseases, *TIBTECH* 12 : 133-141.

Defago, G. 1993, 2,4-Diacetylphloroglucinol, a promising compound in biocontrol, *Pl. Path.* 42 : 311-312.

Duijff, B.J., Gianinazzi-Pearson, V. and Lemanceau, P. 1997, Involvement of the outer membrane lipolysacharides in the endophytic colonization of tomato roots by biocontrol *Pseudomonas fluorescens* strain WCS 417r, *New Phytol.* 135 : 325-334.

Duffy, B.K., Simon, A. and Weller, D.M. 1996, Combination of *Trichoderma koningii* with fluorescent *Pseudomonads* for control of take-all on wheat, *Phytopath.* 86 : 188-194.

Duffy, B.K. and Weller, D.M. 1996, Biological control of take-all of wheat in the pacific northwest of the USA using hypovirulent *Gaeumanomyces graminis* var. *tritici* and fluorescent Pseudomonads, *J. Phytopath.* 144 : 585-590.

Edwards, S.G. and Seddon, B. 1992, *Bacillus brevis* as a biocontrol agent against *Botrytis cinerea* on protected chinese cabbage, In : *Recent Advances in Botrytis Research*, eds. K. Verhoeff, N.E. Malathrakis and B. Williamson, Pudoc Scientific Publishers, Wageningen, pp. 267-271.

Expert, J.M. and Digat, B. 1995, Biocontrol of *Sclerotinia* wilt of sunflower by *Pseudomonas fluorescens* and *Pseudomonas putida* strains, *Can. J. Microbiol.* 41 : 685-691.

Fiddaman, P.J. and Rossall, S. 1993, The production of antifungal volatiles by *Bacillus subtilis*, *J. Appl. Bacteriol.* 74: 119-126.

Fiddaman, P.J. and Rossall, S. 1994, Effect of substrate on the production of antifungal volatiles from *Bacillus subtilis*, *J. Appl. Bacteriol.* 76 : 395-405.

Fiddaman, P.J. and Rossall, S.1995, Selection of bacterial antagonists for the biological control of *Rhizoctonia solani* in oilseed rape (*Brassica napus*), *Pl. Path.* 44 : 695-703.

Forlani,G., Pastorelli, R., Branzoni, M. and Favilli, F. 1995, Root colonization efficacy, plant growth promoting activity and potentially related properties in plant-associated bacteria, *J. Genet. Breed.* 49 : 343-352.

Frandberg, E. and Schnurer, 1994, Chitinolytic properties of *Bacillus pabuli* K 1, *J. Appl. Bacteriol.* 76 : 361-367.

Fravel, D.R., Lumsden, R.D. and Roberts, D.P. 1990, *In situ* visualization of the biocontrol rhizobacterium *Enterobacter cloacae* with bioluminescence, *Plant Soil* 125 : 233-238.

Fridlender, M., Inbar, J. and Chet, I. 1993, Biological control of soil-borne plant pathogens by a beta-1,3 glucanase producing *Pseudomonas cepacia*, *Soil Biol. Biochem.* 25 : 1211-1221.

Fujimoto, D.K., Weller,D.M. and Thomashow, L.S. 1995, Role of secondary metabolites in root disease suppression, In : *Allelopathy : Organisms, Process and Applications*, American Chemical Society, Washington, D.C. pp.330-347.

Gomes, A.M.A., Peixoto, A.R., Mariano, R.L.R. and Michereff, S.J. 1996, Effect of bean seed treatment with fluorescent *Pseudomonas* spp. on *Rhizoctonia solani* control, *Arquivos-de-Biologiae-Technologia* 39 : 537-545.

Handelsman, J., Nesmith, W.C. and Raffel, S.J. 1991, Microassay for biological and chemical control of infection of tobacco by *Phytopthora parasitica* var. *nicotianae*, *Curr. Microbiol.* 22: 317-319.

Handelsman, J, Raffel, S., Mester, E.H., Wunderlich, L. and Grau, C.R. 1990, Biological control of damping-off of alfalfa seedling by *Bacilllus cereus* UW85, *Appl. Environ Microbiol.* 56 : 713-718.

Hebbar, K.P., Davery, A.G. and Dart, P.J. 1992, Rhizobacteria of maize antagonistic to *Fusarium moniliforme*, a soil-borne fungal pathogen : isolation and identification, *Soil Biol. Biochem.* 24 : 979-987.

Hill, D.S., Stein, J.I., Torkewitz, N.R., Morse, A.M., Howell, C.R., Pachlatko, J.D. and Ligon, J.M. 1994, Cloning of genes involved in the synthesis of pyrrolnitrin from *Pseudomonas fluorescens* and role of pyrrolnitrin synthesis in biological control of plant disease, *Appl. Environ. Microbiol.* 60 : 78-85.

Howell,C.R., Beier,R.C. and Stipanovic, R.D. 1988, Production of ammonia by *Enterobacter cloacae* and its possible role in the biological control of *Pythium* pre-emergence damping-off by the bacterium, *Phytopath.* 78 : 1075 - 1078.

Howie, W.J., Cook, R.J. and Weller, D.M. 1987, Effects of soil matric potential and cell motility on wheat root colonization by fluorescent pseudomonads suppressive to take-all, *Phytopath.* 77 : 286-292.

34

Hwang, S.F. 1994, Potential for integerated biological and chemical control of seedling rot and pre-emergence damping-off caused by *Fusarium avenaceum* in lentil with *Bacillus subtilis* and Vitaflo (R)-280, *Zeitschrift fur pflanzenkrankheiten und Pflanzenschutz* (Germany) 101 : 188-199.

Inbar, J and Chet, I. 1991, Evidence that chitinase produced by *Aeromonas caviae* is involved in the biological control of soil borne plant pathogens by this bacterium, *Soil Biol. Biochem.* 23 : 973-978.

Kang, J., Park, C., Kang, J.H. and Park, C.S. 1997, Colonizing pattern of fluorescent pseudomonads on the cucumber seed and rhizoplane, *Korean J. Plant Pathol.* 13 : 160-166.

Keel, C., Schnider, U., Maurhofer, M., Voisard, C., Laville, J., Burger, U.,Wirthner, P., Haas, D. and Defago, G. 1992, Suppression of root diseases by *Paeudomonas fluorescens* CHAO : Importance of the bacterial secondary metabolite 2,4-diacetylphloroglucinol, *Mol. Plant-Microbe Interact.* 5 : 4-13.

Kehlenbeck, H., Krone, C., Oerke, E.C. and Schonbeck, F. 1994, The effectiveness of induced resistance on yield of mildewed barley, *J. Plant Dis. Prot.* 101 : 11-21.

Kerr, A. 1972, Biological contol of crown gall: seed inoculation, *J. Appl. Bacterol.* 35 : 493-497.

Kerr, A. 1980, Biological control of crown gall through the production of agrocin 84, *Plant Dis.* 64 : 25-30.

Kim, Y.S., Son, J.K., Moon, D.C. and Kim, S.D. 1997, Isolation and structure determination of antifungal antibiotics from *Bacillus subtilis* YB-70, a powerful biocontrol agents, *Korean J. Appl. Microbiol. Biotechnol.* 25 : 62-67.

King, E.B. and Parke, J.L. 1993, Biocontrol of *Aphanomyces* root and *Pythium* damping-off by *Pseudomonads cepacia* AMMD on four pea cultivars, *Plant Dis.* 77 : 1185-1188.

Kloepper, J.W., Leong, J., Teintze, M. and Scroth, M.N. 1980, Enhanced plant growth by siderophores produced by plant growth-promoting rhizobacteria, *Nature* (London) 286 : 885-886.

Kloepper, J.W. and Schroth, M.N. 1978, Plant growth promoting rhizobacteria on radish, In : *Station de Pathologie Vegetable et Phytobacteriologie*, Angers, Proc. *4th Int. Conf. Plant Pathogenic Bacteria*, ed. INRA, Vol.2, pp.879-882.

Koby, S., Schickler, H., Chet, I and Oppenheim, A.B. 1994, The chitinase encoding Tn7-based chi A gene endows *Pseudomonas fluorescens* with the capacity to control plant pathogens in soil, *Gene* 147 : 81-83.

Leifert, C., Li, H., Chid buree, S., Hampson, S.,Workman, S., Sigee, D., Epton, H.A.S. and Harbour, A. 1995, Antibiotic production and biocontrol activity by *Bacillus subtilis* CL 27 and *Bacillus pumilus* CL45, *J. Appl. Bacterol.* 78 : 97-108.

Leeman, M., Pelt, J.A-van, Hendrickx, M. J. and Scheffer, R.J. 1995, Biocontrol of *Fusarium* wilt of radish in commercial greenhouse trials by seed treatment with *Pseudomonas fluorescens* WCS374, *Phytopath.* 85 : 1301-1305.

Li, H. and Leifert, C. 1994, Development of resistance in *Botryotinia fuckeliana* (de Barry) whetzel (*Botrytis cinerea*) against the biological control agent *Bacillus subtilis* CL27, *Zeitschrift fuer pflanzenkrankheiten und pflanzenschutz*, 101 : 4l4-418.

Lim, H.S. and Kim,S.D. 1997, Role of siderophores in biocontrol of *Fusarium solani* and enhanced growth response of bean by *Pseudomonas fluorescens* GL20, *Korean J. Microbiol. Biotechnol.* 7 : 13-20.

Lindow, S.E. 1983, Methods of preventing frost injury caused by epiphytic ice nucleation active bacteria, *Plant Dis.* 67 : 327-333.

Liu, Z.L. and Sinclair, J.B. 1992, Population dynamics of *Bacillus megaterium* strain B 153-2-2 in the rhizosphere of soybean, *Phytopath.* 82 : 1297-1301.

Martinetti, G. and Loper, J.E. 1992, Mutational analysis of gene determining antagonism of *Alcaligenes* sp. strain MFA 1 against the phytopathogenic fungus *Fusarium oxysporum*, *Can. J. Microbiol.* 38 : 241-247.

Mathre, D.E., Johnston. R.H., Callan, N.W., Mohan, S.K., Martin, J.M. and Miller, J.B. 1995, Combined biological and chemical seed treatments for control of two seedling diseases of sh2 sweet corn, *Plant Dis.* 79 : 1145-1148.

Mauch, F. and Staehelin, L.A. 1989, Functional implications of the subcellular localization of ethylene-induced chitinase and β-l,3, glucanase in bean leaves, *Plant Cell* 1 : 447-457.

Maurhofer, M., Keel, C., Haas, D. and Defago, G. 1995, Influence of plant species on disease suppression by *Pseudomonas fluorescens* strain CHAO with enhanced antibiotic production, *Pl. Path.* 44 : 40-50.

Maurhofer, M., Keel, C., Schnider, U., Voisard, C., Haas, D. and Defago, G. 1992, Influence of enhanced antibiotic production in *Pseudomonas fluorescens* strain CHAO on its disease suppressive capacity, *Phytopath.* 82 : 190-195.

Mazzola, M. and Cook, R.J. 1991, Effects of fungal root pathogens on the population dynamics of biocontrol strains of fluorescent pseudomonads in the wheat rhizosphere, *Appl. Environ. Microbiol.* 57: 2171-2178.

Mazzola, M., Cook, R.J., Thomashow, L.S., Weller, D.M and Pierson, L.S. 1992, Contribution of phenazine antibiotic synthesis in the ecological competence of fluorescent pseudomonads in soil habitats, *Appl. Environ. Microbiol.* 58 : 2616-2624.

Mazzola, M., Fujimoto, D.K. and Cook, R.J. 1994, Differential sensitivity of *Gaeumannomyces graminis* populations to antibiotics produced by biocontrol fluorescent pseudomonads, *Phytopath.* 84 : 1091.

Mazzola, M., Fujimoto, D.K., Thomashow, L.S. and Cook, R.J. 1995, Variation in sensitivity of *Gaeumannomyces graminis* to antibiotics produced by fluorescent pseudomonads spp. and effect on boilogical control of take-all disease, *Appl. Environ. Microbiol.* 61 : 2554-2559.

McInroy, J.A. and Kloepper, J.W. 1995, Population dynamics of endophytic bacteria in field grown sweet corn and cotton, *Can. J. Microbiol.* 41 : 895-901.

McLoughlin, T.J., Quinn, J.P., Bettermann, A. and Bookland, R. 1992, *Pseudomonas cepacia* suppression of sunflower wilt fungus and role of antifungal compounds in controlling the disease, *Appl. Environ. Microbiol.* 58 : 1760-1763.

Merriman, P.R., Price, R.D., Kollmorgan, J.F., Piggott, T. and Ridge, E.H. 1974, Effect of seed inoculation with *Bacillus subtilis* and *Streptomyces griseus* on growth of cereals and carrots, *Aust. J. Agric. Res.* 25 : 219-226.

Meshram, S.U. and Jager, G. 1983, Antagonism of *Azotobacter chroococcum* isolates to *Rhizoctonia solani, Neth. J. Plant Pathol.* 89 :191-192.

Michereff, S.J., Silveira, N.S.S. and Mariano, R.L.R. 1994a, Antagonism of bacteria to *Colletotrichum graminicola* and potential for biocontrol of sorghum anthracnose, *Fitopatologia Brasileira* 19 : 541-545.

Michereff, S.J., Silveira, N.S.S. and Mariano, R.L.R. 1994b, Epiphytic bacteria antagonistic to *Curvularia* leaf spot of yam, *Microbial Ecol.* 28 : 101-110.

Misaghi, I.L., Olsen, M.W., Billotte, J.M. and Sonoda, R.M. 1992, The importance of rhizobacterial motility in biocontrol of bacterial wilt of tomato, *Soil Biol.Biochem.* 24 : 287-293.

Mitchell, R. and Hurwitz, E. 1965, Suppression of *Pythium debaryanum* by lytic rhizospheric bacteria, *Phytopath.* 55 : 156-158.

Moulin, F., Lemanceau, P. and Alabouvette, C. 1994, Control by fluorescent pseudomonads of *Pythium aphanidermatum* root rot responsible for yield reduction in soilless culture of cucumber, In : *Improving Plant Productivity with Rhizosphere Bacteria*, eds. M.H. Ryder, P.M. Stephens and G.D. Bowen, CSIRO, Adelaide, pp. 47-50.

Moulin, F., Lemanceau, P. and Alabouvette, C.1996, Suppression of *Pythium* root rot of cucumber by a fluorescent pseudomonad is related to reduced root colonization by *Pythium aphanidermatum, J. Phytopath.* 144 : 125-129.

Nautiyal, C.S. 1997, Selection of chickpea-rhizosphere-competent *Pseudomonas fluorescens* NBRI 1303 antagonistic to *Fusarium oxysporum* f.sp. *ciceris, Rhizoctonia bataticola* and *Pythium* sp., *Curr. Microbiol.* 35 : 52-58.

Nelson, E.B. 1988, Biological control of *Pythium* seed rot and pre-emergence damping-off of cotton with *Enterobacter cloacae* and *Erwinia herbicola* applied as seed treatments, *Plant Dis.* 72 : 140-142.

O'Sulivan, D.J. and O'Gara, F. 1992, Traits of fluorescent *Pseudomonas* spp. involved in suppression of plant root pathogens, *Microbiol. Rev.* 56 : 662-676.

Pal, K.K. 1995, Rhizobacteria as biological control agents for soil borne plant pathogens, Ph.D. Thesis, Indian Agricultral Reseaech Institute, New Delhi, India.

Pan, M.J., Rademan, S., Kunert, K. and Hastings, J.W. 1997, Ultrastructural studies on the colonization of banana tissue and *Fusarium oxysporum* f. sp. *cubense* Race 4 by the endophytic bacterium *Burkholderia cepacia, J. Phytopath.* 145 :479-486.

Parke, J.L. 1990, Population dynamics of *Pseudomonas cepacia* in the pea spermosphere in relation to biocontrol of *Pythium, Phytopath.* 80 : 1307-1311.

Paulitz, T.C., Anas, O. and Frenando, D.G. 1992, Biological control of *Pythium* damping-off by seed-treatment with *Pseudomonas putida* : relationship with ethanol production by pea and soybean seeds, *Bio. Sc. Tech.* 2 : 193-201.

Phipps, P.M. 1992, Evaluation of biological agents for control of *Sclerotinia* blight of peanut, 1991, Biol. cultural tests control, *Plant Dis.* 7 : 60.

Pierson, E.A. and Weller, D.M. 1994, Use of mixtures of flourescent pseudomonads to suppress Take-all and improve the growth of wheat, *Phytopath.* 84 : 940-947.

Powell, K.A. 1993, The commercial exploitation of microorganisms in agriculture, In : *Exploitation of Microorganisms*, D.G. ed. Jones, (Ist edition), Chapman and Hall, London, U.K. pp.442-459.

Quadt-Hallmann, A., Hallman, J. and Kloepper, J.W. 1997, Baterial endophytes in cotton :location and interaction with other plant-associated bacteria, *Can. J. Microbiol.* 43 : 254-259.

Raaijmakers, J.M., Leeman, M., Van Oorschot, M.M.P., Van der Sluis, I., Shippers, B. and Baker, P.A.H.M. 1995, Dose-response relationship in biological control of *Fusarium* wilt of radish by *Pseudomonas* spp., *Phytopath.* 85 : 1075-1081.

36

Reis, A., Silveira, N.S.S., Michereff, S.J., Pereira, G.F.A. and Mariana, R.L.R. 1994, *Bacillus subtilis* as a potential biocontrol agent of the northern leaf blight of corn, *Revista de Microbiologia* 25 : 255-260.

Roberts, D.P., Dery, P.D., Hebbar, P.K., Mao, W. and Lumsden, R.D. 1997a, Biological control of damping-off of cucumber caused by *Pythium ultimum* with a root-colonization-deficient strain of *E. coli*, *J. Phytopath.* 145 : 383-388.

Roberts, D.P., Dery, P.D., Mao, W. and Hebbar, P.K. 1997b, Use of a colonization deficient strain of *Esherichia coli* in strain combinations for enhanced biocontrol of cucumber seedling diseases, *J. Phytopath.* 145 : 461-463.

Salt, G.A. 1979, The increasing interest in minor pathogens, In : *Soil-borne Plant Pathogens*, eds. B. Schippers and W. Gams, Academic Press, London, pp. 289-312.

Sariah, M. 1995, Potential of *Bacillus* spp. as a biocontrol agent for anthracnose fruit rot of chilli, *Malaysian Appl. Biol.* (Malaysia) 23 : 53-60.

Shanahan, P., O'Sullivan, D.J., Simpson, P., Glennon, J. and O'Gara, F. 1992, Isolation of 2,4-diacetyl-phloroglucinol from a fluroscent pseudomonad and investigation of physiological parameters influencing its prodction, *Appl. Environ. Microbiol.* 58 : 353-358.

Sharga, B.M. 1997, *Bacillus* isolates as potential biocontrol agents against chocolate spot on faba beans, *Can. J. Microbiol.* 43 : 915-924.

Sholberg, P.L., Marchi, A. and Bechard, J. 1995, Biocontrol of post harvest diseases of apple using *Bacillus* spp. isolated from stored apples, *Can. J. Microbiol.* 41 : 247-252.

Silo-Suh, I.A., Lethbridge, B.J., Raffel, S.J., Clardy, H., He., J. and Handelsman, J. 1994, Biological activities of two fungistatic antibiotics produced by *Bacillus cereus* UW85, *Appl. Environ. Microbiol.* 60 : 2023-2030.

Smith, K.P., Havey, M.J. and Handelsman, J. 1993, Suppression of cottony leak of cucumber with *Bacillus cereus* strain UW85, *Plant Dis.* 77 : 139-142.

Stutz, E.M., Defago, G. and Kerri, H. 1986, Naturally occurring fluorescent pseudomonads involved in suppression of black root rot of tobacco, *Phytopath.* 76 : 181-185.

Tu, J.C. 1980, Incidence of root rot and over wintering of alfalfa as influenced by rhizobia, *Phytopathol. Z.* 97 : 97-108.

Turner, J.T. Jr. 1987, Relationship among plant growth, yield and rhizosphere ecology of peanuts as affected by seed treatment with *Bacillus subtilis*, Ph.D. Dissertation, Auburn Univ., p.108.

Thara, K.V. and Gnanamanickam, S.S. 1994, Biological control of rice sheath blight in India:lack of correlation between chitinase production by bacterial antagonists and sheath blight suppression, *Plant Soil* 160 : 277-280.

Thomson, J.A. 1987, The use of agrocin-producing bacteria in the biological control of crown gall, In: *Innovative Approach to Plant Disease Control*, ed. I. Chet, John Wiley and Sons, New York, USA, pp. 213-228.

Tosi, L. and Zazzerini, A. 1994, Evaluation of some fungi and bacteria for potential control of safflower rust, *J. Phytopath.* 142 : 131-140.

Toyota, K. and Ikeda, K. 1997, Relative importance of motility and antibiosis in the rhizoplane competence of a biocontrol agent *Pseudononas fluorescens* Mel RC2Rif, *Biol. Fertil. Soils* 25 : 416-420.

Vincent, M.N., Harrison, L.A., Brackin,J.M., Kovacevich, P.A., Mukerji, P., Weller, D.M. and Pierson, E.A. 1991, Genetic analysis of the antifungal activity of a soil borne *Pseudomonas aureofaciens* strain, *Appl. Environ. Microbiol.* 57 : 2928-2934.

Walker, R., Emslie, K.A. and Allan, E.J. 1996, Bioassay methods for the detection of antifungal activity by *Pseudomonas antimicrobica* against the grey mould pathogen *Botrytis cinerea, J. Appl. Biotechnol.* 81 : 531-537.

Wei, G., Kloepper, J.W. and Tuzun, S. 1996, Induced systemic resistance to cucumber diseases and increased plant growth by plant growth promoting rhizobacteria under field conditions, *Phytopath.* 86 : 221-224.

Weller, M. 1988, Biological control of soilborne plant pathogens in the rhizosphere with bacteria, *Annu. Rev. Phytopathol.* 26 : 379-407.

Weller, D.M. and Thomashow, L.S. 1994, Current challenges in introducing beneficial microorganisms into the rhizosphere, In : *Molecular Ecology of Rhizosphere Microorganisms*, eds. F. O'Gara, D.N. Dowling and B. Boesten, VCH, Weinheim, pp. 1-18.

Weststeijn, E.A. 1990, Fluorescent *Pseudomonas* isolate E 11.3 as biocontrol agent for *Pythium* root rot in tulip, *Neth. J. Plant Pathol.* 96 : 261-272.

Wu-Ws, 1995, Biocontrol of seedling blight of cucumber and poinsettia, *Pl. Path. Bull.* 4 : 97-105.

Zhang Jin Xu, Howell, C.R. and Starr, J.L. 1996, Suppression of *Fusarium* colonization of cotton roots and *Fusarium* wilt by seed treatments with *Gliocladium virens* and *Bacillus subtilis, Biocont. Sci. Technol.* (UK) 6 : 175-187.

EXPLOITATION OF PROTOPLAST FUSION TECHNOLOGY IN IMPROVING BIOCONTROL POTENTIAL

Sumeet and K.G. Mukerji

Applied Mycology Laborarory
Department of Botany
University of Delhi, Delhi - 110 007, INDIA

1.   INTRODUCTION

There is world wide swing to the use of ecologically safe and environmentally friendly methods of protecting crops from pests and pathogens. Control methods using pesticides are not the long term solutions to the crop plants because single application of pesticides is seldom sufficient therefore repeated applications are required to combat the pathogen. Such an approach demands huge economic cost and has adverse environmental impacts. Continuous applications of chemical pesticides may lead to development of resistant strains of pathogens. Biocontrol which depends on the activity of naturally occurring organisms offers an alternate method to check the pathogen and answer to many persistant problems in agriculture including problems of resource limitation, non-sustainable agriculture system and over-reliance on pesticides (Bagyaraj and Gonvindan 1996; Cook and Baker 1983).

Presence of any antagonistic organisms in rhizosphere and rhizoplane will influence the pathogen adversely and check its abundance. This makes biological control an attractive proposition for controlling plant diseases. Moreover, biocontrol agents can protect the newly formed plant parts to which they are not initially applied (Harman, 1990; Harman *et al.,* 1989).

Despite these beneficial effects, biocontrol methods using beneficial microbes received little or no approval from users as their ability to protect plant parts have been less effective and more variable to the results obtained by chemical pesticides. A principal reason for this is the poor growth of bioprotectants. Competitive microflora that rapidly colonize planted seeds may inhibit the bioprotectant leading to the failure of biocontrol (Harman, 1990; Harman *et al.,*1989; Hubbard *et al.,* 1983). For becoming an important component of plant disease management system, it must be effective and reliable as competitive chemical pesticides. For this reason biocontrol agents must be manipulated and improved for their effective use in the control of plant diseases.

However, the last few years witnessed a dramatic efforts on biocontrol of plant diseases (Nigam *et al.*, 1997). Among the wide variety of techniques used in the improvement of biocontrol agents namely mutation, hybridization, protoplast fusion, transformation; protoplast fusion seems to be an efficient way to induce genetic recombination in whole genomes, even between incompatible strains. This chapter emphasise the role of protoplast fusion in enhancing biocontrol potential.

2. BIOLOGICAL CONTROL

Presently, the control of plant diseases can be done with pesticides but due to development of resistant pathogenic strains, some of them loose their effectiveness against plant pathogens. Therefore, alternate methods of controlling plant diseases are being always explored. Biological methods mainly consist of using microorganism to control harmful microorganisms (by biological destruction) causing plant diseases without disturbing the ecological balance (Mukhopodhyay, 1994). Cook and Baker (1983) defined biocontrol as reduction in the pathogen inoculum or its disease producing capacity by action of one or more organisms accomplished naturally or through manipulation of the environment, host or antagonist or by mass introduction of one or more antagonist.

The main purpose of biocontrol of plant diseases is to suppress the inoculum load of the target pathogen below the level that potentially causes economically significant outbreak of disease (Kumar *et al.*, 1997; Mukerji *et al.*, 1999). Biocontrol aimed directly at the pathogen mediated through adjustments in the host offers unlimited opportunity to reduce losses caused by biotic and abiotic stresses. Biocontrol of plant pathogen seeks a solution in terms of restoring and maintaining the biological balance within the ecosystem and offers a powerful means to improve the health and hence the productivity of plants increased by suppression or destruction of pathogen incoulum, protection of plants against infection or increasing the ability of plants to resist pathogens (Upadhyay *et al.*, 1996).

Basic aspects of strain selection, efficient production of biomass, formulation, storage ability, method of application are some of the main obstacles in the use of biocontrol agents (Kumar *et al.*, 1997). Further, the biotechnological advances in improving effectiveness of biocontrol agents i.e. to create genetically superior strains than the wild type selected from the environment needs to be examined (Lumsden and Lewis, 1989; Upadhyay *et al.*, 1996, 1997, 1998a,b,c). For successful biocontrol of diseases, the necessity to identify the predominant microbial strain which is highly effective against plant pathogenic fungi must be isolated and employed in time to antagonise pathogen thereby preventing infection.

2.1. Biocontrol Agents

Biocontrol agents or antagonists are microorganisms with potential to interfere with the growth or survival of plant pathogens and thereby contribute to biological control. Potential agents for biocontrol activity are rhizosphere competent fungi and bacteria which in addition to their antagonistic properties are capable of inducing growth responses by either controlling pathogens or producing growth stimulating factors (Chet *et al.*,1993). The management of biocontrol agents is much safer as well as its usage is presumed to be less polluting to the environment than chemical pesticides. A wide range of organisms can be employed for the biological control.

40

An ideal biocontrol agent should possess number of characteristics, *viz.*:
(i) high rate of survival in soil in either active or passive form for prolonged period of time.
(ii) high probability of contacting the pathogen
(iii) active under required environmental conditions
(iv) mass multiplication should be simple and inexpensive
(v) should be efficient and economical
(vi) should not be a health hazard

Although large number of biocontrol agents have been reported but much emphasis is given on fungal biocontrol agents. Cook and Baker (1983) listed 44 microorganisms out of which 25 being fungi. Fungal biocontrol agents represent an excellent reservoir of microorganism with biocontrol ability including species of *Aspergillus, Gliocladium, Paecilomyces, Penicillium, Pythium, Talaromyces* and *Trichoderma.*

2.2. Mechanism of Action of Biocontrol Agents

Biocontrol agents of plant diseases are termed as 'Antagonists'. Antagonist is a microorganism that adversely effect another microorganism by utilizing different modes of action *viz.* competition, mycoparasitism and antibiosis. All these mechanisms may operate independently or together and their activities can result in suppression of plant pathogens (Nigam *et al.*, 1997; Singh and Faull, 1988). In order to improve the reliability of biocontrol it will be necessary to understand the mechanisms of antagonism and those environmental factor which effect changes (Renwick and Poole, 1989).

2.2.1. Competition

Biocontrol agents competes for nutrition and space with pathogen. It is an injurious effect of one organism to another because of the utilization or removal of some resources of the environment thereby determining the growth and infection of soil plant pathogens in competition with other microorganisms.

The rhizosphere is of major concern where competition for space and nutrients occurs. The successful rhizosphere biocontrol agent has the ability to maintain a high population on the root surface thereby provide protection to the whole root for the duration of its life. Rhizosphere competent isolate of *Trichoderma* spp. have been obtained by mutagenesis and a correlation between rhizosphere competence and its ability to utilize cellulose substrate associated with the root has been observed (Ahmad and Baker, 1987; Lewis *et al.,* 1989). Competition for space occurs naturally external to the root, particularly in regions enriched by organic amendments. In some soils, this mechanism helps in suppression of pathogen from causing disease, e.g., a greater biomass present in the Chateaurenard soil in France led to greater nutrient competition and consequent inhibition of *Fusarium* spp. (Alabouvette *et al.,* 1986). Specific antagonists showing combative behavior were also active in the suppressive soil but the relative importance of mechanism was not clear.

2.2.2. Antibiosis

Antibiosis is required as one of the important attribute in deciding the competitive saprophytic ability of the fungus. Antibiotics have been isolated from many antagonistic microbes and have often been indirectly implicated in disease control. Antibiotic antibiosis was first reported by Weindling (1934) and Weindling and Emerson (1936). The production of antibiotics required large amount of carbon and other environmental factors that affects its metabolism (Baker, 1968; Nigam, 1997; Thomashow and Weller, 1991; Williams, 1982). Sometimes antibiotics produced by biocontrol agents are inactivated by soil (Howell and Stipanovic, 1980; Renwick and Poole, 1989).

41

Species of *Gliocladium* and *Trichoderma* are well known biocontrol agents that produce a range of antibiotics which are active against pathogens in *in vitro* and consequently, antibiotic production has been suggested as a mode of action for these fungi. The production of volatile and non-volatile compounds by biocontrol agents has direct impact on pathogenic microorganisms resulting in denaturation of cell contents before coming in contract with the myceluim of antagonist. The production of volatile and non-volatile compounds by *Trichoderma* spp. and *Gliocladium* spp. inhibited the growth of broad range of fungi *viz.* species of *Fusarium, Rhizoctonia, Pythium, Phytophthora, Sclerotium* and *Sclerotinia.* These compounds influenced morphological or physiological sequences leading to successful penetration of the pathogen (Dennis and Webster, 1971a,b,c). Yong *et al.* (1985) demonstrated that the amount of volatiles produced by *T. viride* was dependent on the availability of N and C.

In general, the role of antibiotic production in biological control *in vivo* remains largely unproven (Fravel, 1988) but *in vitro* numerous agar plate tests have been developed to detect volatile and non-volatile antibiotic production by putative biocontrol agent and to quantify their effect on pathogen (Lewis *et al.*, 1989; Nelson and Powelson, 1988). Cultural conditions of microbes influenced the production of secondary metbolites and there is little evidence that antibiotics are produced in natural environment. *In vivo* the antibiotics may be rapidly degraded or bound to the substrates such as clay particles in soil preventing detection (Kumar *et al.*, 1997; Lumsden and Lewis, 1989).

2.2.3. Mycoparasitism

Mycoparasitism or hyperparasitism occurs when one fungus exists in intimate association with another from which it derives some or all of its nutrients while conferring no benefits in return (Lewis *et al.*, 1989). Biotrophic mycoparasites have a persistent contact with the living cells whereas necrotrophic mycoparasites kill the host cells, often in advance of contact or penetration. Several hyperparasites of soil-borne pathogens have been exploited in biocontrol. It is necrotrophic (destructive) mycoparasite which have qualified as potential biocontrol agent. The mechanism of hyperparasitism involve different kinds of interaction like coiling of hyphae around the pathogen, penetration , production of haustoria and lysis of hyphae (Nigam *et al.*, 1997). For instance, hyphae of *Trichoderma harzianum* have been found to grow towards hyphae of susceptible fungi before contact is made (Chet *et al.*, 1981), presumably due to chemical signals originating from the host. Subsequently excessive hyphal coiling and short branching on the host occurs, which in some cases may be lectin mediated (Barak *et al.*, 1985; Elad *et al.*, 1983a). Cytoplasmic degradation may occur before contact or penetration (De Oliveira *et al.*, 1984). But penetration takes place immediately, arising directly from parent hyphae or from appresoria, followed by cytoplasmic breakdown and host exploitation (Elad *et al.*, 1983b). Recent investigations have revealed that the mycoparasites like *Trichoderma* spp. produced cell wall degrading enzymes such as chitinase, β-1,3-D-glucanase and proteases which enable the penetration of host hyphae (Ridout *et al.*, 1988). All these biological processes are influenced by changes in environmental conditions resulting in changed biocontrol activity.

Practical application of mycoparasites for biological control of soil borne plant diseases has yet to be exploited. Here, as in all biological system, the physical and nutritional environment must be favourable for the agents to be active. In practice success is rare, but mycoparasitism has potential for eradication of pathogens and remains an attractive strategy in biocontrol.

3. IMPROVEMENT OF BIOCONTROL AGENTS

Biocontrol agents must be manipulated and improved in order to be effectively used for control of plant diseases. Potential fungal biocontrol agents have developed numerous mechanisms for activity and survival in their appropriate ecological niches. Only some of these characteristics may be considered to be beneficial in growth promotion and/or biocontrol, so the challenge is to manipulate the organism so that its performance matches more closely the purpose for which it is to be employed (Baker, 1989). Greater tolerance to environmental stress may increase the effectiveness of an organism and extend its useful range to new crop situations (Wilson and Pusey, 1985). Biocontrol of an organism can be improved by (i) altering the environment to make it more conducive to the biocontrol strain in question (ii) modifying the genetics of an organism to produce superior strain or both. The prime strategy for improvement is by genetic manipulation of biocontrol agents which can enhance their biocontrol activity and expand its spectrum (Harman and Hayes, 1993). It offers a possible approach for improving their potential for plant disease control (Harman and Stasz, 1991; Upadhyay and Rai, 1988).

Fungal biocontrol agents could acquire desirable traits by sexual recombination but in most of them sexual stages are rare or lacking, hence the genetic modification by classical sexual recombination has not been used to great extent in biocontrol for acquisition of desirable traits. Genetic manipulation of fungi can be achieved by mutation, transformation and protoplast fusion.

3.1. Mutation

Selection and mutation has been used to produce improved strains of fungal biocontrol agents. Ultraviolet(UV) radiation induced mutants of *Trichoderma* spp. which were resistant to benomyl were more effective than wild type strain in controlling *Rhizoctonia solani, Pythium ultimum, Sclerotium cepivorum* and *S. rolfsii* (Abd-El Moity et al., 1982; Papavizas, 1987). Benomyl tolerant mutants were rhizosphere competent even in the presence of 10μg of benomyl per g of soil. In the absence of benomyl, these mutants were also found to be competent than the wild type strains. Several UV-induced mutants of *Trichoderma harzianum* produced two antibiotic metabolites in fermentation medium and strains with higher amount of these antibiotics were found to be most effective biocontrol agents against *S. cepivorum* (Papavizas, 1987).

Mutagenesis by chemicals or UV irradiation, resulting in formation of numerous mutants that can be masked while selecting for particular phenotype. Most mutagens are known to cause multiple mutations. Moreover, changes in geotypes cannot be controlled or characterized. Therefore this strategy poses numerous problems.

3.2. Transformation

The recent developments in genetic engineering made it possible to manipulate a biological control agent by inserting or deleting a specific gene from any source. In the presence of benomyl *Gliocladium virens* cannot be used for biocontrol. Benomyl binds to fungal β-tubulin and prevents microtubule polymerisation. Mutation in β-tubulin gene can cause resistant to this fungicide, but induction of benomyl resistance by UV-mutagensis is a difficult method (Bagyaraj and Govindan, 1996). Ossanna and Mischke (1990) developed methodology to transform protplasts of *G.virens* to benomyl resistance with a *Neurospora crassa* β-tubulin gene. Such strains may be of immense use in specific integrated pest management.

An important problem with the potential application of transformed fungi as biocontrol agent is the overall stability of the foreign DNA (Kistler and Benney, 1988). It was suggested that the stability of the introduced DNA was affected by the stress of plant infection (Leslie and Dickman, 1991).

3.3. Protoplast Fusion

The biocontrol ability is under the control of large number of genes, hence the use of mutagenesis or transformation is not always desirable. Therefore the introduction of superior characters may be best achieved by crossing of strains by appropriate characters (Hocart and Peberdy, 1989). Protoplast fusion provides a means for recombination of whole genomes of species in which sexual or parasexual mechanisms are not present or difficult to exploit and even between the incompatible strains/species. Protoplast fusion combines the entire cell contents as well as genomic DNA of two vegetative cells into one. But transfer of isolated nuclei into protoplasts eliminates recombination of extranuclear components. Sivan *et al.* (1990) transformed the auxotrophic strains of *T. harzianum* with the nuclei of another auxotroph of same strain. This intrastrain nuclear transformation gave rise to numerous progenies which were stable, prototrophic and heterokaryotic. This methodology may lead to better understanding of nucleus-cytoplasmic interaction which may lead to better method of transferring genetic information into an organism (Sumeet, 1999; Sumeet and Mukerji, 1999).

4. PROTOPLAST FUSION IN IMPROVING BIOCONTROL POTENTIAL

In large number of fungi, protoplast fusion is now possible but there are few studies related to fungal biocontrol agents, that also with species of *Trichoderma* . Now a days, much emphasis is given on *Trichoderma* spp. as it possess useful traits for production of cellulolytic enzymes and biocontrol ability and it would be possible to combine these attributes into single superior strain using protoplast fusion technology.

Recently few strains with improved biocontrol ability have been generated by using protoplast fusion technology. The diversity created by this process can provide a source of improved biocontrol strains. The protoplast fusion has been used to improve biocontrol ability of *T. harzianum* (Harman, 1991; Harman *et al.,* 1989; Lalithakumari *et al.,* 1996; Migheli *et al.,* 1995; Pe'er and Chet, 1990; Sivan and Harman, 1991; Stasz *et al.,* 1988). Effective biocontrol strains are rare, only 2-4% of the progeny obtained from a fusion between *T.harzianum* strains T12 and T95 were more effective as biological seed treatment than the wild strain (Harman and Stasz, 1991). Similarly, Pe'er and Chet (1990) found only one progeny strain of 14 tested from fusion among various mutant strains of *T.harzianum* ATCC 32173 gave significant biocontrol of *Rhizoctonia solani.*

Harman *et al.* (1989) and Sivan and Harman (1991) reported that the progeny derived from protoplast fusion between T12 and T95 strains of *T.harzianum* (Stasz *et al.,* 1988) exhibitted better biocontrol ability and more effectively colonize the rhizosphere of several crops. Both parental strains have ability to control several plant pathogenic fungi and strains T95 is rhizosphere competent and benomyl resistant (Ahmad and Baker, 1987; Hadar *et al.,* 1984; Harman and Taylor, 1988). Most of the resulting fusants were no better than the parental strain but one strain 1295-22 showed improved biocontrol efficacy. It grows more rapidly than either of the parental strains and effective against an array of pathogenic fungi including *Pythium ultimum, Fusarium graminearum, Rhizoctonia solani, Sclerotium rolfsii, Sclerotinia homoeocarpa* and *Botrytis cinerea* (Harman *et al.,* 1989). The strain is substantially more rhizosphere competent than the parental strain. After application as seed treatment, it was found that T12, T95, 1295-22 increased the root elongation rate in crops *viz.* cotton, maize, pea and cucumber. It was further observed by Sivan and Harman (1991) that the protoplast fusion progeny 1295-22 resulted in longest roots in maize and cotton as compared to T12 and T95. All the strains were present at higher levels in the upper portion and root tip region and decrease in central region whereas strain 1295-22 maintained a constant higher level along the root.

In soils infested with *Pythium ultimum* seed treatment with either parent or fusant strains of *T.harzianum* provided increased stands compared with untreated control and were effective as treatment with thiram. Further the stands were improved when seeds were treated with strain 1295-22 as compared to either of the parental strain. The fusant progency 1295-22 induced robust growth of cucumber seedlings than the parental strain (Harman *et al.*, 1989). This strain is effective on wide range of agronomic and vegetable crops ranging from cotton to corn to beans.

Migheli *et al.* (1995) observed that the parental strain, mutant strain and six fusant strains of *T. harzianum* significantly reduced grey mould incidence on grape caused by *Botrytis cinerea*. They were giving significant or highly significant almost complete control of the disease. A highly significant strain effect of parental strain, mutants and two of the fusants were observed on percentage plant stand and similar results were obtained with the plant fresh weight in comparison with the control of *P. ultimum* on lettuce.

Intraspecies fusion between *T. harzianum* (an efficient antagonist) and *T. longibrachiatum* (tolerant to copper sulphate and carbendazim) resulted in the recovery of numerous fusants. Two fusant F1 and F2 exhibited enhanced antagonistic potential against *Rhizoctonia solani, Fusarium oxysporum* f. sp. *lycopersici, Venturia inaequalis, Curuvlaria lunata* and *Bipolaris oryzae* along with tolerance of copper sulphate and carbendazim (Lalithakumari *et al.*, 1996).

5. CONCLUSIONS

Owing to adverse effects of chemical pesticides and fungicides there is a need to switch over to non-chemical control of plant pathogens. Biological control of plant pathogens with fungal antagonists has been considered as a potential control strategy. Although, biocontrol has not solved agricultural problems so far, but we are at the turning point in technology to make significant approach. To compete with chemicals biological control agents must be enhanced for greater efficiency and reliability. Identification and enhancement of mechanisms involved in biological control is a logical strategy for solution of the problem.

New approaches must be introduced to develop improved production, formulation and shelf life of biocontrol agents. There is a need to determine the effective ways of improving microbial action against pathogens both through manipulation of their biology and genetic constitution. Investigation of new technologies is needed to induce new biotypes for improved performance, adaptation of microbial agents to new disease control requirements and improved compatibility with chemical fungicides so that microbial agents should be economical feasible, safe and above all effective. The promising potentials of the biotechnology have changed the direction of research in the biological sciences.

Protoplast fusion will undoubtedly play an important role in the exploitation of biocontrol fungi. The use of protoplast fusion to overcome incompatibility barriers allows characteristics from even distantly revealed isolates to be combined. Increased cellulolytic and mycolytic activities could enhance the effectiveness of organism for biocontrol purpose.

REFERENCES

Abd-El Moity, T.H., Papavizas, G.C. and Shatla M.N. 1982, Induction of new isolates of *Trichoderma harzianum* tolerant to fungicides and their experimental use for control of white rot of onion, *Phytopath.* 72:396-100.

Ahmad,J.S. and Baker, R.1987, Rhizosphere competence of *Trichoderma harzianum*, *Phytopath.* 77:182-189.

Alabouvette, C., Couteavdier, Y. and Lemanceau, P. 1986, Nature of intrageneric competition between pathogenic and non-pathogenic *Fusarium* in a wilt suppressive soil, In : *Iron, Siderophores and Plant Diseases,* ed.T.R. Swinburne, Plenum Press, New York, pp. 165-178.

Bagyaraj, D.J. and Govindan, M. 1996, Microbial control of fungal root pathogens, In : *Advances in Botany,* eds. K.G.Mukerji, B.Mathur, B.P.Chamola and P.Chitralekha, APH Publishing Corporation, New Delhi, India, pp. 293-321.

Baker, R. 1989, Some perspectives on the application of molecular approaches to biocontrol problems, In: *Biotechnology of Fungi for Improving Plant Growth,* eds. J.M. Whipps and R.D.Lumsden, Cambridge University Press, Cambridge, pp.219-233.

Barak, R., Elad, Y., Mirelman, D.and Chet, I. 1985, Lectins : a possible basis for specific recognition in the interaction of *Trichoderma* and *Sclerotium rolfsii, Phytopath.* 75 : 458-462.

Chet, I., Harman, G.E. and Baker, R.1981, *Trichoderma hamatum:* its hyphal interaction with *Rhizoctonia solani* and *Pythium* spp., *Microb. Ecol.* 7:29-38.

Chet, I.,Barak, Z. and Oppenheim, A. 1993, Genetic engineering of microganisms for improved biocontrol activity, In: *Biotechnology in Plant Disease Control,* ed. I.Chet, Willey-Liss Inc., USA, pp. 211-235.

Cook, R.J. and Baker, K.F. 1983, *The Nature and Practice of Biological Control of Plant Pathogens,* 2nd ed., American Phytopathological Society, St. Paul, Minnesota, USA.

De Oliveira, V., Bellei, M.de M. and Borges, A.C. 1984, Control of white rot of garlic by antagonistic fungi under controlled environmental conditions, *Can. J. Microbiol.* 30: 884-889.

Dennis, C. and Webster, J.1971a, Antagonistic properties of species group of *Trichoderma,* I, Production of non volatile antibiotics, *Trans. Br.Mycol. Soc.* 157:25-39.

Dennis, C. and Webster, J.1971b, Antagonistic properties of species group of *Trichoderma,* II, Production of volatile antibiotics, *Trans. Br.Mycol. Soc.* 157:41-60.

Dennis,C.and Webster, J. 1971c, Antagonistic properties of species group of *Trichoderma,* III, Hyphal interactions, *Trans.Br.Mycol. Soc.* 57:363-369.

Elad, Y., Barak, R. and Chet, I. 1983a, Possible role of lectins in mycoparasitism, *J. Bacteriol.* 154: 1431-1435.

Elad, Y., Chet, I., Boyle, P. and Henis, Y. 1983b, Parasitism of *Trichoderma* spp. on *Rhizoctonia solani* and *Sclerotium rolfsii* - scanning electron microscopy and fluorescence microscopy, *Phytopath.* 73:85-88.

Fravel, D.R.1988, Role of antibiosis in the biocontrol of plant diseases, *Ann. Rev. Phytopath.* 26: 75-91.

Hader Y., Harman G.E. and Taylor, A.G. 1984, Evaluation of *Trichoderma koningii* and *T.harzianum* from New York soils for biological control of seed rot caused by *Pythium* spp., *Phytopath.* 74:106-110.

Harman, G.E. 1990, Deployment tactics for biocontrol agents in plant pathology, In: *New Directions in Biological Control, Alternative for Suppressing Agricultural Pests and Diseases,* eds. R.R. Baker and P.E. Dunn, Alan R.Liss, New York, pp. 779-772.

Harman, G.E. 1991, Seed treatment for biological control of plant diseases, *Crop Protec.* 10:166-171.

Harman, G.E. and Hayes, C.K. 1993, The genetic nature and biocontrol ability of progeny from protoplast fusion in *Trichoderma,* In : *Biotechnology in Plant Disease Control,* ed. I. Chet, Willey Liss, USA, pp. 237-255.

Harman, G.E. and Stasz, T.E. 1991, Protoplast fusion for the production of superior biocontrol fungi, In : *Microbial Control of Weeds,* ed. D.O.TeBeest, Chapman and Hall, New York, pp. 171-186.

Harman, G.E. and Taylor, A.G. 1988, Improved seedling performance by integration of biological control agents at favourable pH levels with solid matrix priming, *Phytopath.* 78: 520-525.

Harman, G.E., Taylor, A.G. and Stasz, T.E. 1989, Combining effective strains of *Trichoderma harzianum* and solid matrix priming to improve biological seed treatment, *Plant Dis.* 73:631-637.

Hocart, M.J. and Peberdy, J.F. 1989, Protoplast technology and strain selection, In: *Biotechnology of Fungi for Improving Plant Growth,* eds. J.M. Whipps and R.D. Lumsden, Cambridge University Press, Cambridge, pp.235-258.

Howell, C.R. and Stipanovic, R.D. 1983, Gliovirin, a new antibiotic from *Gliocladium virens* and its role in the biological control of *Pythium ultimum, Can. J. Microbiol.* 29:321-324.

Hubbard, J.P., Harman, G.E. and Hadar. Y. 1983, Effect of soil-borne *Pseudomonas* spp. on the biocontrol agent, *Trichoderma hamatum,* on the pea seeds, *Phytopath.* 73:655-59.

Kistler, H.C. and Benney, U.K. 1988, Genetic transformation of the fungal plant wilt pathogen, *Fusarium oxysporum, Curr. Genet.* 13:145-149.

Kumar, R.N., Upadhyay, R.K. and Mukerji K.G. 1997, Strategies in biological control of plant diseases, In : *IPM System in Agriculture,* Vol.2, *Biocontrol in Emerging Biotechnology,* eds. R.K. Upadhyay, K.G. Mukerji and R.K. Rajak, Aditya Books Pvt. Ltd., New Delhi, India, pp. 371-422

Lalithakumari, D., Mrinalini, C., Chandra, A.B. and Annamalai, P.1996, Strain improvement by protoplast fusion for enhancement of biocontrol potential integrated with fungicide tolerance in *Trichoderma* spp., *Zeitschrift fur Pflanzenkrankeiten und Pflanzenschutz* 103:206-212.

Leslie,J.F. and Dickman, M.B. 1991, Fate of DNA encoding hygromycin resistance after meiosis in transformed strain of *Gibberella fujikuroi (Fusarium moniliforme), Appl. Environ. Microbiol.* 57:1423-1429.

Lewis, K.,Whipps, J.M. and Cooke, R.C. 1989, Mechanisms of biological disease control with special reference to the case study of *Pythium oligandrum* as an antagonist, In: *Biotechnology of Fungi for Improving Plant Growth,* eds.J.M. Whipps and R.D. Lumsden, Cambridge University Press, Cambridge, pp. 191-194.

Lumsden, R.D. and Lewis, J.A. 1989, Problems and progress in the selection, production, formulation and commercial use of plant disease control fungi, In: *Biotechnology of Fungi for Improving Plant Growth,* eds. J.M. Whipps and R.D. Lumsden, Cambridge University Press, Cambridge, pp. 171-190.

Migheli, Q., Whipps, J.M., Budge, S.P. and Lynch, J.M. 1995, Production of inter-and intra-strain hybrids of *Trichoderma* spp. by protoplast fusion and evaluation of their biocontrol activity against soil borne and foliar pathogens, *J. Phytopath.* 143:91-97.

Mukerji, K.G., Chamola, B.P. and Upadhyay, R.K. (eds.) 1999, *Biotechnological Approaches in Biocontrol of Plant Pathogens,* Kluwer Academic / Plenum Publishers, New York, USA.

Mukhopodhyay, A.N. 1994, Biocontrol of soil borne fungal plant pathogens, *Indian J. Mycol. Plant Pathol.* 17:1-10.

Nelson, M.E. and Powelson, M.L. 1988, Biological control of grey mold of snap beans by *Trichoderma hamatum, Plant Dis.* 72: 727-729.

Nigam, N. 1997, Fungi in biocontrol, In: *New Approaches in Microbial Ecology,* eds. J.P. Tewari, G. Saxena, N.Mittal, I.Tewari and B.P.Chamola, Aditya Books Pvt. Ltd., New Delhi, India, pp. 427-457.

Nigam, N. , Kumar, R.N., Mukerji, K.G. and Upadhyay, R.K. 1997, Fungi - a tool for biocontrol, In: *IPM System in Agriculture,* Vol. II, *Biocontrol in Emerging Biotechnology,* eds. R.K. Upadhyay, K.G. Mukerji and R.L.Rajak, Aditya Books Pvt. Ltd., New Delhi, India, pp. 503 - 526.

Ossanna, N. and Mischke, S. 1990, Genetic transformation of the biocontrol fungus *Gliocladium virens* to benomyl resistance, *Appl.Environ. Microbiol.* 54:3052-3056.

Papavizas, G.C. 1987, Genetic manipulation to improve the effectiveness of biocontrol fungi for plant diseases control, In: *Innovative Approaches to Plant Disease Control,* ed. I.Chet, John Wiley and Sons, New York, pp. 193-211.

Pe'er, S. and Chet, I. 1990, *Trichoderma* protoplast fusion: a tool for improving biocontrol agents, *Can. J.Microbiol.*36:6-9.

Reyes, F., Perez-Leblic, M.I., Martinez, M.J. and Lahoz, R. 1984, Proptoplast production from filamentous fungi with their own autolytic enzymes, *FEMS Microbiol. Lett.* 24:281-283.

Ridout, C.J., Coley-Smith, J.R. and Lynch, J.M. 1988, Fractionation of extracellular enzymes from a mycoparasitic strain of *Trichoderma harzianum, Enzy. Microbiol. Technol.*10: 180-187.

Renwick, A. and Poole, N. 1989, The environmental challenge to biological control of plant pathogen, In: *Biotechnology of Fungi for Improving Plant Growth,* eds. J.M.Whipps and R.D. Lumsden, Cambridge University Press, Cambridge, pp. 277-290.

Singh, J. and Faull, J.L. 1988, Antagonism and biological control, In: *Biocontrol of Plant Diseases,* Vol. II, eds. K.G.Mukerji and K.L. Garg, CRC Press, Boca Raton, Florida, USA, pp. 167-177.

Sivan, A. and Harman, G.E. 1991, Improved rhizosphere competence in a protoplast fusion progency of *Trichoderma harzianum, J.Gen. Microbiol.* 137:23-29.

Sivan, A., Harman, G.E. and Stasz, T.E. 1990, Transfer of isolated nuclei into protoplasts of *Trichoderma harzianum, Appl. Environ. Microbiol.* 56:2404-2409.

Stasz, T.E., Harman, G.E. and Weeden, N.F. 1988, Protoplast preparation and fusion in two biocontrol strains of *Trichoderma harzianum, Mycologia* 80:141-150.

Sumeet 1999, Protplast fusion for improvement of fungal strains, In : *Advances in Micobial Biotechnology,* eds. J.P. Tewari, T.N. Lakhanpal, J. Singh, R. Gupta and B.P. Chamola, APH Publishing Corporation, New Delhi, India, pp. 187-202.

Sumeet and Mukerji, K.G. 1999, Protoplast fusion is disease control, In : *Biotechnological Approaches in Biocontrol of Plant Pathogens,* eds. K.G. Mukerji, B.P. Chamola and R.K. Upadhyay, Kluwer Academic / Plenum Publishers, New York, USA, pp. 177-196.

Thomashow, L.S. and Weller, D.M. 1991, Role of antibiotics and siderophores in biocontrol of take-all disase, In:*The Rhizosphere and Plant Growth,* eds. D.L.Keister and P.B. Cregan, Kluwer Academic Publ. Dordrecht, pp.245-251.

Upadhyay, R.S. and Rai, B. 1988, Biocontrol agents of plant pathogens: their use and practical constraints. In: *Biocontrol of Plant Disease,* Vol. 1, eds. K.G.Mukerji and K.L.Garg, CRC Press Inc., Florida, USA, pp. 15-36.

Upadhyay, R.K., Mukerji, K.G. and Rajak, R.L.(eds.) 1996, *IPM System in Agriculture,* Vol.1, *Principles and Perspective,* Aditya Books Pvt.Ltd., New Delhi, India.

Upadhyay, R.K., Mukerji, K.G. and Rajak, R.L. (eds.) 1997. *IPM System in Agriculture,* Vol.II, *Biocontrol in Emerging Biotechnology,* Aditya Books Pvt.Ltd., New Delhi, India.

Upadhyay, R.K., Mukerji, K.G. and Rajak, R.L. (eds.) 1998a, *IPM System in Agriculture,* Vol.III, *Cereals,* Aditya Books, Pvt. Ltd., New Delhi, India.

Upadhyay, R.K., Mukerji, K.G. and Rajak, R.L. (eds.) 1998b, *IPM System in Agriculture,* Vol.IV, *Pulses,* Aditya Books, Pvt. Ltd., New Delhi, India.

Upadhyay, R.K., Mukerji, K.G., Chamola, B.P. and. Dubey, O.P.(eds.) 1998c, *Integrated Pest and Disease Management,* APH Publishing Corporation, New Delhi, India.

Weindling, R.1934, Studies on a lethal principle effective in the parasitic action of *Trichoderma lignorum* on *Rhizoctonia solani* and other soil fungi, *Phytopath.* 24:1153-1157.

Weindling, R. and Emerson, D.H. 1936, The isolation of a toxic substance from the culture filtrate of *Trichoderma, Phytopath.* 26: 1068-1074.

Williams, S.T. 1982, Are antibiotics produced in the soil ? *Predobiologia* 23: 427-435.

Wilson, C.L. and Pusey, P.L. 1985, Potential for biological control of post harvest plant diseases, *Plant Dis.* 69:375-378.

Znidarsic, P., Pavko, A. and Komel, R. 1992, Laboratory-scale biosynthesis of *Trichoderma* mycolytic enzymes for protoplast release from *Cochliobolus lunatus, J. Ind. Microbiol.*9:115-119.

Yong, F.M., Wong, H.A. and Lim, G. 1985, Effects of nitrogen source on aroma production by *Trichoderma viride, Appl. Microbial. Biotechnol.* 22:146-150.

MICROBIAL IRON CHELATORS: A SUSTAINABLE TOOL FOR THE BIOCONTROL OF PLANT DISEASES

S. B. Chincholkar, B. L. Chaudhari, S. K. Talegaonkar and R. M. Kothari

Department of Microbiology
School of Life Sciences
North Maharashtra University
P.B.No. 80, Jalgaon - 425 001, Maharashtra, INDIA

1. INTRODUCTION

Since prehistoric times, control of plant diseases has remained a challange to mankind. In spite of large scale use of agro-chemicals and pesticides since long, number of plant diseases are still beyond the control. Every year agricultural losses due to pests and diseases amount to Rs. 20,000 crores and exports worth more than Rs. 1,000 crore are restricted owing to poor quality.

Adverse short and long term effects of agro-chemicals and pesticides on the ecosystem have prompted exploration of natural alternatives for the biocontrol of plant diseases on sustainable busis. Based on this theme, use of microbial iron chelators / sidero phores is reviewed. It has started on a positive consideration of healthy plant as an ecological asset and for its preservation the concept of Integruated Pest Management (IPM) has been considered in the light of conventional vis-a-vis emerging strategies for plant protection. Since siderophores appeared eco-friendly and sustainable, significance of iron deficiency has been highlighted.

1.1. Healthy Plant - an Ecological Asset

Apparently healthy appearing plants are usually infected with fungal, bacterial and viral diseases and some times they are afflicted with insect pests and nematodes. Environment (humidity, temperature and light) contributes its might for the spread / control of these infections. Resulting plant diseases cause reduced yield and at times, a complete crop failure and thereby significant economic losses. To ward off from such a calamity, Ignacimuthu (1996) highlighted the benefits of healthy plants and cautioned about the use of plant species which can act as symptomless carriers of potentially devastating pathogens. Therefore, plant health and hygiene in a breeding/tissue culture programme is accorded prime importance.

1.2. Pesticides - a Twin Edged Weapon

The existing methods of controlling pests and diseases caused by them are heavily biased in favour of chemotherapy through dusting/ foliar spray with chemical pesticides, partly due to incessant publicity by their manufacturers/dealers and partly due to ignorance of illiterate farmers. In fact, a tendency not to take any chance in wiping out a pest and hope of fetching higher productivity by recommending/using more pesticide than perhaps necessary, has brought a number of short and long range miseries. Over the last four decades, this practice has rendered fertilie soil non-fertile, deteriorated microflora in the soil, showed pesticide leaching towards hydrosphere, brought incalculable damage to the eco-system and thus created a frightening scenario for future. In retrospect, pesticides have emerged as a twin-edged weapon, providing us food safety and havoc-multiplier in eco-system.

1.3. Evolution of IPM Concept

The above realities led to introspection and evolution of various strategies under the umbrella term, Integrated Pest Management (IPM), and biological control of plant diseases. Its philosophy of sustainability has been discussed on a common platform of multi-disciplinary intellectuals, facilitating its rapid acceptance in international agenda.

Although various definitions of biological control have been proposed by the plant pathologists, only a few of them have stood the test of time. Baker (1985) defined biological control as "Pest suppression with biotic agent(s), excluding the process of breeding for resistance to pests". Nigam and Mukerji (1988) defined biological control as "the direct or indirect manipulation by man of living natural control agents to increase their attack on pest species".

Paroda (1997) has defined sustainability as maintaining an increasing production trend to ensure food security, implicitly with no or negligible damage to aerial, aqueous, soil and underground ecosystem. He claimed that during 1993-1996, India (i) increased its wheat production due to control of wheat rust, (ii) similarly controlled rice blast (*Pyricularia oryzae*), (iii) successfully controlled yellow vein disease of ladies' finger and viral disease of potato and (iv) attained a comfortable food stock position, concomitant with demands of an ever growing population, implicitly through partial/fair application of IPM.

2. STRATEGIES FOR PLANT PROTECTION

2.1. Conventional Strategies

A number of conventional methods of controlling plant pathogens implicitly rely on the inherent fertility of the soil or use of resident antagonists.

2.1.1. Classical approach

It relied on increasing the fertility of the soil through the application of farm yard manure (FYM), composted night soil, cattle dung, soil conditioners, etc. Implicit understanding behind this strategy was that an application of ample organic carbon will improve soil fertility, releasing sustained/ requisite nutrition and moisture to the plants, rendering them inherently healthy, thereby minimizing the risk of pathogen infestation. In nutshell, it was a pre-emptive rather than post-curative approach.

2.1.2. Regular crop rotation

This approach on a soil, pre-enriched in fertility by a preceding leguminous crop was practised so that soil borne pathogens vanished in the absence of their natural/preferred host.

2.1.3. Regular tillage and flooding

These operations were alternated in the soil for every few weeks, as in paddy fields, so that the pathogens were deprived of adequate oxygen and incubation period for colonisation to show viable impact.

2.1.4. Application of chemical pesticides

Amongst various conventional strategies available for plant disease management, chemical(s) based strategies have so far been dominating. However, indiscriminate/ prolonged application of chemicals to control plant diseases has caused serious imbalance in the agro-ecosystem. Strangely, about 70% chemicals sprayed on field crops, do not stick to foliage, leave alone their absorption in the field crops or subsequent biodgradation. Nonsticking property of the protectant and a keen desire for the assured kill of the infestant by the farmer leads to use of enormous quantities of chemicals, adversely affecting microbial life in the soil/ rhizosphere. This practise culminates in to adverse effect on all of non-target species. Therefore, a shift towards non-chemical strategies is likely to correct the imbalance in our approach as known in Integrated Pest Management (IPM) or Integrated Disease Management (IDM), as depicted in Figure1.

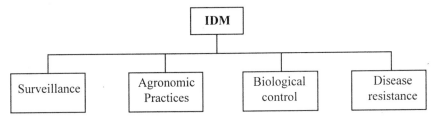

Figure 1. Sub-systems of Integrated Disease Management (IDM)

2.2. Emerging Strategies

A number of strategies are emerging in thought process, at laboratory and green house experiments, in tissue culture laboratories, a few at pilot scale field trials and fewer with potential for commercialisation for the biocontrol of plant pathogens. At a glance their profiles are briefed as follows :

2.2.1. Foliar spray

In this approach, requisite number of sprays are given at pre-determined frequency with phenolic formulations on leaf surfaces, to act as slowly biodegradable antimicrobial agent(s) for biocontrol.

2.2.2. Introduction of known antagonists of pathogens in the soil

Through this approach, control of *Agrobacterium radiobacter* var. *tumefaciens*, a causative agent of crown gall of roses, was achieved using the non-pathogenic strain K-84 of *A. radiobacter* var. *radiobacter*, known to produce antibiotic-like substances.

2.2.3. Spray with *Trichoderma*

Introduction of *Trichorderma roseum*, known to produce antiviral substances, acts indirectly via the host, causing alterations in the host susceptibility to viruses and thereby provide plant protection.

2.2.4. Induction of natural resistance / phytoalexins

Plants are known to produce phytoalexin like compounds under physical, chemical stress conditions. These are plant-specific secondary metabolites whose biogenesis is triggered by pathogen attack. They arrest the growth of pathogens and thereby provide opportunity to the plants to exhibit their optimal output. They help the plants under physical, chemical or biological stress to suppress or resist infectant activity.

2.2.5. Application of siderophore genic microbes/siderophores

This approach of plant protection involves selectively depriving the plant pathogens an availability of soluble Fe^{3+}, which is essential for the growth of plant as well as its pathogens. This is accomplished by the microbes bestowed with strong Fe^{3+} chelation ability through secondary metabolites, recognised as siderophores.

2.2.6. Biological control by gene manipulation

To address a lack of target specificity exhibited by chemical pesticides and eco-hazards arising out of their application, introduction of transgenic varieties of plants has become a reality. In practise it needs engineering the resistance to either pesticides or pests. So also, biological control by gene manipulations is possible by expanding the range of target species, restricting the range of non-pathogenic organisms and improving its durability.

3. MICRONUTRIENTS' DEFICIENCY REFLECTING DISEASE-LIKE SYMPTOMS

3.1. Micronutrients Deficiency

Like vitamins and hormones, micronutrients have also been a vital necessity for plant health and hygiene. Number of times, micronutrient deficiencies in plants lead to symptoms which are described, perhaps rightfully so, as plant disease(s); for instance, red leaf formation in cotton indicates nitrogen deficiency, development of dwarf cotton plants indicates phosphorus deficiency, blossom rot development in tomato indicates Ca^{2+} deficiency, etc. In all such cases, micronutrient deficiencies are manageable by the application of ammonium nitrate or urea, phosphate fertilizers and lime, respectively.

3.2. Iron Deficiency

In the biosphere, all the living creatures, including microorganisms, plants and animals require iron for oxidation-reduction in cellular reactions for living and possess mechanism(s) to acquire it from the environment. Like in animal body, activities of microorganisms affecting plant health are also influenced by the availability of iron. Therefore, its deficiency, like that of other micronutrients, leads to development of iron chlorosis or other unexplored disease-like symptoms in plants. The competitive abilities of microbes to sequester iron and make it available to plants under the stress of pathogen attack through microbial siderophores appears as one of the possible mechanisms to control pathogens.

Iron chlorosis occurs even in soils which are rich in iron but are not available to the plants. High calcium carbonate in the soil reduces the available iron content and whatever little iron that is absorbed by the palnts accumulates in the top most part of the stem. Moreover, iron chlorosis is also linked to poor aeration, high mangnese content, excessive application of phosphate, undecomposed compost and fresh cowdung.

3.3. Iron - A Significant Micronutrient

The history of iron in the nutrition of microorganisms dates back to evolutionary period when anaerobic life alone was into existence and it probably preferred ferrous ions (Fe^{2+}). In due course of evolution, with the emergence of aerobic environment and aerobes, majority of the microbes opted to use ferric form (Fe^{3+}) for their nutrition. Thus, iron came into existence in two oxidation states (Fe^{2+} and Fe^{3+}), which qualifies it to work as an electron carrier.

Today, iron is the best known and perhaps the most important inorganic micronutrient, being involved at a number of stages in the growth of plants and microorganisms. It is abundant (1-6%) in soils, ranking fourth among all the elements on the surface of the earth and yet it is not available to microorganisms and plants, the ferric form (Fe^{3+}) being insoluble at physiological pH. Its acquisition by microorganisms is difficult because it forms hydroxides which have very low solubility at this pH. Unlike organic compounds, minerals can not be synthesized; at best they could be relieved from one sphere/ site of action to another, depending upon the metabolic priorities and hence must be acquired from the environment.

In intermediary metabolic processes, iron is required in nitrogen fixation, photosynthesis, TCA cycle, electron transport chain and oxidative phosphorylation. Besides, iron is also known to regulate biosynthesis of porphyrin, vitamins, antibiotics, cytochromes, pigments, toxins and a number of aromatic compounds. In all these processes, at molecular level, iron is needed as a cofactor by various enzymes and proteins such as peroxidase, superoxide dismutase, nitrogenase, several hydrogenases, glutamate synthase, ribonucleotide diphosphate reductase, aconitase, DAHP synthetase, cytochromes, ferridoxin, flavoproteins, etc.

In view of indispensability of iron and its versatility in the core metabolic processes, many aerobic and facultative anaerobic microorganisms have overcome the difficulty in its solubilization by synthesizing siderophores. They are capable of solubilising Fe^{3+} and thus making it available in adequate amount to the plants. This siderophore mediated iron sufficiency appears to protect the plant(s) from pathogen(s) and thereby lead to development of healthy crop(s). The first few reports of microbial siderophores were perceived as growth factors for mycobacteria and fungi (Hesseltine et al., 1952).

3.4. Historical Progress of Siderophores to Present Status

Since 1952, therefore, iron nutrition of microbes has been an intense area of research. For the first time, Neilands (1952) isolated siderophores in crystalline form and thus opened a virgin area of research for iron chelating new compounds. In the last two decades, a number of siderophores from bacterial and fungal sources have been reported. Light was thrown on their biosynthesis at molecular level and genetically controlled regulatory mechanisms of iron transport was by Guerinot (1994). Knowledge about their isolation and identification slowly developed and compiled in a review (Jalal and van der Helm, 1991). Soon, details on siderophore receptors, their journey from external environs through the membrane barrier into the cytoplasm, specificity with respect to transport etc., have been focused (Chaudhari and Chincholkar, 1998; Guerinot, 1994; Winkelmann, 1991). Recently, Winkelmann and Drechsel (1997) have compiled the available data on bacterial and fungal siderophores, as also highlighted mechanism(s) of siderophore-mediated iron transport.

4. GENESIS, TYPES, STRUCTURES AND PROPERTIES OF SIDEROPHORES

4.1. Genesis of the Term Siderophore

Siderophores are ferric ion-specific binding compounds, secreted by bacteria and fungi, growing under low iron stress (Lankford, 1973). In its earlier literature, designations like siderochromes, sideramines and sideromycins had been given to iron complexed molecules like ferrichrome, ferrioxamine and coprogen, respectively. These are now replaced by an umbrella term "siderophores"" (meaning in Greek: sideros = iron and phores = bearer). Subsequently, a necessity was realized to further clarify the term siderophore, if it was reserved for the iron-containing or iron-free ligands (Neilands, 1981). However, it is now settled that the iron-free form of siderophores is prefixed by desferri- or deferri- (e.g. desferricoprogen), while the corresponding iron-containing forms (enterobactin, aerobactin, staphyloferrin, rhizoferrin, etc.) are prefixed with ferri- or ferric-siderophores.

4.2. Types and Structures of Siderophore

A large number of siderophores have been reported from different microorganisms, varying in structures. Generally, these are categorized into two groups: (i) hydroxamates and (ii) catecholates. Recently, Winkelmann and Drechsel (1997) have classified bacterial siderophores into 5 types namely (i)catecholate, (ii)hydroxymates, (iii) peptide siderophores, (iv)mycobactin and related siderophores and (v)citrate hydroxamate siderophores. Fungal siderophores has been classified in to 5 families as (i) ferrichromes, (ii) coprogens, (iii) rhodotorulic acid, (iv) fusarinines (fusigen) and (v) rhizoferrins. A common feature of the eukaryotic organisms, especially fungi, is to produce preponderantly hydroxamate type of siderophores, while prokaryotic organisms produce both hydroxymates and catecholates. Their productivity seems to be governed by stressful conditions imposed by the iron content, competitive ability of pathogens, numerical superiority of both the competitors, ease of receptor compatibility, growth conditions, pH, temperature, etc.

4.2.1. Hydroxamates

This type of siderophores are produced by both bacteria and fungi. Most of them contain three secondary hydroxamate groups, (C= O), N-(OH) and R, where R is an amino acid or its derivative. Each hydroxamate group provides 2 oxygen atoms which form a bidentate ligand with iron. Therefore, each siderophore forms a hexadentate octahedral complex with Fe^{9+}. e.g. ferribactin produced by *Pseudomonas fluorescens,* aerobactin by *Erwinia carotovora,* francobactin by *Frankia* sp., etc.

4.2.2. Catecholates (phenolates)

This type of siderophores are usually produced by bacteria. Chemically, they are derivatives of 2,3-dihydroxybenzoic acid. Each catecholate group provides two oxygen atoms for chelation with iron so that a hexadentate octahedral complex is formed. e.g. enterochelin, the cyclic triester of 2,3-dihydroxybenzoyl serine produced by *E. coli,* linear catecholate like agrobactin by *Agrobacterium tumifaciens,* parabactin by *Paracoccus denitrificans* etc.

Above cited examples in both categories nearly represents a tip of the iceberg. As depicted in Tables 1 and 2, a wide variety of siderophore(s) are produced by bacteria and fungi and every year, their number is increasing as new siderophores are identified.

Table 1. Profiles of bacterial siderophores

S.No.	Siderophore	Producing organism	Reference
1	Schizokinen	*Bacillus megaterium*	Mullis *et al.*, 1971
		Ralstonia solanocearum	Budzikiewicz *et al.*, 1997a
2	Azotobactin	*Azotobacter vinelandii*	Fusakawa *et al.*, 1972
3	Pseudobactin	*Pseudomonas putida* B 10	Teintz *et al.*, 1981
4	Rhizobactin	*Rhizobium meliloti*	Smith *et al.*, 1985
5	Anguibactin	*Vibrio anguillarum* 775 (PJM)	Actis *et al.*, 1986
6	Pyoverdin	*Pseudomonas aeruginosa*	Meyer *et al.*, 198
		Pseudomonas chlororaphis	Hohlneicher *et al.*, 1995
7	Cepabactin	*Pseudomonas cepacia*	Meyer *et al.*, 1989
8	Chrysobactin	*Erwinia chrysanthemi*	Persmark *et al.*, 1989
9	Staphyloferrin A	*Staphylococcus hyicus*	Meiwes *et al.*, 1990
10	Ferribactin	*Pseudomonas fluorescens*	Linget *et al.*, 1992
11	Ornibactin	*Pseudomonas cepacia*	Holger *et al.*, 1993
12	Yersiniabactin	*Yersinia enterocolitica*	Haag *et al.*, 1993
13	Yersiniophore	*Yersinia enterocolitica*	Chambers *et al.*, 1996
14	Amonabactin	*Aeromonas hydrophilla*	Telford *et al.*, 1994
15	Alcaligin E	*Alcaligenus eutrophus*	Khan *et al.*, 1994
16	Alcaligin	*Bordetella pertussis,*	Brickman *et al.*, 1996
		B. bronchiseptica	
17	Vulnibactin	*Vibrio vulnificus*	Okujo *et al.*, 1994
18	Acinetobactin	*Acinetobacter baumanii*	Yamamoto *et al.*, 1994
19	Desferrioxamine B & E	*Streptomyces viridosporus*	Imbert *et al.*, 1995
20	Protochelin	*Azotobacter vinelandii*	Cornish and Page, 1995
21	Exochelin MS	*Mycobacterium smegmatis*	Sharman *et al.*, 1995
22	Corynebactin	*Corynebacterium glutamicum*	Budzikiewicz *et al.*, 1997b

Table 2. Profiles of fungal siderophores

S.No.	Siderophore	Producing organism	Reference
1	Fusarinine A & B	*Fusarium roseum*	Sayer and Emery,1968
2	Ferrichrome A	*Ustilago sphaerogena*	Emery, 1971
3	Ferrichrome C	*Neurospora crassa*	Horowitz *et al.*, 1976
4	Malionichrome	*Fusarium roseum*	Emery, 1980
5	Ferrichrome	*Penicillium parvum*	Winkelmann and Braun, 1981
6	Ferricrocin	*Microsporum canis*	Bentley *et al.*, 1986
7	Dimerum acid	*Stemphylium botyrosum*	Manulis *et al.*, 1986
8	Coprogen and Neocoprogen I & II	*Curvularia lunata*	Hossain *et al.*, 1987; Chincholkar *et al.*, 1995
9	Rhizoferrin	*Rhizopus microsporus,*	Drechsel *et al.*,1991
		R. arrhizus	Shenker *et al.*,1995
10	Alterobactin	*Alteromonas luteoviolaces*	Reid *et al.*,1993

4.3. Tailor-made Properties

By virtue of their inherent physical necessity of easy solubilization, transport and storage of iron via specific membrane receptors, siderophores are low molecular weight compounds, produced under iron starvation conditions by the microorganisms. To fulfill the demand of iron, microorganisms have evolved a specific and high affinity versatile mechanism of producing siderophores which becomes operative only in the iron deficient environment (Bagg and Neilands, 1987).

Siderophores exhibit novel structures with two types of ligands and containing modified amino acids, not found elsewhere in nature, with variation from one species to another. Although significance of modified amino acids associated with soderophores is not known presently, it is believed to have larger role than apparently perceived in chelation, solubilization and transport. While ferri-siderophores are intensely coloured, iron-free (desferri) forms are usually colourless, fascilatating their easy identification.

They exhibit requisite (i) hydrophilic properties for chelating iron in extracellular aqueous environment, (ii) lipophilic properties for entering through the lipoproteinous membrane receptors of the cell and (iii) hydro-/ lipo- philic properties, depending upon the aqueous or fatty environment under which they are destined to function (Jalal et al., 1984).

4.4 Applications of Siderophores

Several monographs have emphasized on the applications of siderophores in (i) agriculture in plant disease management and plant growth improvement, (ii) health care in new drugs development for combating diseases products and (iii) environment for balancing iron availability and elimination of pathogens (Messenger and Ratledge, 1986).

5. MECHANISMS OF BIOLOGICAL CONTROL

Although real mechanism(s) of biological control of plant pathogen(s) is not known, evidences are available suggestive of possible operative mechanisms. These have been proposed on the basis of antagonistic activities of microorganisms in soil.

5.1. Antibiosis

While screening potential biocontrol agent, it is customary to select these organisms which produce antibiotics, secondary metabolites, which are directly inhibitory to the growth of the pathogen. An outstanding example of successful development of an antagonist as a biocontrol agent is the use of *Agrobacterium tumefaciens* (strain K-84). This strain is sold commercially in USA under the trade name GALLTROL, meaning an effective controller of crown gall, particularly in the nursery industry. Unlike the virulent strain of *A. tumefaciens,* strain K-84 lacks the Ti-plasmid, but contains another plasmid that codes for the production of an antibiotic, agrocin. Therefore, plants pretreated with strain K-84 are protected against the virulent forms (Sequeira, 1987).

Certain strains of bacteria produce bactericidal proteins known as bacteriocins. They have been used against foliar bacterial pathogens (Vidaver, 1976). Similarly, under experimental conditions, there has been moderate success in the control of brown spot on beans by bacteriocin producing non-pathogenic strains of *Pseudomonas syringae* (Smidt and Vidaver, 1982). Because of their high specificity and lack of persistence in the environment, they have been considered as ideal biocontrol agents for many years.

There exists a controversy over the actual secondary metabolite responsible for suppression of plant pathogen. For instance, a fluorescent *Pseudomonas* isolate from iron-rich acidic soil used for tea plantation, was capable of inhibiting the growth of *Fusarium oxysporum, F. lycopersici, F. ciceri, F. udum , F. monoliformi, F. semitectum, F. solani,*

56

Ustulina zonata and *Fomes lamonensis* under *in vitro* conditions. Their inhibition was attributed to an antibiotic, 2-phenaziniamine, and not due to siderophores (Dileepkumar and Bezbaruha, 1997)

Although many organisms produce antibiotics in culture under optimised nutritional medium and physical parameters, there is little evidence that antibiotics are produced in natural environments, where inputs of optimised organic ingredients lack, unlike the culturing conditions. Within bacterial biocontrol agents, several species of the genus *Pseudomonas* produce antibiotics/ antibiotic-like effectors (secondary metabolites), as depicted in Table 3.

Table 3. Antibiotics/ antibiotics-like effectors produced by *Pseudomonas* sp. (Johri *et al.*, 1997).

Species	Secondary metabolite
P. fluorescens	Pyoluteorin
	Pyrrolnitrin
	Alginate
	IAA
	HCN
	Pseudomonic acid
	Oomycin A
	2,4-Diacetyl phloroglucinol
	2-Hydroxy-2,4,6-cyclohepta-triene-1-pseudomonic acid
	Ovafluorin
	Fluopsin C & F
	Sorbisitin A1 & B
	Salicylic acid
P. aeruginosa	Pyoluteorin
	HCN
P. cepacia	Pyrrolnitrin
P. putida	IAA

From Table 3, *P. fluorescens* has emerged as a versatile *Pseudomonas* sp., secreting a number of secondary metabolites, which have a killing/ controlling (antibiosis) effect on several pathogens. It is amazing that *P. fluoresens* alone secrets a number of antibiosis catalysing metabolites as a function of its environment. It is further noteworthy that inspite of vast differences in their structures and physico-chemical properties, their biocontrol effect was unaltered. It is, therefore, tempting to postulate that their site of action at molecular level must also be as varied. This leaves a totally virgin area for research as to why a particular chemical was selectively secreted, when each one of them could catalyze antibiosis. Its fluorescence affords an additional handle in the hands of researchers to identify the site of antibiosis, so that light could be thrown on the mechanism of biocontrol at a molecular level. These studies could exploit fluoroscopic examination and spare the creation of elaborate radioactive facilities.

5.2. Competition by Rhizobacteria

Competitive edge of the biocontrol agent over its opponent(s) is a major prerequisite for the success of biocontrol strategy. This is possible only when the biocontrol agent is a rapid colonizer and proliferates at a rate outnumbering that of the pathogen. Earlier workers have postulated that competition for a particular nutrient is an operating mechanism accounting for biocontrol and the addition of such nutrient must eliminate biocontrol activity.

An emphasis was always accorded on competition for carbon, nitrogen, phosphate and moisture between soil, root and bacteria inhabiting in them. Now competition for iron has gained multi-fold awareness and recognition after some light was thrown on the role of siderophores. Recent emphasis on the selection of rhizobacteria (bacteria which are highly adapted for survival on the root surface) in general and their secondary metabolites (siderophores) in particular, as biocontrol agents is due to their aggressive colonization on roots, presumably by displacing pathogens from infecting.

Application of rhizobacteria (mostly *Pseudomonas* sp.) in several field trials on potato seeds, for example, has significantly reduced the root zone population of *Erwinia carotovora* (Kloepper, 1983). The ability (effect) of these bacteria to suppress the colonization by root pathogens seems to have been traced to the production of siderophores, which have higher affinity for Fe^{3+} vis-à-vis pathogenic species. This postulate has gained ground by establishing a direct cause and effect relationship between the amount of siderophore produced by various fluorescent pseudomonads and their ability to inhibit *Fusarium* chlamydospore germination in soil. Since rhizobacteria also enhanced the growth of treated plants, there has been a great deal of interest in their commercial development as seed inoculants.

5.2.1. Siderophores in plant growth promotion and disease suppression

That iron plays an important role in the crop development was demonstrated by Powell *et al.* (1980), who reported the presence of siderophores, particularly hydroxamate siderophores (10^{-7} -10^{-8} M), in 67 different soils of the United States, at concentrations high enough to be useful to plant roots. Akers (1983) also succeeded in detecting siderophores (schizokinen) in paddy crop, suggesting that insoluble form of iron present in the soil, was made available to the plants, to a larger extent by microbial siderophores. Further evidences also suggested that the plants have an ability to incorporate Fe^{3+} of siderophores into their biomass (Backer *et al.*, 1985; Reid *et al.*, 1984). This reality was confirmed at molecular level by Castignetti and Smarrelli (1986) that at least one plant enzyme, NADH: nitrate reductase, functioned as ferri-siderophore reductase also, indicating that plants accept iron available through iron-siderophore chelates.

Since modern agricultural practices require eco-friendly and sustainable technologies, Integrated Pest Management (IPM) strategies find a potential in siderophores as a new dimension to biocontrol. On this premise, plant disease control and elimination of harmful diseases of some cash crops through siderophore producing bacteria is summarized in Table 4.

5.2.2. Siderophore secreting microbes for assured control of pathogens

Painstaking work of Loper (1988) suggested siderophore producing microbes as biocontrol agents, by apparently limiting the amount of iron available to potential plant pathogens. Thus, protection imparted against plant pathogens, reflected in increasing the yield of potato tuber (Bakker *et al.*,1986). Freitas and Pizzinatto (1997) also reported inhibition of *Colletotrichum gossypii* by rhizobacteria and thereby promoting the growth of cotton seedlings. Similarly, De-Meyer and Hofte (1997) also observed that iron regulated and siderophore-induced resistance was imparted by *Pseudomonas aeruginosa*, capable of producing pyoverdin, pyochelin and salicylic acid, to bean leaf infection by *Botrytis cinerea*. In due course it was found that low concentration of iron, temperature and catabolic repression were regulating the synthesis of pyoverdin in fluorescent *Pseudomonas* sp. That salicylic acid was essential for induction of resistance indicated a regulatory role of iron.

Table 4. Biocontrol agents for arresting pathogenic/ infestation on cash crops (Sindhu *et al.*, 1997)

Biocontrol agent	Crop affected	Pathogen controlled
Pseudomonas fluorescence	Potato	*Erwinia carotovora*
	Wheat	*Gaeumannomyces graminis*
	Cotton, Peanut	*Rhizoctonia solani*
	Flax	*Fusarium oxysporum*
	Rice	*Sarocladium oryzae*
	Soybean	*Fusarium glycinea*
	Tobacco	*Thielaviopsis basicola*
Pseudomonas putida	Potato	*Erwinia carotovora*
	Tomato, Flax	*Fusarium oxysporum*
	Soybean	*Fusarium glycinea*
	Apple	*Fusarium dianthi*
Pseudomonas cepacia	Onion	*Fusarium oxysporum*
Bacillus subtilis	Corn	*Fusarium roseum*
Rhizobium	Soybean	*Macrophomina phaseolina*
Bradyrhizobium	Sunflower	*Fusarium solani*
	Mungbean	*Rhizoctonia solani*

Work on rhizobacteria leading to plant pathogens' inhibition has been reviewed by Sindhu *et al.*(1997). Accordingly, rhizobacteria are a consortium of fluorescent pseudomonads alongwith other soil bacteria viz. *Alcaligenes, Azotobacter, Agrobacterium, Arthrobacter, Rhizobium, Clostridium, Bacillus* and *Serratia*. Most *Pseudomonas* strains tested were antagonistic *in vitro* to several fungi on King's B medium but not on potato dextrose agar (PDA) medium, indicating the possibility of siderophore mediated inhibition, whereas all *Bacillus* and two *Pseudomonas* isolates were antagonistic on PDA, probably by antibiotic mediated inhibition.

Johri *et al.*(1997) have also reviewed the role of fluorescent pseudomonads strain RBT 13, isolated from rhizoplane, in tomato plant disease management. It produced siderophores, which exhibited *in vitro* antagonism (disease suppression) against several bacterial and fungal pathogens and simultaneously increased the growth of four crops.

Erwinia carotovora, a member of Enterobacteriaceae, causative agent of soft rot diseases in potato tubers, was controlled by the application of *Pseudomonas* sp. prior to planting or storage. Biological control of decay in potato seeds was therefore, thought to be siderophore mediated iron competition between *Pseudomonas* sp. and *E. carotovora*. The fluorescent siderophores (pyoverdins or pseudobactins) produced by *Pseudomonas* sp. presumably deplete the pathogen's micro-environment for available iron by sequestering Fe^{3+} as ferric-pyoverdin complexes, which are utilized exclusively by *Pseudomonas* sp. (Loper and Buyer, 1991).

Thus, evidences are replete, according the elite position to fluorescent pseudomonads as versatile biocontrol agents.

5.2.3. Multiple biological functions bring more versatility to biocontrol

That *Pseudomonas* sp. promoted (i) symbiotic nitrogen fixation in clover under controlled conditions, (ii) growth of different plants under field conditions and (iii) secretion of sizable amount of B group vitamins renders pathogen control mechanism far from being clear. Their significance in biocontrol, especially if they are alleviative Vs curative, or coincidental Vs a cause or an effect, need to be examined.

Saikia and Bezbaruah (1995) showed that a nitrogen fixing *Azotobacter chroococcum* (strain RRLJ 203) produced hydroxamate type siderophores in both, iron-free and iron-fortified medium. Fekete *et al.* (1989) also made a similar observation on *Azotobacter chroococcum* (strain B-8). *Rhizobium* species too produced water soluble catecholates under low iron conditions. By efficient fixation of nitrogen in the soil, it promoted the acquisition of iron and inhibited several fungal pathogens, including *Fusarium oxysporum, F. udum, F. solani, F. moniliforme, F. semitectum, Ustulina zonata* and *Fomes lamonensis,* in the presence and absence of soluble iron.

Association of iron and molybdenum (Mo) with nitrogenase and an ability of mycobactin to bind both, Mo as well as Fe, (Messenger and Ratledge, 1986) appears to have complicated the mechanism of biocontrol. A question needs to be addressed whether nitrogen fixation precedes to impart strength to the plant to overcome pathogen effect or whether siderophores-mediated plant protection precedes to neutralize pathogen effect and thereby facilitates nitrogen fixation? That, the deferri form of siderophores was capable of promoting plant growth (Dileepkumar and Bezbaruah, 1997) indicated hitherto unknown role of iron in nitrogen metabolism. This dilemma is due to absolute requirement of high complement of iron for nitrogenase and other enzymatic machinery involved in nitrogen fixation, as also respiratory system responsible for protecting nitrogenase against oxygen inactivation.

Mulya *et al.,* (1996) found that *Pseudomonas fluorescens* (strain PfG32) isolated from the rhizosphere of onion was capable of actively suppressing the bacterial wilt disease of tomato (BWT) in vermiculite amended soil. Since the strain was capable of producing antibiotics and siderophores, it was worthwhile to establish if antibiotics or siderophores were responsible for its biological control.

5.2.4. Utilization of heterologous siderophores

It is interesting to note that many microorganisms are able to utilize the ferric siderophores which they themselves did not synthesize. Plessner *et al.* (1993) have reported the utilization of 3 types of siderophores (ferrichrome, rhodotorrulate and ferric citrate) by *Bradyrhizobium japonicum* (strains USDA 110 and 61A152), to overcome iron starvation. Strain 61A152 afforded consistently high yields on a variety of soybean cultivars in field trials. These strains could also utilize pseudobactin St3 (pyoverdin type siderophore). Their ability to utilize another organisms' siderophores certainly confers upon them an additional iron scavenging ability and versatility in the rhizosphere.

It is postulated that the ability to utilize a variety of siderophores may improve ecological fitness of *Pseudomonas* as a biocontrol inoculant in the rhizosphere. In order to utilize different ferric siderophores, an attempt to improve competence of fluorescent Pseudomonads in rhizosphere was made by Moenne *et al.*(1996) through genetic engineering approach. Plasmid pCUP2 carries a copy of the gene pbuA, coding for the membrane receptor of ferric pseudobactin M114. *Pseudomonas* sp. B24 Rif, containing pCUP2, can utilize ferric pseudobactin of *P. fluorescens* M114, in addition to its own siderophores. However, the ability to utilize ferric pseudobactin M114 did not improve the ecological fitness of B24 Rif in the rhizosphere of sugarbeet, although a larger fraction of the culturable resident fluorescent pseudomonads could supply pseudobactin M 114 - complexed iron to B24 Rif (pCUP2) rather than to B24 Rif.

Similarly, utilization of heterologous siderophores and competence of fluorescent *Pseudomonas* sp. in the rhizosphere of radish was investigated by Raaijmakers *et al.* (1995). For this purpose, interactions between *Pseudomonas putida* WCS358 and WCS374 and between WCS358 and indigenous 8 strains of *Pseudomonas*, capable of utilizing pseudobactin 358 were investigated. Siderophore mediated competition for iron was found to be the major determinant of interaction between WCS358 and WCS374 in the rhizosphere.

From the above evidences, it is clear that the design of the microbial cell by the Nature is concerned as an integrtaed concept. By extrapolation, various functions are also the part of the intergation. This meant that microbes discharging multiple integrated functions are complementory in Nature and would be ideal biocontrol agents, once little known intricacies are addressed squarely and understood in the spirit of sustainable application.

5.3. Parasitism and Lysis

Parasitism, another approach to pathogen control, involves the destruction of pathogen(s) by the action of enzyme(s) capable of lysing their cell wall. Boosalis (1964) has referred mycoparasitism, hyperparasitism, direct parasitism and inter-fungus parasitism to the phenomenon of one fungus parasiting on another, deriving nutrition in the biotropic (from the living cells of host) or necrotropic (from the dead host cells) manner, depending upon its mode of parasitism. Illustrative examples of parasitism and lysis are given in Table 5.

Table 5. Parasitism- and lysis-based biocontrol of pathogens (Johri *et al.*, 1997)

Host plant	Biocontrol agent	Pathogen controlled
Carnation	*Arthrobacter* sp.	*Fusarium* sp.
Tomato	*Arthrobacter* sp.	*Pythium aphanidermatum*
Grasses	*Bacillus* sp.	*Gaeumannomyces graminis*
Peas & Sugarbeet	*Enterobacter cloaceae*	*Pythium* sp.
Beans	*Serratia marcescens*	*Sclerotium rolfsii*
		Rhizoctonia solani

Mycoparasitism occurs when one fungus is intimately associated with another, from which it derives some or all of its nutrition. For example, *Sporidesmium sclerotivorum* trapping *Sclerotinia minor* (Ayers and Adams, 1979). Many illustrative examples to explain parasitism and lysis could be cited, the most published one being an attack by *Trichoderma* sp. on a variety of phytopathogenic fungi, responsible for the most dreadful diseases suffered by the crops of worldwide economic significance (Esterlla and Chet, 1998).

5.3.1. *Trichoderma as* a biocontrol agent

Weindling (1932) was the first to report *Trichoderma* as a mycoparasite, which exhibited both, antibiosis and competition, as the mechanism(s) of disease control. Albeit, antibiotic antibiosis was reported due to *Gliocladium*. After three decades, Godtfredsen and Vangedal (1965) reported trichodermin, a sesquiterpene, as an antibiotic produced by *Trichoderma*. Subsequently, Pyke and Dietz (1966) reported the secretion of dermadine, a monobasic acid, by *Trichoderma*, active against a wide range of fungi as well as Gram positive and Gram negative bacteria. This system of non-antibiotic antibiosis by *Trichoderma* is due to its array of lytic enzymes, chitinase, cellulase, β-1, 3-glucanase, and proteases, which are optimally active at neutral pH (Rodriguez-Kabana *et al.*, 1968, 1969, 1978). Through them, *Trichoderma* lyses the host cell wall, thrives on its sap and comes out as an efficient and broad spectrum biocontrol agent.

Omnipresence of *Trichoderma* in agriculture and natural soils in all geo-climatic conditions throughout the world is an evidence of its superior ability to compete for nutritional resources and survival strategy. Chemical treatment of soil to sublethal dose imparted further competitive advantage to *Trichoderma* for disease suppression. In many cases, fungi responsible for wood rotting have been controlled by *Trichoderma*. Like Toole (1951), Hulme and Shields (1970) were able to preserve wood logs upto 7 months due to the ability of *Trichoderma* to utilize non-structured carbohydrates, thereby preventing destructive basidiomycetes from getting established.

Wells (1988) found *Trichoderma* as an exceptionally good biocontrol agent due to its ubiquitous and non-pathogenic nature for higher plants, ease of cultivation , rapid growth on a number of substrates and secretion of antibiotics and enzymes, which are capable of attacking a wide range of plant pathogens. However, its competitive ability and biocontrol potential is affected by soil pH and availability of iron required for growth.On the contrary, fluorescent pseudomonads seem to be versatile as judged from their suppressing *Trichoderma hamatum* by secreting siderophores and thus competing for iron in soil (Hubbard *et al.,* 1983). They demonstrated that seed colonizing pseudomonads are responsible for the failure of *T. hamatum* as a biocontrol agent in low iron soil. Addition of iron to this soil permitted *T. hamatum* to protect pea seeds, thus suggesting that pseudomonads inhibit *T. hamatum* through iron competition by siderophore production. Siderophores produced *in vitro* also inhibited *T. hamatum* mycelial growth; however, its inhibition was abolished by the addition of iron.

5.4. Suppressiveness

Some soils are referred to as suppressive soils because of their disease suppression activity against a variety of plant pathogens. The main factor(s) contributing to suppression of plant pathogens is the presence of beneficial microorganisms. Additionally, physical factors such as type of clay minerals, pH and tropical climate also contribute to their suppression.

Kloepper *et al.* (1980) revealed that siderophores of a symbiont plant arrested plant pathogenesis by sequestering available Fe^{3+} and thus denying plant pathogens iron adequate enough to establish themselves in the rhizosphere to cause disease. They characterized nature of such soils as disease suppressive. Simeoni *et al.* (1987) have gone a step further by focusing on critical iron level required for the biocontrol of *Fusarium* wilt. These efforts not only eliminated the pathogen, but also promoted height of plants by 25% and yield by 50% under field conditions.

Baker and Chet (1982) and Elad and Baker (1985) have studied the suppressivity of *Fusarium* wilt pathogens by Metz soils. They showed inhibition of germination of chlamydospores related with the production of siderophores by pseudomonads and reversal of this inhibition, totally or partially, by the addition of Fe chelators. The suppressivity by some soils of "take-all disease" of wheat appeared induced by continuous monoculture of wheat, apparently by an increased antagonism between the gradually reduced pathogen and numerically stronger fluorescent pseudomonads in the rhizosphere (Cook, 1985, Weller and Cook, 1983). It was then postulated that either antibiotics and/ or siderophores were involved in the antagonistic effect of the pseudomonads. However, it is now clear that denial of Fe^{3+} to pathogens by pseudomonads was responsible for their control.

5.5. Induced Resistance

It is evident that host population responds to microbial products or potential pathogens in a number of ways. Many of these responses lead to induction of disease resistance. The efficacy of biocontrol agents is, therefore, dependent upon neutralizing the induced resistance. One of the examples of induced resistance is protection of tomatoes against tobacco mosaic virus (Fulton, 1986).

Induced resistance can be (i) localized, detectable in the area immediately adjacent to the administration of an inducer, or (ii) systemic, judged from the resistance throughout the plant. During localized resistance, a series of biochemical events are catalysed by the host such as (i) induction of peroxidase activity to disable the attacker (pathogen), (ii) induction of protease inhibitors to arrest puncturing of the cell wall, (iii) incorporation of hydroxyproline-rich glycoproteins in the cell wall and reinforcing the cell through lignification and (iv) finally the secretion of phytoalexin and increased suberine, cumulatively imparting the underdog (attacked plant) maximum avenues of self-defence (Dean and Kuc, 1987; Hammerschmidt *et al.*, 1984; Roby *et al.*, 1987). In a systematically protected tobacco or cucumber, an increase in pathogenesis retarding (PR) nascent proteins, such as chitinase, glucanase, osmotin, etc. have been detected (Fritig *et al.*, 1987; Gianinazzi *et al.*, 1980; Metraux *et al.*, 1988).

In nutshell, Nature has provided the pathogen with certain strategies for attack and the living system under the attack with counter strategies of self-defence. It is their degree of efficiency and persistence that determines the outcome, pathogenesis or pathogenolysis.

6. MERITS AND LIMITATIONS OF BIOCONTROL STRATEGY

6.1. Merits of Biological Control

6.1.1. Microbes for biocontrol
Withdrawl of umpteen number of chemical pesticides from the market due to mounting evidence of toxicity problems by them in short and long range, would not have been feasible in the absence of alternatives. Fluorescent *Pseudomonas* and *Trichoderma* are already accepted as the versatile members of the microbial comity for biocontrol. Their use *per se* or siderophores derived from them, apparently appear innocuous today, both chemically and biologically. As biological control alone is a logical path forward to sustainable ecosystem, there is no viable alternative except to give it a fare chance to prove its utility. *Pseudomonas* and *Trichoderma* have the potential to succeed as their propagation technology is an accomplished task, are simple and economical and free from energy-intensive, capital-intensive, pollution-intensive and cost-intensive ills.

6.1.2. Siderophores for biocontrol
Although low solubility of Fe^{3+} at neutral and alkaline pH rendered it unavailable to both plants and microorganisms, hydroxamate siderophores alleviate this problem by maintaining chelated iron in solution. Similarly, Reid *et al.*(1984) showed that ferri- / desferri-oxamine-B (FeDFOB) siderophores reduced chlorosis and enhanced the growth of sorghum and sunflower plants in iron-deficient situation.

Thus, genesis of siderophores would be *in situ* at site, being necessity- induced. Depending on the characteristics of soil, whenever catecholates could not succeed, hydroxamates will step in or *vice-versa*. An army of siderophores, differing in their structures, make them more versatile by logically differing in their functional aspects, hopefully to provide a broad spectrum handle for combating wide spectrum of pathogens. Furthermore, in our opinion, Nature's provision of desferri-siderophores is articulated to ensure rapidity in response for a sudden pathogen attack, merely by conversion from desferri to ferric-siderophores. Thus, desferri-siderophores are strategically a standby arrangement for any eventuality. Their genesis, storage, transport and use appears to be auto-regulated and hence poses no danger(s) to the ecosystem.

Being biodergadable, siderophores will be metabolisable and hence there is no reason for them to cause pollution or resistance in the target pathogen. Also its accumulation in food, feed and fodder is inconceivable to pose danger to wild life or ecology. In fact they are anticipated to leave micro- and macro- environment unaffected, after accomplishing their designated task. Therefore, siderophores are hoped to (i) provide sustainability in agriculture, horticulture, floriculture and forestry, (ii) the effective and eco-friendly and (iii) leave environment cleaner and safer for future generations.

6.2. Limitations of Biological Control

6.2.1. Colonization : A pre-requisite for the success of biocontrol agents

Root colonization is an ongoing process to ensure numerical rise in the strength of biocontrol agents required for their success, starting on the seeds in the soil, multiplication in the spermosphere on seed exudates, which are rich in carbohydrates and amino acids, attachment to the root surface and subsequent proliferation on the root system in soil, containing indigenous microorganisms. For this purpose, the biocontrol strain should be able to integrate itself in the micro-environs of soil for features like soil type, soil texture, minerals and trace elements, composition of root exudates, local microflora, water potential, pH, temperature, etc. Acclimatization with these factors facilitates their colonization, which in turn assures pathogen control. However, if acclimatization and colonization by the biocontrol agent is delayed, it could pose danger to the crop, resulting in less or little yield.

6.2.2. Limitations of microbes

Biocontrol agents have been found effective in *in vitro* studies, reproducible in pot scale studies in the laboratory/ green house, but their success in the field leaves much desired. In farm trials, it is difficult to test their effectiveness and long range impact in a reasonable period, while satisfying Governmental regulations.

Secondly, their micro-habitats, rhizosphere and phyllosphere, are complex and therefore changes in population dynamics and thereby their altered behavior, cannot be predicted/ assured. Daily periodic fluctuations in humidity and temperature, as also effect of seasonal changes bringing extreme (low/ high) temperature, 100% humidity by flooding, reducing the availability of oxygen to aerobes, effect of routine agro-inputs by farmers do not allow a desired control over their equlibration with local environs, colonization and efficacy.

Further, a number of problems are associated with their formulation, storage, transport, stability and application, so also environmental impact after application before their efficacy is observed beyond doubt.

In countries like United States, because of the strict regulations by federal agencies, testing of biocontrol agents has to be within the set parameters. Even a slight variation invites hindrance in testing. Sequeira (1987) cited that for testing genetically engineered strains of *Pseudomonas syringae* for biocontrol, Dr. Lindow, University of California, Berkley was prohibited by federal agencies.

Biocontrol agents are new microorganisms added to the alien soil. Their equilibration with the microflora of the local site needs a sustained effort/ strategy so that they maintain themselves for longer period.

A systematic appraisal of the present knowledge of biocontrol agents, statistical significance of their efficacy, remedial strategy to overcome uncertainties posed by environment, protocol on its stability during storage and rapidity during colonization etc. would bring clarity on their drawbacks and time schedule to rectify them. As of today, *Agrobacterium tumefaciens*, *Trichoderma* and fluorescent *Pseudomonas* appear as the probable candidates to sort out known/ anticipated limitations in their applications.

6.2.3. Limitations of siderophores

Large scale application of siderophores is not conceivable at present in view of problems in developing its production technology. It is mired in a number of knitty-gritty problems such as selection of strain, optimization of its medium, optimization of physical parameters for its maximum productivity, methods of extraction, stabilization, quality control and problems pertaining to its stability during storage. Secondly, one or few preparations of siderophores may not be varsatile to control different pathogens under different atmospheric conditions. Thirdly, problems in its formulation, potency, application mode need to be sorted out. Unexpected problems due to variable bioefficacy, strategies of passing through cell membrane barrier, their sensitivity to pH and biodegradability by the local microflora remain to be addressed. The effect of siderophores on other microbes, especially *Trichoderma* and other rhizobacteria needs careful study so that eco-sustainability desired does not become its first casualty. Finally, in the absence of any semi-commercial trials, their economic availability can not be determined.

7.   COMMERCIAL FEASIBILITY

Schroth and Hancock (1981) observed that sound quantitative data and examples that have been thoroughly tested in the field under natural environments were very few. Only two well recognized cases, where a biological control agent had been registered for use on a commercial scale in the United States, could not offer reliable data for statistically meaningful inferences. Underlying reasons were (i) absence of funding of this area of research, (ii) lower priority accorded by the granting agencies, (iii) poor problem selection, (iv) lack of sustained efforts etc. It is merely hoped (Sequeira, 1987) that the situation will change as new approaches and new techniques are exploited for biocontrol.

Laboratory and field trials on the development of a commercially viable technology for the production of biocontrol agent(s) (siderophores) has a strong potential. In fact, *Agrobacterium tumefaciens* is already commercially sold in USA under the trade name Galltrol, to control crown gall. Table 6 summerises some of the commercial successes in this field. Therefore, siderophores have a fair chance to emerge as a sustainable technology for 21st century.

Table 6. Worldwide application of biocontrol agents (Elad and Chet, 1995)

Country	Biocontrol agent	Pathogen/ Disease suppressed
USA	*Agrobacterium radiobacter*	Crown gall
	Pseudomonas fluorescence	Seedling disease
	T.harzianum/polysporum	Wood decay
	Trichoderma sp.	Root disease
USSR	*Pythium oligandrum*	*Pythium* sp.
	Glioacdium virens	Seedling disease
Europe	*Trichoderma viride*	Timber pathogen
UK	*Peniophora gigantea*	Fommes annosus
Japan	*Fusarium oxysporum*	*Fusarium oxysporum*
Australia	*Agrobacterium radiobacter*	Crown gall
New Zealand	*Agrobacterium radiobacter*	Crown gall

In our laboratory, a variety of *Pseudomonas* strains from the recognized Culture Collection Centers and a few local isolates have shown *Fusarium oxysporum* and *Aspergillus flavus* suppression (Chincholkar, 1997). These were confirmed by *in vitro* studies where, in the presence of siderophores of fluorescent pseudomonads, these pathogens failed to grow. Bioefficacy studies on wheat have indicated its beneficial role in promoting plant growth, evidenced by an increase in the biomass and chlorophyll content of wheat crop.

For the release of biocontrol agents in the field, press mud is being explored as a carrier (matrix) to afford sustained release of nutrients for preserving initial number even after six months' storage under ambient conditions. Towards this objective, a laboratory scale experiment simulating commercial scale conditions is underway in our laboratory for deciding its usefulness. However, these studies are far away from commercialization.

A final word on the commercialization depends upon its success, simplicity in production, favourable consumer responce, lack of viable competing product(s), presence/absence of competing technology, assured returns, future prospects and possibility of backward integration for profit optimization. Unless the above stated each aspect is critically assessed, it is premature at this stage to venture for commercialization. However, semi-commercial operations are the need of the time to gain experience in the right direction.

8. CONCLUSION

In view of different structures of siderophores in different microbes, vast possibilities exist for their varying applications, efficacy and significance. Nature seems to have ingeniously devised an integrated strategy of plant protection, healthcare, environment remediation, etc. A focus in understanding the Nature will help in preserving it.

Although use of biocontrol agents is the solution in controlling many plant diseases, new plant diseases have arisen like red rot of sugarcane in South India, wilt of sugarcane in Gujarat, karnal bunt in North-West India and bacterial blight of rice in Punjab. These emerging diseases and contingencies invite continuous development of novel biocontrol agents in an eco-friendly, easy to apply, sure to work, free from hazzards and economically viable way.

REFERENCES

Actis, L.A., Fish, W., Crosa, J. H., Kellarman, K., Ellenberger, S. R., Hauser, F. M. and Sanders-Lohr, J. 1986, Characterization of anguibactin, a novel siderophore from *Vibrio anguillarum* 775 (pJM), *J. Bacteriol.* 167 : 57-65.

Akers, H. A. 1983, Multiple hydroxamic acid microbial iron chelators (siderophores) in soil, *Soil Sci.* 135 : 156-159.

Ayers, W.A. and Adams, P.B. 1979, Mycoparasitism of sclerotia of *Sclerotinia* and *Sclerotium* sp. by *Sporidesmium sclerotivorum, Can. J. Microbiol.* 25 : 17-32.

Backer, J. O., Messens, E. and Hedges, R. W. 1985, The influence of agrobactin on the uptake of ferric iron by plants, *FEMS Microbiol. Ecol.* 31 : 171-175.

Bagg, A. and Neilands, J. B. 1987, Molecular mechanism of regulation of siderophore-mediated iron assimilation, *Microbiol. Rev.* 51: 509-518.

Baker, R. and Chet, I. 1982, Introduction of suppressiveness, In: *Suppressive Soils and Plant Disease,* ed. R.W. Schneider, Amer. Phytopathol. Soc., St. Paul, MN, USA, pp. 88.

Baker, R. 1985, Biological control of plant pathogens : definitions, In: *Biological Control in Agricultural IPM Systems,* eds. D.C. Herzog and. M.A. Hoy, Academic Press, New York, USA pp. 25-39.

Bakker, P. A., Lamers, J. G., Bakker, A. W., Marngg, J. D., Weisbeek, P. J. and Schippers, B. 1986, The role of siderophores in potato tuber yield increase by *Pseudomonas putida* in a short rotation of potato, *Net. J. Plant Pathol.* 92 : 249-256.

Bentley, M. D., Aderegg, R. J., Szaniszlo, P. and Davenport, R. F. 1986, Isolation and identification of the principal siderophore of the dermatophyte *Microsporum gypseum*, *Biochem.* 25: 1455-1459.

Boosalis, M. G. 1964, Hyperparasitism, *Ann. Rev. Phytopathol.* 2: 363-373.

Brickman, T. J., Hansel, J. G., Miller, M. J.and Armstrong, S. K. 1996, Purification, spectroscopic analysis and biological activity of the macrocyclic dihydroxamate siderophore alcaligin produced by *Bordetella pertussis* and *B. bronchiseptica*, *BioMetals* 9 : 191-203.

Budzikiewicz, H., Muuenzinger, M., Taraz, K. and Meyer, J. M. 1997a, Schizokinen the siderophore of plant deleterious bacterium *Ralstonia (Pseudomonas) solanacearum* ATCC 11696, Zeitschriftfhr Naturforschung Section C, *J. Biosciences* 52 : 496-503.

Budzikiewicz, H., Boessenkamp, A., Taraz, K., Pandey, A. and Meyer, J. M. 1997b, Corynebactin, a cyclic catecholate siderophore from *Corynebacterium glutamicum* ATCC 14067 (*Brevibacterium* sp. DSM 20411), Zeitschrift fhr Naturfordchung Section C, *J. Biosciences* 52 : 551-554.

Castignetti, D. and Smarrelli, J. 1986, Siderophores, the iron nutrition of plants and nitrate reductase, *FEBS Lett.* 209 : 147-151.

Chambers, C. E., McIntyre, D. D., Mouck, M. and Sokol, T.A.1996, Physical and structural characterisation of yersiniophore, a siderophore proudced by clinical isolate of *Yersinia enterocolitica*, *BioMetals* 9 : 157-167.

Chaudhari, B. L. and Chincholkar, S. B. 1998, Microbial siderophores, In: *Selected Topics in Biotechnology*, ed. A. M. Deshmukh, Pama Publications, Karad, India, pp. 32-42.

Chincholkar, S. B., Chaudhari, B. L. and Wakharkar, R. D. 1995, Siderophore production by *Curvularia lunata*, Paper presented at 2nd International Conference on "Pharmaceutical Biotechnology", Ghent, Belgium.

Chincholkar, S. B. 1997, Role of microbial siderophores in plant disease management, Paper presented at Int. Conf. on Plant Disease Management, IARI, New Delhi.

Cook, R.J. 1985, Biological control of plant pathogens : theory to application, *Phytopathol.* 75: 25-29.

Cornish, A.S. and Page, W.J. 1995, Production of the tricatecholate siderophore protochelin by *Azotobacter vinelandii*, *BioMetals* 8 : 332-338.

Dean, R.A. and Kuc, J. 1987, Rapid lignification in response to wounding and infection as a mechanism for induced systemic protection in cucumber, *Physiol. Mol. Plant Pathol.* 31: 69-81.

De-Meyer, G. and Hofte, M. 1997, Salicylic acid produced by the rhizobacterium *Pseudomonas aeruginosa* 7NSK2 induces resistance to leaf infection by *Botrytis cinerea* on bean, *Phytopathol.* 87 : 588-593.

Dileepkumar, B. S. and Bezbaruah, B. 1997, Plant growth promotion and fungal pest control through an antibiotic and siderophore producing fluorescent Pseudomonas strain from tea [*Camellia sinensis* (L) O kuntze] plantations. *Ind. J. Microbiol.* 35: 289-292.

Drechsel, H., Metzger, J., Freund, S., Jung, G., Boelaert, J. R. and Winkelmann, G. 1991, Rhizoferrin- A novel siderophore from the fungus *Rhizopus microsporus* var. *rhizopodiformis*, *Biol. Met.* 4 : 238-243.

Elad, Y. and Baker, R. 1985, Influence of trace amounts of cations and siderophore-producing pseudomonads on chlamydospore germination of *Fusarium oxysporum*, *Phytopathol.* 75: 1047-1052.

Elad, Y. and Chet, I. 1985, Practical approaches for biocontrol implementation, In : *Novel Approaches to Integrated Pest Management*, ed. R. Reuveni, CRC Press, Boca Raton FL, USA, 1995, pp. 369-378.

Emery, T. 1971, Role of ferrichrome as a ferric ionophore in *Ustilago sphaerogena*, *Biochem.* 10 : 1483-1488.

Emery, T. 1980, Malionichrome, a new iron chelate from *Fusarium roseum*, *Biochim. Biophys. Acta* 629 : 382-390.

Estrella, A. H. and Chet, I. (ed.) 1998, Biocontrol of bacteria and phyto-pathogenic fungi, In : *Altman Agricultural Biotechnology*, Marcel Dekker Inc., New York, USA, pp. 263-282.

Fekete, F. A., Lanzi, R. A., Beaulieu, J. B., Longcope, D. C., Sulya, A. W., Hayes, R. N. and Mabboti, G.A. 1989, Isolation and preliminary characterisation of hydroxamic acids formed by nitrogen -fixing *Azotobacter chroococcum* B-8, *Appl. Environ. Microbiol.* 55 : 298-305.

Freitas, S. D. S. and Pizzinatto, M. A. 1997, Action of rhizobacteria on the *Colletotrichum gossypii* incidence and growth promotion in cotton seedlings, *Summa Phytopathologica* 23 : 36-41.

Fritig, B., Kauffmann, S., Dumas, B., Geoffrey, P., Kopp, M. and Legrand, M. 1987, Mechanism of the hypersensitivity reaction of plants, In: *Plant Resistance to Viruses*, Ciba Foundation Symp. 133, Wiley, Chichester, pp. 92-108.

Fulton, R.W. 1986, Practices and precautions in the use of cross protection for plant virus disease control, *Ann. Rev. Phytopathol.* 24: 67-81

Fusakawa, K., Goto, M., Sasaki, K., Hirata, Y. and Sato, S. 1972, Structure of the yellow-green fluorescent peptide produced by iron-deficient *Azotobacter vinelandii* strain 0, *Tetrahedron* 28 : 5359-5365.

67

Gianinazzi, S., Ahl, P., Cronu, A. and Scalla, R. 1980, First report of host β-protein appearance in response to a fungal infection in tobacco, *Physiol. Plant Pathol.* 16: 337-342.

Godtfredsen, W.O. and Vangedal, S. 1965, Trichodermin, a new sequiternpene antibiotic, *Acta. Chem. Scand.* 19: 1088.

Guerinot, M. L. 1994, Microbial iron transport, *Ann. Rev. Microbiol.* 48 : 743-772.

Haag, H., Hantke, K., Drechsel, H., Strojilkovic, I., Jung, G. and Zahner, H. 1993, Purification of yersiniabactin: A siderophore and possible virulence factor of *Yersinia enterocolitica, J. Gen. Microbiol.* 139 : 2159-2165.

Hammerschmidt, R., Lamport, D. T. A. and Muldoon, E. 1984, Cell wall hydroxyproline enhancement and lignin deposition as an early event in the resistance of cucumber to *Cladosporium cucumerinum, Physiol. Plant Pathol.* 24: 43-47.

Hesseltine, C. W., Pidacks, C., Whiteihall, A. R., Bohonos, N., Hutchings, B. L. and Williams, J. H. 1952, Coprogen, a new growth factor for coprophilic fungi, *J. Am. Chem . Soc.* 74 : 1362-1368.

Hohlneicher, U., Hartmann, R., Taraz, K. and Budzikiewicz, H. 1995, Pyoverdin, ferribactin, azotobactin - a new triad of siderophores from *Pseudomonas chlororaphis* ATCC 9446 and its relation to *Pseudomonas fluorescens* ATCC 13525, Zeitschrift fhr Naturforschung Section. C, *J.Biosciences* 50 : 337-344.

Holger, S., Freund, S., Werner, B., Jung, G., Meyer, J. M. and Winkelmann, G. 1993, Ornibactins : A new family of siderophores from *Pseudomonas, BioMetals* 6 : 93-100.

Horowitz, N.A., Charlang, G., Horn, G. and Williams, N.P. 1976, Isolation and identification of the conidial growth factor of *Neurospora crassa, J. Bacteriol.* 127 : 135-140.

Hossain, M. B., Jalal, M. A. F., Benson, B. A., Barnes, C. L.and van der Helm, D. 1987, Structure and confirmation of two coprogen type siderophores: Neocoprogen I and Neocoprogen II, *J. Am. Chem. Soc.* 109 : 4948-4954.

Hubbard, J. E., Harman, G.E. and Hadar, Y. 1983, Effect of soil born *Pseudomonas* sp. on the biological control agent *Trichoderma hamatum* on pea seeds, *Phytopathol.* 73: 655-659.

Hulme, M.A. and Shields, J. K. 1970, Biocontrol of decay fungi in wood by competition for non-structured carbohydrates, *Nature* 227: 300-305.

Ignacimuthu, S. (ed.) 1996, Agriculture and biotechnology, In: *Applied Plant Biotechnology*, Tata Mcgrew Hill Publishing Company Ltd., New Delhi, India, pp. 65-130.

Imbert, M., Bechet, M. and Blondeau, R. 1995, Comparison of the main siderophores produced by some species of Streptomyces, *Curr. Microbiol.* 31: 129-133.

Jalal, M. A. F., Mocharla, R. and van der Helm, D. 1984, Separation of ferrichromes and other hydroxamate siderophores of fungal origin by reversed-phase chromatography, *J. Chromatogra* 301 : 247-252.

Jalal, M. A. F. and van der Helm, D. 1991, Isolation and spectroscopic identification of fungal siderophores, In : *Handbook of Microbial Iron Chelates*, ed. G. Winkelmann, CRC Press, Boca Raton, USA, pp. 235-269.

Johri , B. N., Rao, C. V.S. and Goel, R. 1997, Fluorescent pseudomonads in plant disease management, In : *Biotechnological Approaches in Soil Microorganisms for Sustainable Crop Production*, ed. K.R. Dadarwal, Scientific Publishers, Jodhpur, India, pp. 193-223.

Khan, M. A., Lelie, D. Van Der, Cornelis, P. S. and Mergeay, M. 1994, Purification and characterisation of alcaligin E, a hydroxamate type siderophore produced by *Alcaligenus eutrophus,* Ch 34, Colloq. INRA, 66 (Plant pathogenic bacteria), pp. 591-597.

Kloepper, J. W., Leong, J.,Teintze, M. and Schroth, M. N. 1980, *Pseudomonas* siderophores: a mechanism explaning disease-suppressive soils, *Curr. Microbiol.* 4 : 317-320.

Kloepper, J.W. 1983, Effect of seed piece inoculation with plant growth promoting rhizobacteria on populations of *Erwinia carovora* on potato roots and in daughter tubers, *Phytopathol.* 73: 217-219.

Lankford, C. E. 1973, Bacterial assimilation of iron, *Crit. Rev. Microbiol.* 2 : 273-331.

Linget, C., Stylianou, D. G., Dell, A., Wolff, R. E., Piemont, Y. and Abdallah, M. 1992, Bacterial siderophores: the structure of a desferribactin produced by a *Pseudomonas fluorescens* ATCC 13525, *Tetrahedron Lett.* 33: 3851-3854.

Loper, J. E. 1988, Role of fluorescent siderophore production in biological control of *Pythium ultimum* by a *Pseudomonas fluorescence* strain, *Phytopathol.* 78 : 166-172.

Loper, E. and Buyer, J. S. 1991, Siderophores in microbial interactions on plant surfaces, *Mol. Plant Microbe Interact.* 4: 5-13.

Manulis, S., Kashman, Y. and Barash, I. 1986, Identification of siderophores and siderophore mediated uptake of iron in *Stemphylium botyrosum, Phytochem.* 26: 1317-1321.

Meiwes, J., Fiedler, H., Haag, H., Zahner, H., Konetschny-rapp, S. and Jung, G. 1990, Isolation and characterization of Staphyloferrin A, a compound with siderophore activity from *Staphylococcus hyicus* DSM 20459, *FEMS Microbiol. Lett.* 67 : 201-206.

Messenger, A.J. M. and Ratledge, C. 1986, Siderophores, In: *Comprehensive Biotechnology,* Vol. III, ed. Murray Moo Young, Pergamon Press, London pp. 275-295.

Metraux, J. P., Streit, L. and Staub, T. 1988, A pathogenesis-related protein in cucumber is a chitinase, *Physiol. Mol. Plant Pathol.* 33: 1-9.

Meyer, J. M., Halle, F., Hohnadel, D., Lemanceau, P. and Ratfiarivelo, H. 1987, Siderophores of *Pseudomonas*-biological properties, In : *Iron Transport in Microbes, Plants and Animals,* eds. G. Winkelmann, D. van der Helm and J.B. Neilands, VCH Verlagsgesellshaft, Weinheim, pp. 188-205.

Meyer, J. M., Hohnadel, D. and Halle, F. 1989, Cepabactin from *Pseudomonas cepacia,* a new type of siderophore, *J. Gen. Microbiol.* 135 : 1479-1487.

Moenne, L. Y., McHugh, B., Stephens, P. M., McConnell, F. I., Glennon, J.D., Dowling, D. M. and O'Gara, F. 1996, Rhizosphere competence of fluorescent *Pseudomonas* sp. B24 genetically modified to utilise additional ferric siderophores, *FEMS Microbiol. Ecol.* 19 : 215-225.

Mullis, K. B., Pollack, J. R. and Neilands, J. B. 1971, Structure of schizokinen, an iron transport compound from *Bacillus megaterium, Biochem.* 10 : 4894-4898.

Mulya, K., Watanabe, M., Goto, M., Takikawa, Y. and Tsuyumu, S. 1996, Suppression of bacterial wilt disease in tomato by root dipping with *Pseudomonas fluorescence* PfG32: The role of antibiotic substances and siderophore production, *Ann. Phytopath. Soc. Japan* 62 : 134-140.

Neilands, J. B. 1981, Microbial iron compounds, *Ann. Rev. Biochem.* 50 : 715-731.

Nigam, N. and Mukerji, K.G. 1988, Biological control - concepts and practices, In : *Biocontrol of Plant Diseases,* eds. K.G. Mukerji and K.L. Gerg, CRC Press Inc., Boca Raton, Florida, USA, pp. 1-14.

Okujo, N., Saito, M., Yamamoto, S., Yoshida, T., Miyoshi, S. and Shinoda, S. 1994, Structure of vulnibactin, a new polyamine-containing siderophore from *Vibrio vulnificus,* BioMetals 7 : 109-116.

Paroda, R.S. 1997, Valedictory lecture delivered at International Conference on Integrated Plant Disease Management For Sustainable Agriculture, New Delhi, India.

Persmark, M., Expert, D. and Neilands, J.B. 1989, Isolation, characterization and synthesis of chrysobactin, a compound with siderophore activity from *Erwinia chrysanthemi, J. Biol. Chem.* 264 : 3187-3193.

Plessner, O., Klapatch, T. and Guerinot, M. L. 1993, Siderophore utilization by *Bradyrhizobium japonicum, Appl. Environ. Microbiol.* 59 : 1688-1690.

Powell, P. E., Cline, G. R., Reid, C. P. and Szaniszlo, P. J. 1980, Occurrence of hydroxamate siderophore iron chelators in soils, *Nature* 287 : 833-834.

Pyke, T.R. and Dietz, A. 1966, U-21, 963, A new antibiotic I, Discovery and biological activity, *Appl. Microbiol.* 14: 506-509.

Raaijmakers, J. M., Bitter, W., Punte, H. L. M., Bakker, P. H. A. M., Weisbeek, P. J. and Schippers, B. 1995, Siderophore receptor PupA as a marker to monitor wild-type *Pseudomonas putida* WCS358 in natural environments, *Appl. Environ. Microbiol.* 60: 1184-1190.

Reid, C.P.P., Crowley, D.E., Powell, P.E., Kim, H.J. and Szaniszlo, P.J. 1984, Utilization of iron by oat when supplied as ferrated hydroxamate siderophore or as ferrated synthetic chelate, *J. Plant. Nutr.* 7: 437-447.

Reid, R. T., Live, D. H., Faulkner, D. J. and Buttler, A. 1993, A siderophore from a marine bacterium with an exceptional ferric ion affinity constant, *Nature* 366: 455-458.

Roby, D., Toppan, A. and Esquerre-Tugaye, M.T. 1987, Cell surfaces in plant microorganisms interactions VIII, Increased proteinase inhibitor activity in melon plants in response to infection by *Colletotrichum lagenarium* or to treatment with an elicitor fraction from this fungus, *Physiol. Mol. Plant Pathol.* 30: 453-460.

Rodriguez-Kabana, R. and Curl, E.A. 1968, Saccharase activity of *Sclerotium rolfsii* in soil and the mechanism of antagonistic action by *Trichoderma viride, Phytopath.* 58: 985-990.

Rodriguez-Kabana, R. 1969, Enzymatic interactions of *Sclerotium rolfsii* in soil and the mechanism of antagonistic action by *Trichoderma viridie, Phytopath.* 58: 487-491.

Rodriguez-Kabana, R., Kelly, W. D. and Curl, E.A. 1978, Proteolytic activity of *Trichoderma viride* in mixed culture with *Sclerotium rolfsii* in soil, *Can. J. Microbiol.* 24: 487-492.

Saikia, N. and Bezbaruah, B. 1995, Iron dependent plant pathogen inhibition through *Azotobacter* RRLJ203 isolated from iron rich acid soil, *Ind. J. Expt. Biol.* 35: 571-575.

Sayer, J. M. and Emery, T. 1968, Structures of naturally occurring hydroxamic acids, fusarinines A and B, *Biochem.* 7 : 184-190.

Schroth, M.N and Hancock, J.G. 1981, Selected topics in biological control, *Ann. Rev. Microbiol.* 35: 453-476.

Sequeira, L. 1987, Biological control of plant pathogens: Present status and future prospects, In : *Integrated Pest Management,* ed. V. Delucchi, Parasitis, Geneva, Switzerland, pp. 217-235.

Sharman, G. J., Williams, D. H., Ewing, D. F. and Ratledge, C. 1995, Isolation, purification and structure of exochelin MS, the extracellular siderophore from *Mycobacterium smegmatis, Biochem. J.* 305: 187-196.

Shenker, M., Ghirlando, R., Oliver, I., Helmann, Y., Hadar, Y. and Chen, Y. 1995, Chemical structure and biological activity of a siderophore produced by *Rhizopus arrhizus*, *Soil Sci. Soc. Am. J.* 59 : 837-843.

Simeoni, L. A., Lindsay, W. L. and Baker, R. 1987, Critical iron level associated biological control of *Fusarium* wilt, *Phytopath.* 77 : 1057-1061.

Sindhu, S. S., Suneja, S. and Dadarwal, K. R. 1997, Plant growth promoting rhizobacteria and their role in crop productivity, In : *Biotechnological Approaches in Soil Microorganisms for Sustainable Crop Production,* ed. K.R. Dadarwal, Scientific Publishers, Jodhpur, India pp. 149-193.

Smidt, M.L and Vidaver, A. 1982, Bacteriocin production by *Pseudomonas syringae* Ps W-1 in plant tissue, *Can. J. Microbiol.* 28: 600-604.

Smith, M. J., Shoolery, J. N., Schwyn, B., Holden, I. and Neilands, J. B. 1985, Rhizobactin, a structurally novel siderophore from *Rhizobium meliloti*, *J. Am. Chem. Soc.* 107 : 1739-1743.

Teintz, M., Hossain, M. B., Barnes, C. L., Leong, J. and van der Helm, D. 1981, Structure of ferric pseudobactin, a siderophore from a plant growth promoting *Pseudomonas*, *Biochem.* 20: 6446-6457.

Telford, J.R., Leary, J. A., Tunstad, L. M. G., Byers, B. R. and Raymond, K. N. 1994, Amonabactin: Characterisation of a series of siderophore from *Aeromonas hydrophilla*, *J. Am. Chem. Soc.* 116 : 4499-4500.

Toole, E.R. 1971, Interaction of mold and decay fungi on wood in laboratory tests, *Phytopathol.* 61: 124-129.

Vidaver, A. 1976, Prospects for control of phytopathogenic bacteria by bacteriophages and bacteriocins, *Ann. Rev. Phytopath.* 14: 451-465.

Weindling, R. 1932, *Trichoderma lignorum* as a parasite of other soil fungi, *Phytopath.* 22: 837-842.

Weller, D.M. and Cook, R.J. 1983, Suppression of take-all of wheat by seed treatments with fluorescent pseudomonads, *Phytopath.* 73: 463-469.

Wells, H.D. 1988, *Trichoderma* as a biocontrol agent, In : *Biocontrol of Plant Diseases,* eds. K.G. Mukerji and K.L. Gerg, CRC Press Inc., Boca Raton, Florida, USA, pp. 71-82.

Winkelmann, G. and Braun, V. 1981, Stereoselective recognition of ferrichrome by fungi and bacteria, *FEMS Microbiol. Lett.* 11: 237-241.

Winkelmann, G. 1991, Specificity of iron transport in bacteria and fungi, In : *Handbook of Microbial Iron Chelators,* ed. G. Winkelmann, CRC Press, Boca Raton, Florida, USA, pp. 65-105.

Winkelmann, G. and Drechsel, H. 1997, Microbial siderophores, In : *Biotechnology,* (2nd Edn.), Vol VII, eds. H.J. Rehm and G. Reed, VCH Publishers, Weinheim, pp. 199-245.

Yamamoto, S., Okujo, N. and Sakakibara, Y. 1994, Isolation and structure elucidation of acinetobactin, a novel siderophore from *Acinetobacter baumanii, Arc. Microbiol.* 162: 249-254.

ANTIVIRAL PHYTOPROTEINS AS BIOCONTROL AGENTS FOR EFFICIENT MANAGEMENT OF PLANT VIRUS DISEASES

V. K Baranwal[1] and H. N. Verma[2]

[1]National Centre for Integrated Pest Management (ICAR)
Pusa Campus,
New Delhi - 110012, INDIA

[2]Department of Botany
Lucknow University
Lucknow - 226 001, U.P., INDIA

1. INTRODUCTION

Plant virus diseases have always been of great concern to farmers, researchers and policy makers because they cause enormous yield loss in many cereal, vegetable, fruit, legume and cash crops like cotton. Plant virus diseases are very difficult to control and are much problematic than commonly occurring fungal diseases. Virus infection causes loss by reducing the productive life of the crop e.g. potatoes infected with leaf roll , by adverse effect on vegetative propagation, on germination of seeds and growth of seedlings, by reduction in quality of fruits, by loss of vigour and by number of other usual and unusual ways (Waterworth and Hadidi, 1998). The recent outbreak of cotton leaf curl virus disease in cotton in northern cotton growing region of India has led to huge yield loss of cotton fibre. Approximately 12,000 ha of area under cotton was affected by leaf curl virus disease during 1996 in Rajasthan alone. An annual loss of US $ 300 million is caused by MYMV by reducing the yield of black gram, mungbean and soybean (Varma *et al.,* 1992). It is therefore, important to develop management strategy for virus diseases of important crops so that the losses can be minimised. Viral parasitism is unique, in contrast to fungi and bacteria. Viruses do not attack the structural integrity of their host tissues, but instead they subvert the synthetic machinery of the host cell, acting as molecular pirates. Therefore, the management of virus diseases is a difficult task. Control strategies are mainly aimed at reducing or eliminating existing sources of infection, prevention of virus transmission, etc. The development of methods of virus disease control, continue to be vital elements in the drive to improve crop productivity. The use of molecular genetic techniques has provided new insights into how plants defend themselves against pest attack. This new understanding is suggesting novel ways in which plant disease might be controlled. Most of the current researches on management of plant viruses are centred around

71

to confer resistance to host plants against virus infection. Usually the resistance is derived from the host only and it is termed as host derived resistance (HDR). It is achieved either through conventional breeding programme or by genetic engineering. The conventional breeding programme have provided limited successes of durable resistance but overall the process of incorporation of resistance by crossings and repeated back crossing takes a number of years and quite expensive. The use of recent biotechnological tools now facilitate identification of natural genes for resistance and their incorporation in host plants. However, genetic engineering is being used more and more for transgressing of genes from pathogen often termed as pathogen derived resistance (PDR) which includes resistance derived from coat protein, replicase protein, movement protein, helper component, satellite RNA, sense and antisense RNA. Here, we shall be discussing on resistance derived from non- host plants i.e. non -host derived resistance (NHDR) or resistance mediated by antiviral phytoproteins derived from non - host healthy plants and their potential use as biocontrol agents. Plants defence systems can be classified as passive (constitutive) or active (inducible). It is the inducible defence system which is switched on after treatment with phytoproteins from non-host plants. These proteins are active inducers of plant defence responses against viruses. The induced resistance by phytoproteins against plant viruses is a new, environmentally safe means of disease management. The study of induced resistance is rapidly expanding and there is great interest in this area now a days because of practical application of induced resistance in virus disease management. This paper is concerned with induced resistance towards viruses. Systemic induced resistance attracts the greatest interest by providing potential for its use in new methods of crop protection. The use of systemic induced resistance as part of programme for disease management is an exciting prospect.

2. ANTIVIRAL PHYTOPROTEINS

2.1. Characteristics

Though the studies on antiviral property of non - host plants dates back to 1918, it is only during the last 20 years some very interesting and valuable informations on molecular level characterisation, mode of action and their possible use in management of plant viruses have been generated , however, antiviral substances occurring in these plants have been characterised only from a very few plants. Most of characterised antiviral substances are basic proteins. Their molecular weight ranges between Mr. 20,000 to Mr. 32000. A few of the antiviral substances have been also characterized as polysaccharide, phenolic, alkaloid, quinone and salt. However, detailed studies have only been carried out for proteinaceous antiviral proteins (Verma *et al.,* 1995 a,b; 1998). Another property associated with many of these antiviral proteins is that they are ribosome inactivating proteins(RIPs) (Table 1). The RIPs like abrin, ricin, modeccin, gelonin and momordin inhibited local lesions produced by tobacco mosaic virus (TMV) in the range of 42 to 85 % but *Phytolacca* antiviral protein and crude extract from *Bryonia* seeds and *Dianthus caryophyllus* caused 100% inhibition (Stevens *et al,* 1981). Similarly antiviral protein from *Clerodendrum aculeatum* shown to be RIP also produced 99.8 % inhibition (Kumar *et al.,* 1997; Verma *et al.,* 1996). Thus it appears that these antiviral proteins though show ribosome inactivating property but they vary in their virus inhibitory activity. Many of RIPs are also inhibitory to animal viruses (Battelli and Stripe, 1995; Gerbes *et al.,* 1996). There are many other applications of these plant antiviral RIPs which is beyond the scope of this paper.

The antiviral proteins isolated and characterised from *Boerhaavia diffusa, Chenopodium amaranticolor, Spinacea oleracea, Lentinus edodes, Clerodendrum inerme* and *Celosia cristata* have not been studied for depurination of ribosome as yet. The antiviral proteins from *Phytolacca americana* (PAP), *Mirabilis jalapa* (MAP) and *Dianthus caryophyllus*

Table 1. Some virus inhibitory plant proteins possessing property of host ribosome inactivation.

Antiviral* protein	Source	Virus-host system	References
Abrin	*Abrus precatorius* seeds	TMV/*N. glutinosa*	Stevens *et al.*, 1981
Agrostins	*Agrostemma githago* seeds	TMV/*N. glutinosa*	Stirpe *et al.*, 1983
Bryodin	*Bryonia dioica* leaves and roots	TMV/*N. glutinosa*	Stirpe *et al.*, 1986
CA-SRI	*Clerodendrum aculeatum* leaves	TMV/*N.glutinosa* TMV/*N.tabacum* Samsun NN SRV/*Cyamopsis tetragonoloba*	Verma *et al.*, 1996, Kumar *et al.*, 1997
Dianthins	*Dianthus caryophyllus* leaves	TMV/*N.glutinosa*	Stirpe *et al.*, 1981
EHL (*Eranthis hyemalis* lectin)	*Eranthis hyemalis* bulb	AMV/*Vigna vigna*	Kumar *et al.*, 1993
Gelonin	*Gelonium multiflorum* seeds	TMV/*N. glutinosa*	Stevens *et al.*, 1981
MAP	*Mirabilis jalapa* roots	TMV/ *N. tabacum*	Kubo *et al.*, 1990
Modeccins	*Adenia digitata* roots	TMV/*N. glutinosa*	Stevens *et al.*, 1981
Momordin	*Momordica charantia* seeds	TMV/*N. glutinosa*	Stevens *et al.*, 1981
PAP	*Phytolacca americana* seeds, leave, roots	TMV/*N.glutinosa* SBMV/*Phaseolus vulgaris*, TMV/ *P. Vulgaris*, AIMV, CMV,PVY,TMV/ *Chenopodium amaranticolor*, PVX/*Gomphrena globosa*, ACMV/*N. benthamiana*, CaMV/*Brassica campestris*	Batelli and Stripe 1995
Ricin	*Ricinus communis* seeds	TMV/*N. glutinosa*	Stevens *et al.*, 1981
Saporins	*Saponaria officinalis* leaves, roots, seeds	TMV/*N. glutinosa*	Stirpe *et al.*, 1983
YLP (*Yucca recurvifolia* protein)	*Yucca recurvifolia* leaves	TMV/*C. amaranticolor*	Ito *et al.*, 1993*

* When these proteins are mixed with viruses prior to inoculation, local lesions are reduced or completely suppressed.

(Dianthins) and systemic resistance inducer protein from *Clerodendrum aculeatum* (CA-SRI) have been studied in detail with regard to their characterisation, cloning and transgressing of genes are concerned. They have potential for developing non-host and non-pathogen derived transgenics for prevention/ reduction of virus diseases. We shall also be focusing on other antiviral plants which have not been fully characterised but at the same time have shown good promise in reducing virus infection.

2.2 Mechanism of Antiviral Action

It is indeed very difficult to explain the mechanism of action of the antiviral proteins. Polysaccharide inhibitors from *Physarum polycehalum* and *Abutilon striatum* have been shown to act on the surface of virus articles (Mayhew and Ford, 1971; Moraes *et al.*, 1974) but none of the proteinaceous antiviral substances has been shown to do so. Ragetli (1975) indicated that inhibitors from a number of plants from Centrospermae inhibited the virus establishment by blocking the receptor sites situated near the leaf surface. This was only circumstantial and no experimental evidence was provided. Studies on antiviral proteins from a number of plants showed that they were also having the property of ribosome inactivation and virus inhibition by these plant proteins was explained differently. Most of these works were centred around the phytoproteins from the genus *Phytolacca* known for a number of proteins having antiviral properties. These antiviral proteins inhibited a number of viruses (Table 1). They have the property of inactivating the ribosomes and enzymatically function as ribosome - specific N- glycosidases by specifically removing a single adenine from the larger RNA component of the ribosomes (Irvin, 1995). *Phytolacca* antiviral protein (PAP), a single chain ribosome inactivating protein (RIP) was initially thought to prevent viral infection by inhibiting the synthesis of viral proteins (Owens *et al.*, 1973). It was hypothesised that any invasion of a cell by a virus that causes rupture of the cell wall and membrane (as caused by abrasives like carborundum at the time of virus inoculation) would allow PAP to enter the cytoplasm and inactivate the ribosomes which would prevent the virus from replication. However, subsequent report that PAP inhibits the virus infection when applied on the lower surface while virus infection accomplished on upper surface (Chen *et al.*, 1991) did not prove this hypothesis. Indeed in 1997, it was shown that depurination of tobacco ribosomes and toxicity could be dissociated from virus inhibiting activity of PAP when Tumer *et al.* (1997) demonstrated that non - toxic C-terminal deletion mutant of PAP inhibited viral infection but did not depurinate host ribosomes. Earlier gene for PAP was cloned and was transgressed into tobacco (*Nicotiana tabacum*). The transgenics showed resistance to infection by TMV and PVX but the growth of transgenics was reduced due to toxicity of PAP caused by RIP property (Lodge *et al.*, 1993). To circumvent the problem of cytotoxicity of antiviral proteins, Hong *et al.* (1996) developed a novel strategy for engineering virus resistance in transgenic plants inducing a virus - inducible promoter from African cassava mosaic virus (ACMV) to regulate the expression of dianthin , an antiviral protein from *Dianthus caryophyllus*. They demonstrated that *Nicotiana benthamiana* plants transformed with the dianthin coding sequence under the control of the ACMV coat protein promoter were less susceptible to ACMV infection. This study indicates that even infection by gemini viruses can be reduced by antiviral phytoproteins. However, the role of RIPs implicated in reducing the virus infection has not been conclusively shown. To remove the cytotoxicity, random mutagenesis has been used and a number of non toxic PAP have been isolated and transgenics have been developed (Smirnov *et al.*, 1997). After having shown that RIP activity of PAP is not associated with antiviral property, it appears that PAP is interacting and activating the host system for antiviral activity. Smirnov *et al.* (1997) demonstrated that PAP is responsible for generating signal in the host plants for inhibition of virus infection. They also indicated that PAP expression in transgenic rootstocks of grafted plants induces resistance to viral infection in wild type scions in the absence of

74

detectable levels of PAP, salicylic acid (SA) accumulation , and PR-protein synthesis. This suggests that PAP expression induces a pathway different from the classical systemic acquired resistance which is normally associated with SA accumulation and PR-protein synthesis. Thus, it seems antiviral proteins expressed in transgenic plants trigger the signal transduction to prevent the virus infection and thus can be very useful if genes of these signal transducing molecule can be transgressed into economic plants prone to virus infections and could be activated at the time of virus invasion.

Most of the antiviral phytoproteins have been reported to prevent virus infection only locally in the host plants reacting hypersensitively. *Phytolacca* antiviral proteins (PAP) from *Phytolacca americana*, carnation antiviral proteins from *Dianthus caryophyllus*, *Mirabilis* antiviral proteins (MAP) from *Mirabilis jalapa* were initially reported to prevent local lesions production by tobacco mosaic virus in *Nicotiana glutinosa*, *N. tabacum* cv. *xanthi* - nc. Subsequent studies showed that PAP could inhibit virus in systemic hosts (*Brassica campestris/* cauliflower mosaic virus; *N. benthamiana* / African cassava mosaic virus) also when PAP was mixed with the virus inoculum (Chen *et al.*, 1991). Carnation antiviral protein and *Mirabilis* antiviral protein were later on shown to induce systemic resistance in host plant against TMV infection (Kubo *et al.*, 1990; Plobner and Leiser, 1990). These proteins are however, not widely distributed in the plant kingdom, but they share some homology, suggesting that they all have evolved from an ancestral gene with a common defensive function. Systemic resistance by phytoproteins has been discussed later in this chapter. A number of plants showing localised resistance have been reported from India (Baranwal, 1988; Baranwal and Verma, 1993). A few of them like *Celosia cristata*, *Chenopodium ambrosioides* have shown complete inhibition of TMV in *N.glutinosa*, sunnhemp rosette virus in *Cyamopsis tetragonoloba* and potato virus X in *Capsicum pendulum*. An interesting observation was made when it was found that proteinaceous antiviral substances from *Celosia cristata* and *Chenopodiun ambrosiodes* inhibited local lesion formation by TMV completely in *N. glutinosa* and other solanaceous hosts but not in *Chenopodium amaranticolor* (Baranwal and Verma, 1992). Thus it seems reasonable to assume that the response to virus inhibitor might be due to incompatibility between the host and the source of inhibitor as inhibitor when introduced in the same or related species causes little or no physiological disturbance which may be in the form of signal transduction. These plants could be considered for cloning of genes for antiviral proteins and development of transgenics of economically important crops.

2.2.1. Systemic Induced Resistance

Partially purified leaf extract from *Aerva sanguinolenta*, *Bougainvillea spectabilis*, *Pseuderanthemum atropurpureum tricolor*, *Clerodendrum aculeatum* and *C. inerme* and root extract from *Boerhaavia diffusa* were shown to produce pronounced systemic effect in *N. glutinosa* against TMV and also in other host virus system (Table 2). The other plants showing systemic induced resistance against virus infection are *Cuscuta reflexa*, *Capsicum fruitesence*, *Datura metel*, *Dianthus caryophyllus*, *Mirabilis jalapa*, and *Solanum melongena* . However, many of these plants show low order of systemic induced resistance. It has also been observed that the concentration of proteinaceous resistance inducers in *C. aculeatum* varies in different months of a year and correspondingly the extent of resistance induced in test plants also varies.

The systemic resistance inducer from *Clerodendrum aculeatum* and *Clerodendrum inerme* have been purified and characterised. A 34 kda basic protein was isolated from the leaves of *Clerodendrum aculeatum*. 64µg/ml of protein provided complete protection of untreated leaves against TMV infection in *N. tabacum* Samsun NN (Verma *et al.*, 1996). Two basic proteins of Mr 29 and 34 kda (CIP-29 and CIP- 34) were isolated from the leaves of *Clerodendrum inerme*. The minimum amount of purified proteins required to induce systemic resistance varied from 16 ig/ml for CIP-29 to 800 ig/ml for CIP-34 (Prasad *et al.*, 1995). While systemic resistance inducer (SRI) from *Clerodendrum aculeatum* (CA-SRI) could induce systemic resistance and provide protection to whole plants within 30 min of its application, the SRI from *Clerodendrum inerme* took more than 2 h to induce systemic resistance. It appears that CA-SRI is a better candidate for its utilisation in management of virus diseases. The systemic virus inhibitory activity of these proteins is due to host mediated phenomenon of formation / accumulation of a new virus inhibitory protein in the treated plants showing systemic resistance. When TMV is mixed with the induced virus inhibitory protein, the virus gets completely inhibited. However, it is not yet clear how these induced virus inhibitory protein in CA-SRI treated plants inhibits virus infection. The induction of systemic resistance triggered by proteins from *Clerodendrum* spp. was sensitive to actinomycin D indicating thereby that protein synthesis is involved in this mechanism. The fact that CA-SRI also behaves as RIP as it inhibited *in vitro* protein synthesis in rabbit reticulocyte lysate and wheat germ lysate (Kumar *et al.*, 1997), strongly confirm this belief that antiviral proteins generally have the property of RIP. The deduced amino acid sequence CA-SRI protein showed varying homology (11-54%) to the RIP's from other plant species (Kumar *et al.*, 1997). However the absence of hybridisation between CA-SRI gene and the DNA/RNA of *Mirabilis, Bougainvillea*, rice , pea and tobacco shows that the virus inhibitory genes do not share any significant sequence homology. Antibodies to CA-SRI protein also did not react with leaf proteins from *Mirabilis, Bougainvillea, Boerhaavia*, rice , pea and tobacco. This only indicate that antiviral proteins from different plants may behave differently and can be specifically used in particular host-virus system.

Table 2. Plants that confer systemic induced resistance in different hosts reacting hypersensitively to viruses.

Plant	Host virus systems	% Inhibition at treated leaf	% Inhibition at untreated leaf
Aerva	SRV/*C. tetragonoloba*	99	90
sanguinolenta	TMV/*C.amaranticolor*	nil	nil
	TMV/*D. stramonium*	100	99
	TMV/*N.glutinosa*	68	41
	TMV/*N.rustica*	80	76
	TMV/*N.tabacum* SamsunNN	68	65
Boerhaavia diffusa	TMV/ *N.glutinosa*	100	100
Bougainvillea spectabilis	TMV/ *N.glutinosa*	100	100
Clerodendrum aculeatum	SRV/*C. tetragonoloba*	98	96
	TMV/*N.glutinosa*	100	98
	TMV/*N.tabacum* SamsunNN	96	93
C.inerme	SRV/*C. tetragonoloba*	100	100
	TMV/*N.glutinosa*	98	98
Pseuderanthemum atropurpureum	PVX/*G.globosa*	100	100
	SRV/*C. tetragonoloba*	100	100
	TMV/*Datura stramonium*	100	100

3. APPLICATION OF PHYTO PROTEINS AS BIOCONTROL AGENTS

It is interesting to note that much of the work on phytoantivirals has been done on their characterisation and possible mode of action using local lesion hosts in the glass house conditions. However, the ultimate objective of these studies is to make use of phytoantivirals as biocontrol agents at field levels. The pre-inoculation spray of SRI's from *Clerodendrum aculeatum* , *Clerodendrum inerme*, *Boerhaavia diffusa*, *Bougainvillea spectabilis* and *Pseuderanthemum bicolor* has been shown to modify the susceptibility of several host plants like tomato, tobacco, mungbean, urdbean, bhendi and sunnhemp against subsequent infection by viruses producing local lesions as well as systemic symptoms. The treatment helps to protect the susceptible hosts during the vulnerable early stages of development. Induced resistance appears to be a universal process in all susceptible hosts and can be used to advantage in plant protection. Pre-inoculation sprays (4 sprays) of *Bougainvillea spectabilis* leaf extract protected plants of *Cucumis melo* against cucumber green mottle mosaic virus (CGMMV), *Crotalaria juncea* against sunnhemp rosette virus (SRV) and *Lycopersicon esculentum* against TMV for 6 days (Verma and Dwivedi, 1983). Since the duration of resistance conferred by SRI's was only up to 6 days, it was realised that the durability of resistance needed to be prolonged in field condition for getting better protection against virus infection. It has been shown that proteinaceous modifiers like papain enhanced the activity of CA - SRI and also prolonged the durability of induced resistance against sunnhemp rosette tobamo virus in *Crotalaria juncea* up to 12 days (Verma and Versha, 1995). In another study, 5 weekly sprays of CA-SRI in potted plants of tomato in open fields protected the plants from natural virus infection for more than two months (unpublished). Yet in another study it was shown that weekly sprays of leaf extract of *Clerodendrum aculeatum* delayed the symptom appearance of leaf curl virus in tomato and promoted the growth of the plants (Baranwal and Ahmad, 1997). Natural plant compounds called NS-83 was demonstrated to reduce and delay the disease symptoms by TMV, PVX and PVY in tobacco and tomato plants under field condition. And the fruit yield in tomato was increased by 23.4 %. (Xin-yun *et al.,* 1988). It appears that antiviral phytoproteins trigger the host defence mechanism in a specific manner either by signal transduction as demonstrated in case of *Phytolacca* antiviral protein (PAP) and /or by increased synthesis of antiviral proteins in the host plants treated by systemic resistance inducers like *Clerodendrum aculeatum*. While success has been achieved in developing transgenics having genes for mutant PAP without RIP activity (Smirnov *et al.,* 1997), it remains to be seen how these transgenics are utilised under field condition. On the other hand systemic resistance inducers like CA-SRI do show a potential for their utilisation under field condition but a larger quantity of SRI's would be required for their wider application. Cloning of genes of CA-SRI and its expression has been achieved in *Escherichia coli* expression vector. The expressed protein has been shown to inhibit protein synthesis (Kumar *et al.,* 1997). A detailed study is still required whether transgenics with native and mutant CA-SRI gene would be able to show systemic protection against virus infection or not?. Although *Clerodendrum aculeatum* protein has been demonstrated to have very high antiviral activity and to be extremely useful as plant immunizing agent, yet it has not been commercialized so far. The proteins from a few other non-host plants have also been recognized as good defence stimulants, but they have not been developed into products for disease control, since industry finds it easier to patent newly synthesized compounds than natural plant products.

4. CONCLUSIONS

With the precipitous withdrawal of some of the toxic plant protectants, it may be profitable to explore natural plant products as alternatives particularly against virus diseases where all other methods fail. The phytoproteins or their smaller peptides may prove valuable as 'lead structures' for the development of synthetic compounds. It behaves us to explore more intensely this rich source of antivirals. The value of these antiviral proteins is unlimited because they are quite safe, non - toxic even after repeated and prolonged use, enhance substantially the plant growth and yield. Antiviral phytoproteins may be more useful if they are integrated with other strategies of virus disease management.

REFERENCES

Baranwal, V.K. and Verma, H.N. 1992, Localized resistance against virus infection induced by leaf extract of *Celosia cristata, Pl. Path*. 41 : 633-638

Baranwal, V.K. and Verma, H.N. 1993, Virus inhibitory activity of leaf extracts from different taxonomic group of higher plants, *Indian Phytopath*. 46: 402-403 .

Baranwal, V.K. and Verma, H.N. 1997, Characteristics of a virus inhibitor from the leaf extract of *Celosia cristata, Pl. Path*. 46: 523-529.

Baranwal, V. K. and Ahmad, N. 1997, Effect of *Clerodendrum aculeatum* L. leaf extract on tomato leaf curl virus, *Indian Phytopath*. 50 : 297-299.

Batelli , M. G. and Stirpe, F. 1995, Ribosome inactivating proteins from plants. In: *Antiviral Proteins in Higher Plants,* eds. M. Chessin, D. DeBorde and A. Zipf, CRC Press, USA, pp 39-64.

Chen, Z. C. , White, R. F., Antoniw, J. F. and Lin, Q. 1991, Effect of pokeweed antiviral protein (PAP) on the infection of plant viruses, *Pl. Path*. 40:612-620.

Gerbes, T., Ferreras, J.M., Iglesias, R., Citores,L., DeTorre,C., Carbajales, M.L., Jimenez,P., De Benito, F.M. and Munoz, R. 1996, Recent advances in the uses and application of ribosome inactivating proteins from plants, *Cell Mol. Biol*. 42: 461-471

Hong, Y., Saunders, K., Hartley, M.R. and Stanley, J. 1996, Resistance to gemini virus infection by virus - induced expression of dianthin in transgenic plants, *Virol*. 220:119-127.

Irvin, J.D. 1995, Antiviral proteins from *Phytolacca*. In: *Antiviral Proteins in Higher Plants,* eds. M. Chessin, D. DeBorde and A. Zipf, CRC Press, USA, pp. 65-94.

Ito, Y., Seki, I., Tanifuji, S. and Hiramatsu, A. 1993, Inhibition of protein synthesis by antiviral protein from *Yucca recurvifolia* leaves, *Biosci. Biotech.Biochem*. 7 : 518-519.

Kubo, S., Ikeda, T., Imaizumi, S., Takanami, Y. and Mikami, Y. 1990, A potent plant virus inhibitor found in *Mirabilis jalapa* L., *Ann. Phytopathol.Soc. Japan* 56: 481-487.

Kumar, D., Verma, H.N., Tuteja, N. and Tewari, K.K. 1997, Cloning and characterization of a gene encoding an antiviral protein from *Clerodendrum aculeatum* L., *Pl. Mol. Biol*. 33: 745-751.

Kumar, M.A., Timm,D.E., Neet, K.E., Owen, W.G., Peumans,W.J. and Roa, A.G. 1993, Characterization of the lectin from the bulbs of *Eranthis hyemalis* (winter aconite) as an inhibitor of protein synthesis, *J.Biol. Chem*. 33:25176-25183.

Lodge, J. K., Kaniewski, W.K. and Tumer, N.E. 1993, Broad spectrum virus resistance in transgenic plants expressing pokeweed antiviral protein, *Proc. Natl. Acad. Sci*. USA 90: 7089-7093.

Mayhew, D.E. and Ford, R. F 1971. An inhibitor of tobacco mosaic virus produced by *Physarum polycephalum, Phytopath*., 61: 636-640.

Moraes, W.B.C., July, J. R., Alba, A.B.C. and Oliveira, A.R. 1974, The inhibitory activity of extracts of *Abutilon striatum* leaves on lant virus infection, *Phytopath. Z*. 81: 240-253

Owens, R.A., Bruening, G. and Shepherd, J.R. 1973, A possible mechanism for the inhibition of plant viruses by a peptide from *Phytolacca americana, Virol*. 56: 390-393.

Plobner, L. and Leiser, R.M. 1990, Induction of virus resistance by carnation proteins, In : *Proc. International Congress on Virology,* Berlin 26-21.

Prasad, V., Srivastava, S., Varsha and Verma, H.N. 1995, Two basic proteins isolated from *Clerodendrum inerme* Gaertn. are inducers of systemic antiviral resistance in susceptible plants, *Plant Sci*. 110:73-82.

Ragetli, H.W. J. 1975, The mode of action of natural plant virus inhibitors, *Curr. Adv.Pl. Sci*. 19: 321-334

Smirnov, S., Shulaev, V. and Tumer, N. E. 1997, Expression of pokeweed antiviral protein in transgenic plants induces virus resistance in grafted wild -type plants independently of salicylic acid accumulation and pathogenesis-related protein synthesis, *Plant Physiol*. 114:1113-1121.

Stevens , W. A. , Spurdon, C., Onyon, L.J. and Stirpe, F. 1981, Effect of inhibitors of protein synthesis from plants on tobacco mosaic virus infection, *Exeperientia* 37: 257-259.

Stirpe, F., Williams, D.G., Onyon, L.J., Legg, R. F. and Stevens , W. A. 1981, Dianthins, ribosome inactivating proteins with antiviral properties from *Dianthus caryophyllus* L. (Carnation), *Biochem. J*. 195: 399-405.

Stirpe, F. , Gaseri-Campani, A., Barbieri, L., Falasca, A.I., Abdondanza,A. and Stevens , W. A. 1983, Ribosome inactivating protein from the seeds of *Saponaria officinalis* L. (Soapwart), of *Agrostemma githago* L. (Corn cockle) and *Asparagus officinalis* L. (Asparagus) and from the latex of *Hura crepitans* L. (sandbox tree), *Biochem. J.* 216:617-625.

Stirpe, F., Barbieri, L., Batelli , M. G., Falasca, A.I., Abdondanza, A., Lorenzoni, E. and Stevens, W. A. 1986, Bryodin, a ribososme inactivating protein from the roots of *Bryonia dioica* L. (White bryony), *Biochem. J.* 240: 659-665.

Tumer, N. E., Hwang, Duk-Ju and Bonness, M. 1997, C-terminal deletion mutant of pokeweed antiviral protein inhibits viral infection but does not depurinate host ribosomes, *Proc. Natl. Acad. Sci.* USA 94:3866-3871.

Verma, H.N. and Baranwal, V.K. 1983, Antiviral Activity and the physical properties of the leaf extract of *Chenopodium ambrosiodes* L., *Proc. Indian Acad. Sci.* 92: 461-465.

Verma, H.N., Baranwal, V.K. and Srivastava, S 1998, Antiviral substances of plant origin, In: *Plant Virus Disease Control*, eds. A. Hadidi, R.K. Khetarpal and H. Koganezawa, APS Press, Minnesota, USA, pp. 154-162.

Verma, H.N., Srivastava, S., Varsha and Kumar, D. 1996, Induction of systemic resistance in plant viruses by a basic protein from *Clerodendrum aculeatum* leaves, *Phytopath.* 86:485-492.

Verma, H.N. and Dwivedi, S. D. 1983, Prevention of plant virus diseases in some economically important plants by *Bougainvillea* leaf extract, *Indian J.Plant Pathol.* 1:97-100.

Verma, H.N. and Varsha 1995, Induction of durable resistance by primed *Clerodendrum aculeatum* leaf extract, *Indian Phytopath.* 47:19-22.

Verma, H.N., Varsha and Baranwal, V.K. 1995a, Endogenous virus inhibitors from plants, their physical and biological properties, In: *Antiviral Proteins in Higher Plants,* eds. M. Chessin, D. DeBorde and A. Zipf, CRC Press, USA pp. 1-21.

Verma, H.N., Varsha and Baranwal, V.K. 1995b, Agricultural role of endogenous antiviral substances of plant origin, In: *Antiviral Proteins in Higher Plants,* eds. M. Chessin, D. DeBorde and A. Zipf, CRC Press, USA pp. 23-37.

Varma, A., Dhar, A. K. and Mondal, B. 1992, MYMV transmission and control in India, In: *Proc. Int. Workshop,* Bangkok, Thailand, eds. S.K. Geen and D.Kim, Asian Vegetable Research and Development Centre, Taipei, Taiwan, pp. 8-27.

Waterworth, H. E. and Hadidi, A. 1998, Economic losses due to plant viruses, In: *Plant Virus Disease Control,* eds. A. Hadidi, R.K. Khetarpal and H. Koganezawa, APS Press, Minnesota, USA, pp. 1-13.

Xin-yun, L., Huang-fang, L. and Wei-fan, C. 1988, Studies on the application of virus tolerance inducer- NS-83, In: *Proc. International Symposium of Plant Pathology,* Beijing, China.

MYCOVIRUSES : A NOVEL OPTION FOR MANAGING SOME PLANT DISEASES

A. K. Chowdhury and C. Sen

Department of Plant Pathology
Bidhan Chandra Krishi Viswavidyalaya
Mohanpur - 741252, West Bengal, INDIA

1. INTRODUCTION

The term 'mycovirus' was coined using two words i.e *mykos* and virus literally meaning fungal viruses. These were not known before 1950 although their presence was suspected in some pathogenic fungi (Linderberge, 1959) Presence of virus particles in fungal body was first scientifically established with the dieback disease of cultivated mushrooms in 1962 (Gandy and Hollings, 1962 ; Hollings, 1962). Since then a large number of mycoviruses or similar agents have been isolated, purified and characterised from many species of fungi belonging to all major classes (Bozarath, 1972; Hollings and Stone, 1971; Lemke, 1980; Lemke and Nash, 1974). Mycoviruses have economic significance as many of the fungi in which they were found caused serious diseases in plants or used in the industry for the production of many essential commodities like mushrooms, toxins, antibiotics and medicine.

The present day agriculture is basically yield oriented, necessiated by the need to feed the several billions of population projected for the twenty first century. Potentiality of yield has a limit and all attempts are being made to reach the staggering targets through genetic manipulation and with adequate supply of yield attributing inputs like fertilizers, irrigation water and plant protection chemicals. On an average 26 per cent crop loss has been estimated to be due to insects, diseases and weed problems and another 15 per cent encountered during storage and transit. Such huge losses could be ameliorated to some extent by adopting planned plant protection measures with an eye on sustainability. Overuse of plant protection chemicals in intensive agriculture system has invited newer problems leading to environmental damage and reducing yield or reduced rate of increase in yields. All these factors at the present moment are serious concerns to the Scientists, Administrators, and the farmers who are looking for alternatives to live in harmony with the microbes in a pollution free environment. Some progress has already been achieved in these lines through intensification of fundamental and adaptive researches on Integrated

Pest Management (IPM), Biological Control (BC), Cultural Management (CM), etc. interfaced with the integrating components like soil, water and natural environment but the end points are as yet far from our reach.

In BC, a living agent is used either to kill or to suppress a pathogenic agent doing little, if any harm to the living flora and fauna. The history of BC started as early as 1892 but it remained suppressed until the middle of twentieth century when the hazards of chemical control became more evident. The control of fungal diseases of plants is now one of the great challenges to plant pathologist inspite of some striking successes reported under specific situation (Baker and Cook, 1974) or are in the process of implementation. The viruses mostly recognized as pathogenic agents of plants, human or animals have also been identified as one of the potential biocontrol agents and have already been used successfully to control a number of insect pests (Van Driesche and Bellows, 1996), fungal pathogens (Fulbright *et al.,* 1988) and pathogenic viruses (Sen *et al.,* 1986). Still there is a long way to go before one is able to exploit the viruses as biocontrol agents. Use of viruses for the control of plant diseases and pest is comparatively a recent one although possibilities were apparent for a fairly long time. Presently a number of viruses have been identified from all classes of fungi that are responsible for reducing the power of pathogenicity (Boland, 1992; Ghabrial, 1980; Heiniger and Rigling, 1994; Hollings and Stone, 1971; Smart and Dennis, 1997). Control of viral / bacterial pathogens by agrochemicals is very rare and the present day management are based mostly on CM and/or host resistance. On the contrary, fungicides are most widely used to control the fungal diseases in all major and minor crops. Viruses as a biocontrol agent has been successfully used to control a few fungal diseases where conventional fungicides failed to achieve the desired results. There are many advantages of using viruses as biocontrol agents over chemical pesticides which can be formulated or to be inoculated in the fungal culture without having any pollution problem in nature. Risk factors are equally important. The success of the application will depend largely on our knowledge of viruses in their host, their identification, mode of action and on safe application, affecting the target (pathogen) without affecting the commodity.

2. HISTORY AND ECONOMIC IMPORTANCE

During the 1950's mushroom growers of different countries experienced a disorder of mushroom which deformed the shape and reduced yield of mushrooms. The reasons for such a malady was not known but it was found to rapidly spread in the stock of different mushrooms and such abnormalities were uncurable with available technologies. Various names were given to express this disorder that included X–diseases, die-back and *La France,* etc. and the losses due to this disease was recorded to be upto 95 per cent (Dielman Van Zaayen, 1970; Hollings *et al.,* 1993; Rasmussen *et al.,* 1968). Cream and off-white type of mushrooms were usually found to suffer less than pure white types.

Association of a virus in the mushroom was first scientifically established by Hollings in the year 1962 and subsequently a number of viruses have been isolated from time to time in different classes of fungi (Lemke, 1977; Nigam and Mukerji, 1991). In 1967 a virus was detected in ascomycetous fungi like *Penicillium* and *Aspergillus* (Banks *et al.,* 1969a,b; 1970). Within short period many viruses were identified from fungi belonging to different classes of pathogenic fungi causing diseases of cultivated crop plants (Buck, 1986; Ghabrial, 1980; Hollings and Stone, 1971). Presently comprehensive information on the physiological properties, replication, transmission, epidemiology, Killer system, hypovirulence and on molecular characterization of the phenomenon are available and have been reviewed (Fulbright *et al.*, 1980; Nuss, 1992; Nuss and Kottin, 1990; Smart *et al.*, 1996).

This article summarise the available information on mycoviruses with a view to enrich the knowledge on the applicability of viruses as biocontrol agents in managing pathogenic fungi.

A highly classical example of biocontrol has been established for the fungal pathogen *Cryphonectria parasitica* (Morr) Barr. (formerly *Endothia parasitica*) causing blight in chestnut (*Castanea dentata*) (Marsh) (Bork). It is a very serious disease in North America and in early 1900 it killed over 3 billion trees by a pathogen believed to be introduced from Asia (McDonald and Fulbright 1991). American chestnut trees were found to be extremely susceptible and this blight first recorded in New York city in 1904. The disease rapidly spreads in to chestnut populations. The disease was reported to Italy in 1930_s (Anagnostakis, 1982) which attacked European chestnut trees (*Castanea sativa*). In both the continents the fungus killed large number of trees and only a few of them escaped from infection. The fungus isolated from resistant trees was found to be less virulent and named as hypovirulent strain of *C. parasitica* (Grente *et al.*, 1969). Day *et al.* (1977) observed the presence of hypovirulent strain of fungus which had the capacity to suppress the biological activity of a virulant strain and it contained a double stranded RNA (dsRNA). Fulbright *et al.* (1983) and Heiniger and Rigling (1994) made a survey and collected hypovirulent isolates of *C. parasitica* from France, Italy and also from several locations in America and they found recovering plants or fungus collected from plants resistant to blight contained a ds RNA. Such hypovirulent strains of *C. parasitica* also contained a virus like particles (VLPs) in the cytoplasm.

After the establishment of association of dsRNA in chestnut blight fungus several reports were published indicating the hypovirulence mechanism to be operative among a wide range of pathogenic fungi. These included *Rhizoctonia solani, Gaumanomyces graminis.* var *tritici., Ophistroma ulmi, Sclerotinia sclerotiorum* and *Magnaporthe grisea* (Boland, 1992; Brasier, 1983; Chumley and Valant, 1990; Dunn and Boland, 1993; Fulbriught *et al.*, 1983; Naiki and Cook, 1983).

3. THE BIOCONTROL COMPONENTS

It will be pertinent to analyse atleast briefly the biocontrol components under consideration- the virus as a parasite, the fungi as pathogens and viruses that parasite fungi.

3.1. Viruses as Plant Pathogens

A virus is distinctly different from fungi or with other disease producing organisms in many aspects. A virus can be defined as a set of one or more nueleic acid templated molecules, normally encased in a protective coat of protein or lipoprotein that is able to organise its own replication only in suitable host cells (Matthews, 1991). In susceptible hosts under favourable conditions viruses multiply and trigger metabolic abnormalities which may lead to development of disease syndrome. A virus is active only in its living hosts which range from single celled microbes to flowering plants or most developed animal cells. Viruses on plants cause a number of economically important diseases and the losses have been assessed as high as US $ 60 billion per year (Klausner, 1987).

A virus lacks its own machinary for spread and hence it has to depend on other agents (vector), if not transmitted mechanically, asexually or through sexually propagating materials.

3.2. Fungi as Plant Pathogens

Fungi are at the top of the list amongst the pathogenic organisms inducing diseases of crop plants. They are characterised by chlorophyllless, nucleated. unicellular or multicellular filamentous bodies reproducing both asexually and sextually. Micheli was' the first to scientifically study the fungi in 1729 and in France Tillet in 1755 performed some experiments with the bunt pathogen of wheat. (Stakman, 1967). It took a very long time before Anton de Bary, the German scientist to establish that the fungus *Phytophthora infestans* was the cause of late blight disease of potato in 1861 and thus establish the germ theory. There are many classical examples of fungal diseases of crop plants which caused famine, migration of man from one continent to another, ruination of national economy and even changed the social customs (Agrios, 1997). Presently available disease management technologies certainly minimize the chances of outbreak to cause epidemic but no one can give any guarantee for prevention of reccurrence in near future. Some again, are not currently managed through available technologies. These include many soil-borne pathogens causing diseases of perennial trees. A continuous research is necessary to understand the pathogen and their behaviour as a function of changes in time and agricultural scenario.

3.3. Mycoviruses

The viruses whose host are distributed in the plants and animal kingdom have been extensively studied since its discovery by Adolf Mayer in the year 1886. Presence of virus in the fungi was speculated during 1950's but its presence was scientifically established in the year 1962 (Hollings, 1962; Gandy and Hollings, 1962). Subsequently VLPs containing dsRNA were isolated from two fungi *Penicillium* and *Aspergillus* spp. widely used for the production of antibiotic, toxins and other chemical compounds. Since then mycoviruses received the attention of many scientists to study the biological activities of the virus associated fungi and their utilization for the management of some economically important diseases. In behaviour, mycoviruses are related to the viruses causing infection to plants and animals and in many instances they are infectious to their fungal hosts. Unlike the plant viruses, a mycovirus does not produce any visible symptoms in their host but inhibits the normal growth.

Presence of VLPs or dsRNAs are now widespread in all classes of fungi including plant pathogens (Nuss, 1990). Such dsRNAs in their fungal host remain as naked uncapsidated or encapsidated molecules (Seroussi *et al.,* 1989). Presently viruses having dsRNA are placed into six taxonomic groups (Francki *et al.,* 1991) of which two are assiciated mostly with fungi and belongs to the family Totiviridae and Partiviridae (Zhang *et al.,*1994).

The Totiviridae are characterized by dsRNA with 4-7 Kbp encapsdated in isometric perticks with diameter of 40 nm and the Partiviridae having many unrelated segments of dsRNA of 1-3 Kbp encapsidated with 35 nm particles. Presence of dsRNA in uncapsidated form have been detected in the mitochondria, vesicles, cytoplasm and are transmitted during cell division or hyphal anastomoses.

3.4. Hypovirulence

The term hypovirulence is most widely used with some BC systems and also applicable to some mycoviruses. Hypovirulence has also been variously described as a form of induced resistance, antibiosis and hyperparasitism. Nuss (1992) indicated "hypovirulence as simply a pathogen phenotype where virulence is reduced".

Virus or virus like agents are common in many genera of fungi and in a few cases their presence showed hypovirulence usually through its dsRNA. The most thoroughly studied case of hypovirulence is related to chestnut blight caused by the fungus *C. parasitica*. Mycoviruses and hypovirulent strains of fungi could be used to control plant diseases where conventional chemicals control failed to give any encouraging results. But before going into that it is necessary to understand the nature and biological behaviour of these viruses.

In this chapter, the exhaustive information available on mycoviruses have been critically evaluated in a comprehensive manner emphasising the possibility of their use and applicability as biological control agents (Fulbright *et al.*, 1988; Ghabrial, 1980; Hollings and Stone, 1971; Nuss, 1990; Saksena and Lemke, 1978; Samrat and Fulbright, 1988; Zhang *et al.*, 1994). The following section will focus on the general, biological, physiochemical and molecular characterization of mycoviruses followed by their distribution in different classes of fungi. In subsequent part emphasis will be paid to discuss on the possibility of use of mycovirus as an agent for BC.

4. CHARACTERISTICS OF MICROVIRUSES

4.1. Nomenclature

The term mycovirus has been used in this review to denote the fungal viruses although several other terms like mycophage (Nash *et al.*, 1973) ds RNA plasmid (Wikner, 1976), Virus- like particles (VLPs) (Saksena and Lemke, 1978) have been used as near synonyms in the literature.

4.2. Nature of Mycoviruses

Majority of the viruses isolated from different species of fungi are isometric with an average diameter of 25-28nm. The other morphologically different rod shaped, or baclliform viruses are not uncommon.

Biophysical charactersties of mycoviruses are still considered as one of the important criteria for their identification and differentiation. The genome of most of the mycoviruses are composed of dsRNA which are usually segmented. The uncapsided dsRNA have been reported from a large number of plant pathogenic fungi. Segmentation of genome is an important creteria by which mycoviruses could be separated from plant and animal dsRNA viruses. Multiple infection by different viruses is not uncommon in fungi. Virus preparation from fungi are likely to exhibit heterogenous characters in respect to size, sedimentation properties, electrophoretic mobility, RNA content, etc. and such heterogenecity have been well documented with *Penicillium stolonifer* (Buck *et al.*, 1977) and *Aspergillus foetidus* (Buck and Rattix, 1977).

4.3. Cultural and Transmission Characteristics

Hypovirulent strains of many mycoviruses have been characterized by their cultural morphology, dsRNA content and virulence (Elliston, 1985). In general the mostly studied dsRNA free culture of *C. parasitica* produced similar colony in the culture medium and developed large sized cankers in the field bioassay. Changes in cultural morphology and growth rate of the fungus due to the infection of the virus have been thoroughly investigated among the viruses of mushrooms, *Penicillium* and other fungi (Nuss, 1992

Rogers *et al*., 1987). Pigmentation and sporulation characters are used to differentiate the North- American and European strains of chestnut blight pathogens. The North – American strains are fully pigmented but European ones are mostly lacking in pigmentation.

Mycoviruses have no vector for transmission as found with viruses of plants and animals. Available information indicates that the presence of a mycovirus inside the host is controlled by a combination function encoded by the dsRNA and nuclear and mitocondrial genes. Transmission usually taken place through the progeny host through asexual reproduction or hyphal or hybrid anastomosis or by cell fusion between genetically compatible host strains (Lemke, 1980). Fungal spores could play a most significant role in the spread of mycoviruses. Attempts for mechanical transmission has been made since very long (Hollings, 1978) but with limited success. This may be due to the rigitdity of fungal cell wall or due to other barriers causing obstacles in its entry or establishment in host cell.

4.4. Distribution of Mycoviruses

Since the first observation of the transmissible disorders of fungi and involvement of VLP of de RNA in the middle of twentieth century many such agents have been identified from a large number of fungal species. These fungal species include both plant pathogenic, biologically and economically important genera of fungi.

Lemke (1977, 1979) recorded the presence of VLPs in 40 species of plant pathogenic fungi distributed in 26 genera belonging to all classes of fungi. In comparison members of identified viruses are much common more in class Basidiomycetes and Fungi Imperfecti.

5. BIOLOGICAL FEATURE OF SOME MYCOVIRUSES

It is apparent from the distribution of mycoviruses in fungi that a considerable number of viruses or segmented dsRNA have been identified from a wide range of fungal species. Nature of these agents are either pathogenic or nonpathogenic and found both in economically important fungi like *Agaricus, Penicillium* and *Aspergillus* spp. or in many pathogenic fungi causing plant diseases. In pathogenic fungi the effect of virus (es) has been successfully utilized to control a few such diseases of international importance in an economic and efficient manner. In this section fungal -virus dsRNA interaction are summarized.

5.1. ds RNA and *Cryptonectria parasitica* Association

Chestnut (*C. dentata*) is widely growin in Europe and in many other countries. The large size of the mature trees with a height of more than 100 feet and a basal diameter of 12 feet or more serves as a very useful timber. Due to the high demand of timber chestnut trees were planted all over North America, British Columbia and Canada (Fulbright *et al*., 1988). The fungus *C. parasitica* was found to cause chestnut blight disease extensively and only in North America during early part of 1990 over 3 billion trees were destructed. The disease was incurable, in North America through all available non-conventional plant protection measures. Attempt were made to breed resistant of American chestnut to blight pathogen by interspecific hybridization (Jaynes, 1964), irradiation of seedlings (Dietz, 1978) but none of the methods gave adequate protection to the destructive pathogen.

The important findings on hypovirulence in Europe were followed in the United States and results showed that French and Italian isolats cured the cankers caused by virulent Americal isolats (Anagnostakis and Jaynes, 1973).

The disease spread to Europe and through England, France, Netherland, Germany, etc. It reached Italy in 1938 probably from Japan. The most important discovery of fungal hypovirulence was first observed in Italy fifteen years after the appearence of epidemic. Surprisingly it was noted that some Italian trees survived inspite of the infection of C. parasitica. Further it was observed that the fungus from such infected trees were less virulent than the isolate causing North American blight and was subsequently recognized as a hypovirutent variant (Grente and Sauret, 1969). It was further observed that the hypovirulent character could be cytoplasmically transmited from one strain to another by hyphal fusion. The authors subsequently established the presence of a dsRNA in such hypovirulent strains. After the detection of hypovirulent strain in European chestnut blight pathogen, similar of hypovirulent strain were identified from American chestnut in Michigan in 1976 which contained a dsRNA (Paul and Fulbright, 1988). Hypouirulent of American type was first found 200 miles West of the natural range of chestnut in Michigan, Wisconsin and Minnesota. A considerable number of chestnut trees from such areas escaped infection suggesting the presence of hypovirulent strains. Virulent and hypovirulent isolates showed much variation in growth, colony characteristics and fungal morphology. The most important variation between North American and European hypovirulent strain were based on pigmentation and sporulation. North American isolates were pigmented while European hypovirulent strains were white. Nevertheless both the strain from Europe and North America have some similarities like cytoplasmic transmissibility, gradual declining the rate of virulence and presence of membrane bound particles (Nuss, 1992). The hypovirulent strain have the power to altered the colony morphology, reduced sporulation, oxalate accumulation, Laccase production and pigmentation (Kazmierczak et al., 1996; Nuss and Koltin, 1990; Rigling, 1995). While the disease is being managed successfully in Europe through spray application of the hypovirulent strain, similar approach is in a process of development in USA.

5.2. Molecular Characterization of dsRNA in *Cryptonectiria parasitica*

The evidence of dsRNA hypovirulent of C. parasitica was first demonstrated by Day et al. (1977). Later Dodds et al. (1980) characterised the segments of dsRNA which were found to vary in size and number between the strains. On purification the North American type (GH_2) and European (EP 713) ds RNAs showed some common structural properties. The dsRNAs from both strains consisted of number of species. European type hypovirulent strains EP 713 from *C. parasitica* has been characterized in detail and designated as *Cryptonectria* hypovirus (CHVI – 713) and given a family status Hypoviridae (Hillman et al., 1992). The dsRNAs from CHVI have been classified into three groups, a large (12.7 kbp), a medium (8-10.0 kbp) and a small (0.6 – 1.7 kbp) in length. Shapira et al .(1991b) showed that medium and small sized dsRNA were due to deletion from large dsRNA.

Characterization of North American or Michigan type hypovirulent ds RNA have also been made and designated as CHV3 – GH2 and it contained three ds RNA segments of more or less 9.5, 3.5 and 0.8 kbp in length. Similarly the ds RNA of New Jersey hypovirulent strain NB 58 was completely sequenced and designated as hypovirus CHV2 – NB58. It is similar to CHVI – 713 with a size of 12.5 kbp. In Italy Maresi et al. (1995) tested six hypovirulent strains for ds RNA content, virulence and conversion ability. On inoculation of hypovirulent strain these as well as the converted strain showed similar banding pattern of dsRNA. On analysis of dsRNA in the termini of isolate GH2 and EP713 revealed each dsRNA species to contain a stretch of polyadenylic acid (Poly A) at 3'- terminus of one strand that was base paired with a stretch of polyuridylic acid (Poly LL) at 5' - terminus of the complementary strand. The presence of 3' Poly A, 5' Poly U structure distinguished hypovirulence associated dsRNAs from genomic RNAs of mycoviruses and dsRNA viruses (Bozarath *et al.*, 1979). The hypovirus encoded protein P29 was found to have a significant impact on both sporulation and pigmentation in different fungal host (Chen *et al.*, 1996). Recently progress in molecular biology of hypovirus have been discussed by Nuss (1996) where he critically assesed the mechanisms by which hypoviruses alter fungal gene expression and virulence. Enebak *et al.* (1994a) obtained many isolate of *C. parasitica* which contained dsRNA from central Appalachians of USA. They obtained an icosahedral particles of 60 nm, diameter in isolate C-18. It contained 11 segments of dsRNA ranging from 1-5 kbp which reduced the virulence and altered the cultural morphology. The dsRNA molecules in *C. parasatica* have been associated with VLP in the cytoplasm of hypovirulent strain of the fungus. Further they (Enebak, *et al.*, 1994b) determined the relationship of dsRNA found in hypovirulent isolation from Europe and North America using cDNA libraries for 3 different ds RNA. The first type designated as SR-2 containing 1 segment of dsRNA (12 kb). Second type as D2 containing 2 segments ds RNA (5 and 12 kb) and the third type designated C–18 contained 1 segment of dsRNA (1-5 kb). The classes SR-2, D-2 and C-18 are named as Maryland, Penesylvania and West Virginia respectively. They could not be cross hybridized thus suggesting that these dsRNA neither have close aftinity   to one another nor to the dsRNA from New Jersey (NB 58) on from Europe (EP 713 and EP747).

6.   HYPOVIRULENCE WITH SOME OTHER FUNGI

6.1. *Ophiostoma ulmi*

The ascomycetous fungus *O. ulmi* caused a devastating disease popularly known as Dutch elm disaease commnly found in Europe, North America and South West Asia. This fungus has both aggressive and nonaggressive strains (Rogers *et al.,* 1987). A third strain known as *O.novo-ulmi* has also been identified (Brasier, 1991). The disease isolates indicated the presence of dsRNA and hypovirulence factor called d–factor are cytoplasmically transmissible and the dsRNA has also been delected to be associated with the mitocondria. The possible use of dsRNA-du-factor as a biocontrol agent is very promising but the dsRNA is not transmitted through sexual cycles of the fungus. A high percentage of naturally occuring strain contained the power of hypovirulence which could be used as one of the potential biocontrol agents (Buck, 1988).

6.2. *Gaeumannomyces graminis* var. *tritici*

The causal fungus of wheat take-all disease lost its pathogenicity in storage within a short period and a search for the cause led to discovery of dsRNA in *G. graminis* where. VLP have been detected in both virulent and weakly virulent strains. Later Stanway (1985) indicated the presence of dsRNA in some strains of *G. graminis*. The virulent strains were free from viral dsRNA but it was present in the hypovirulent strain. The hypovirulent strain also lost the sexual, reproductive capacity. This hypovirulence is yet to be exploited.

6.3. *Helminthosporium victoriae*

Victoria blight of oat was an important disease which caused the epidemic in the United States during 1947 and 1948. Abnormality in growth and deviation in morphological characters in some of the isolates indicated the presence of two dsRNA viruses (Sanderlin and Ghabrial, 1978). They were serologically distinct and designated as 190S and 145S on the basis of sedimentation coefficient. The infected isolate contained both the viral genome where as healthy isolate contained only 190S genome. The 190S genome had 4.5 kb segments of dsRNA where as in 145S segments number ranged over 3.3 – 3.6 kb. The pathogenesis of the fungus is related with the production of a toxin known as "Victorin". The hypovirulent strain produced a low level of toxin and possibly that virus ecnoded protein might have influenced on the biosynthetic pathway related with the toxin production has been postulated.

6.4. *Phytophthora infestans*

Tooley *et al.* (1989) isolated a dsRNA from Mexican strain of late blight pathogen. Styer (1978) first suspected the presence of VLP in *P. infestans* through cytological evidences but failed to purify the virus(es). On the contrary presence of dsRNA was found to be transmissible to zoospores. Virulence of the Mexican strain having deRNA was more than dsRNA free strain with exceptions, showing thereby that presence of a dsRNA in a fungal pathogen does not necessarily mean that a natural biological control process is 'on'.

6.5. *Rhizoctonia solani*

Like *Phytophthora* an enhancement of virulence have been observed with the association of dsRNA in *R.solani* an important pathogen of many tropical and subtropical cultivated crop plants (Finkler *et al.*,1988) although earlier work by Castanho and Bulter (1978) indicated a negative virulence with a patricular strain and the phenomenon was recognized as *Rhizoctonia* decline. A survey was conducted by Kinsey and Helber (1989) where they recovered 139 natural isolates from the uninfected plant in Isreal, out of which 23% of the population were considered hypovirulent. Upon subculturing the hyphal types of hypovirulent strains, remained hypovirulent, while the subculture obtained from virulent strains were either virulent or hypovirulent. Molecular basis for lower levels of pathogenesis for hypovirulent strains was not clearly understood but the possible involvement of pectolytic enzymes (Mancus *et al.*,1986) has been suggested. Hypovirulent strains of a few *Rhizoctonia* spp were also obtained from 13 locations in Gifu perfecture of Japan (Juan *et al.*, 1996).

6.6. *Sclerotinia homoecarpa*

S. homoecarpa is the causal agent of dollar spot disease of turfgrass were evaluated for the presence of dsRNA (Zhou *et al.*,1997). At least 4 isolates were associated with the presence of dsRNA which could be transmitted to virulent strain to make it hypovirulent. Another transmissible hypovirulence fungi *S. minor* was also detected in Holland (Melzer and Boland, 1996). When a mycelial suspension of a hypovirulent strain was sprayed on lesions on lettuce leaves caused by a virulent strain, there was reduction in lesion size and number of sclerotia produced. The virulent isolates lost their virulence with the association of hypovirulence and this character was retained after repeated subculturing.

7. DOUBLE STRANDED RNA IN BASIDIOMYCETOUS FUNGI

Association of dsRNA either in encapsidated or unencapsidated form is very widespread among the rust and smut fungi. Isometric virus particles were observed from a number of species of *Puccinia* (Rawlinson and Maclean, 1973). Newton *et al.* (1985) first recognized the presence of dsRNA from different isolates of *P. striiformis*, *P. recondita* and *P. hordei*. A large number of rust fungi have been analysed and found that most dsRNA have a size of 4-5 kb pairs. The three *Puccinia* species causing rust of wheat phenotypically varied among themselves, which could be distinguished on the basis of banding patterns (Zhang *et al.*,1994).

The ds genome from *Puccinia* species were encapsidated and contained VLP of 38 to 40 nm diameter composed of polypeptides of 72–96 Kd. Presence of a large number of encapsidated dsRNA of different sizes in *Puccinia* rust raised the question whether these were variants or resulted from the fragmentation of larger dsRNA molecules. Alternatively they could be independently encoded with definite polypeptides. Critical investigations with PCR and hybridization techniques by Shapira *et al.* (1991a) suggested that medium and small dsRNAs have resulted from deletions of the larger 12.7 kb segments. In the Uiridinales presence of dsRNA in encapsidated isometric particles of 30-40nm in diameter and also rod shaped particles have been isolated from *Uromyces vignal* causing cowpea rust. On separation 7 or more dsRNA segments were identified with four large (5-8 kb), one medium (2-3kb) and two small fractions of less than 0.7 kb (Zang *et al.*, 1994).

Presence of dsRNA in *Tilletia indica* causing Karnal bunt disease of wheat was identified in 6 isolates (Beak *et al.*, 1994). The average size of the genome ranged from 2.8 to 12.3 kb and are similar to other dsRNAs of pathogenic fungi.

Melampsora lini is the causal fungus of linseed (*Linum utitatissimum* L.) rust contained ds RNA and there are evidences of sexual transmission of this dsRNA. The inheritance pattern of dsRNA through sexual crosses illustrate that the dsRNA is organised into groups or units that are transmitted to progeny. In strain I of *M. lini* there are 11 dsRNA genomes of which the large one has 5.2 kb followed by 6 dsRNA of medium size ranging from 2- 2.7 kb, while the four small dsRNA are within the range of 1.1 to 1.5 kb (Lawrence *et al.*,1988). Asexual or somatic transmission of dsRNA was also recorded when two strains of rust were grown in a single host. The virus particles associated with linseed rust pathogen are isometric, 40 nm in diameter and remain as uncapsidated.

Ustilago maydis causing smuts of corn exhibit an inhibition property during the interaction between two natural isolates. Such inhibition was associated with a dsRNA virus (Koltin, 1988). The virus induced the host *Ustilago* to produce some kind of toxins. Three different types of viruses have been isolated and each encode three different toxin. The toxins found in *Ustilago* have some similarities with the toxin encoded by the virus present in *Saccharomyces cerevisiae* and it is very specefic to U. *maydis*. Seroussi *et al.* (1989) identified both unencapsidated and encapsidated dsRNA from natural isolates of *U. maydis*. The toxin has a very adverse effect on the growth of the pathogens which indicated the possibility of using this "Killer" system in management of plant diseases.

8. CONCLUSION

Biocontrol is now most widely used term which consist of various methods for suppressing the pathogens, utilizing non pathogenic or other agents. Control of fungal pathogen is one of the most chalenging areas of sustainable agriculture. Use of chemicals to control the fungal pathogens is most common but it has its own limilations. The opportunities to control fungal diseases using microbes is well known since a long time past. In nature the biocontrol is operative as a regular system and failure or disruption of such a system usually favour the outbreak of diseases.

Presence of viruses in the fungi is of common occurrence and they have been detected from all the major groups of pathogenic fungi. A large number of mycoviruses have been found that are less virulent and have the ability to suppress the growth of the aggressive fungi. Fungal viruses which induce hypovirulence could be used as biocontrol agents and thus offer the potentially of an excellent option for controlling many major diseases that are not effectively controlled by any of the existing traditional strategies. However, the presence of dsRNA is not only an indicator of the available tool for management through hypovirulence. In depth probe is needed to this too as an effective biocontrol option through understanding this molecular basis of interactions and potentiality of natural spread.

REFERENCES

Agrios,G.N. 1997, *Plant pathology*, (Fourth edition), Academic Press, New York, USA, pp. 635.

Anagnostakis, S.L. 1982, Biological control of chestnut blight, *Science* 215 : 466-471.

Anagnostakis, S.L and Janes, R.A. 1973, Chestnut blight control, Use of hypovirulent culture, *Plant . Dis. Rep.* 57 : 225-226.

Baker, K.F and Cook R.J. 1974, *Biological Control of Plant Pathogens*, Freeman San Fransisco, California, New York, USA.

Banks, G.T., Buck, K.W., Chain, E.B., Darbyshire; J.E. and Himmelweit, F. 1969a, Virus like particles in *Penicillium chrysogenum*, *Nature* 222 : 89-90.

Banks, G.T., Buck, K.W., Chain, E.B., Darbyshire. J.E. and Himmelweit F. 1969b, *Penicillum cyane ofuluum* virus and interferon stimulation, *Nature* 223 : 155-158.

Banks, G.T. 1970, Antiviral activity of double stranded RNA from a virus isolated from *Aspergillus foetidus*, *Nature* 227 : 505-507.

Barros, G. and Labarere, J. 1990, Evidences for viral and naked double-stranded RNAs in the basidiomycete *Agrocube aegerita*, *Curr. Genet.* 18: 231-237.

Beek R.J., Smith, O.P. Tooley, P.W. Peterson, G.L. and Bonde, M.R. 1994, Characterization of double stranded RNA from *Tilletia indica*, *Mycologia* 86 : 656-659.

Boland, G.J. 1992, Hypovirulence and double stranded RNA is *Sclerotinia sclerotiorum*, *Can J. Plant Pathol.* 14: 10.

Bozarath, R.F. 1972, Mycoviruses : A new dimenson in microbiology, In *Experimental Health Perspective*, US Dep. Health Educ. & Welfare, Washington, D.C., pp. 23-29.

Bozarath R. F. 1979, Physiochemical properties of mycroviruses : An over view, In : *Fungal Viruses*, eds. H.P. Molitoris., M. Hollings and H.A. Wood, Springler Verlag, Berlin German, pp. 48-61.

Brasier, C.M. 1991, *Ophiostoma novo-ulmi* sp. Nov. causative agent of current Dutch elm disease pandemics, *Mycopathologia* 115 : 151.

Brasier, C.W. 1983, A cytoplasmically transmitted disease of *Ceratocystis ulmi*, Nature 305 : 220-222.

Buck, K.W. 1988, Control of plant pathogens with viruses and related agents, *Phil. Trans, R. Soc. Lond.* B 318 : 295-317.

Buck, K.W. 1986, Fungal virology – An overview In : *Fungal Virology* ed. K.W. Buck, CRC Press, Boca Paton, USA, pp. 1-84.

Buck, K.W. and Kempson-Jones, G.F. 1973, Biophysical properties of *Penicillium stoloniferum* virus, *J. Gen. Virol.* 18 : 223-235.

Buck, K.W. and Ratti, G. 1977, Molecular weight of double-stranded RNA, A re- examination of *Aspergillus foetidus* virus SRNA components, *J. Gen. virol.* 37 : 215–219.

Castanho, B and Butter, E.E. 1978, *Rhizoctonia* decline : Studies on hypovirulence and potential use in biological control, *Phytopath.* 68 : 1511-1514.

Chen–Boashan, Chen-cheinhwa, Bowman, B.H., Nuss, D.L. and Chen, B.S. 1996, Phenotypic changes associated with wild type and mutant hypovirus RNA transfection of plant pathogenic fungi phylogenitically related to *Cryphoneceteria parasitica* , *Phytopath.* 3 : 301-310

Chumley, F.G and Valent, B. 1990, Genetic analysis of lnelanin deficient, nonpathogenic mutant of *Magnoparthe grisea, Mol. Plant. Microb. Interact.* 3: 135.

Day, P.R., Dodds, J.A., Elliston, J.E., Jaynes, R.A. and Anagonostakis, S.L. 1977, Double stranded RNA in *Endothia parasitica, Phytopath.* 67 : 1393.

Dieleman – Van Zaayen, A. 1967, Virus like particles in a weed mould growing on mushroom trays, *Nature* 216 : 595-596.

Dielman - Van Zaayen, A. 1970, Means by which, virus diseases in cultivated mushroom is spread and methods to prevent and control it, *Mushroom Growers Assoc. Bull* 244 : 358-378.

Dietz, A. 1979, The use of ionizing radiation to develop a blight resistant American chestnut, *Castanea dentata* through induced mutation, In : *Proc. Americal Chestnut Symp.*, W.V. Univ. Press, Morgantown, USA.

Dodds, J.A. 1980, Association of type l viral like ds RNA with club shaped particles in hypovirulent strains of *Endothia parasitica, Virol.* 107 : 1-12.

Dunn., M.M and Boland, G.J. 1993, Hypovirulent isolates of *Cryphonecteria parasitica* in southern Ontaris, *Can. J. Plant. Pathol.* 15 : 345.

Elliston, J.E. 1985, Characteristics of ds RNA free and ds RNA containing strains of *Endothia parasitica* in relation to hypovirulence, *Phytopath.* 75 : 151.

Enebak, S.A., Mac Donald, W.L. and Hillman, B. I. 1994, Effect of ds RNA associated with isolates of *Cryphonectria parasitica* from the central Appalachians and their relatedness to other ds RNA from North America and Europe, *Phytopath.* 84 : 528-534.

Enebak, S.A., Hillman, B.I. and Mac Donald, W.A. 1994, A hypovirulent isolate of *Cryphonecteria parasitica* with multiple genetically unique ds segments, *Mol. Plant Interac.* 7 : 590-595.

Finkler, A., Benzvi, B.S and Koltin, Y. 1988, In ds RNA virus of *Rhizoctonia solani*, In : *Viruses of Fungi and Simple Eukaryotes*, eds. Y. Koltin and Leibowitz, Marcel Dekker Inc., USA, pp. 387-409.

Fulbright, D.W., Paul, C.P and Garrod, S. W. 1988, Hypovirulence a natural control of chestnut blight, In : *Biocontrol of Plant Diseases*, Vol. II, eds. K.G. Mukherjee and K.L. Gorg, CRC Press, Boca Ratan, USA, pp. 121-139.

Francki, RIB, Fauquet, C.M., Knudson, D.L. and Brown, F. 1991, Classification and Nomenctature of viruses, Fifth Report of the International Committee on Taxonomy of Viruses, Springer – Verlag. USA, pp. 450.

Gandy, D.G. and Hollings, M. 1962, Dieback of mushroom :a disease associated with a virus, *Rep. Glasshouse Crops Res. Inst.* 1961 : 103-108.

Ghabrid, S.A. 1980, Effect of fungal viruses on their host, *Ann. Rev. Phytopath.* 18 : 441-461.

Grente, J. and Sauret, S. 1969, Hypovirulence expulsive phenomenon originale on pathologia vegetable, *C.R. Acad. Sci. Ser.* D 268 : 2347-2350.

Hansen, D.R., Van Alfen, N.K., Gillies, K. and Powell, W. A. 1985, Naked ds RNA associated with hypovirulence of *Endothia parasitica* packaged in fungal vesicles, *J. Gen. Virol.* 66 : 2605.

Heiniger, U. and Rigling 1994, Biological control of chestnut blight in Europe, *Ann. Rev. Phytopath.* 32 : 581.

Hillman, B.I., Tian, Y., Bedker, P.J. and Brown, M.P. 1992, American hypovirulent isolate of the chestnut blight fungus with Eropean isolate related ds RNA, *J. Gen Virol.* 73 : 681.

Hollings, M. 1962, Viruses associated with a dieback disease of cultivated mushrooms, *Nature* 196 : 962-965.

Hollings, M. 1978, Mycoviruses : Viruses that infect fungi., *Adv. Virus. Res.* 22 : 1-53.

Hollings, M., Gandy, D.C. and Last, F.T. 1963, A virus diseases of a fungus; dieback of mushroom, *Endeavour* 22 : 112-117.

Hollings, M. and Stone, O.M. 1971, Viruses that infect fungi, *Ann. Rev. Phytopath.* 94 : 118.

Jaynes, R.A. 1964, Interspecific crosses in genus *Castania silvae, Genet.* 13 : 146-154.

Juan, Abgona R.V., Katsuno, N., Kageyama, K. and Hyakumachi, M. 1996, Isolation and identification of hypovirulent *Rhizortonia* spp. from soil, *Pl. Path.* 45 : 896 – 904.

Kazmierezak, P., Pfeiffer, P., Zhang, Lei., Alfen, N.K., Van zhang, L. and Van- Alfen- N.K. 1996, Transcriptional repression of specific host genes by the mycovirus *Cryphonecteria* hypovirus, *J. Virol.* 70 : 1137-1142.

Kinsey, J.A. and Helber, J. 1989, Isolation of a transposable elements from *Neurospora crasa., Proc. Natl. Acad. Sci.* USA 86 : 1929-1933.

Klausner, A. 1987, Immunoassays flourish in new markets, *Biotech.* 5 : 551-556.

Koltin, Y. 1988, The killer system of *Ustilago maydis* : Seereted polypeptides encoded by viruses, In : *Viruses of Fungi and Simple Eukaryotes*, eds. M. Koltin and Leibowitz, Marcel Dekkar Inc., USA, pp. 434.

Lemke, P.A. 1977, Fungal viruses in agriculture, In : *Virology in Agriculture,* ed. J.A. Rosenberger, R. Anderson and L. Powell Marcel Decker Inc. New York, USA, pp 159-75.

Lemke, P.A. 1979, *Viruses and Plasmids in Fungi*, Marcel Dekkar Inc. USA, pp. 653.

Lemke, P.A. 1980, Viruses of conidial fungi, In : *The Conidial Fungi*, ed., G. Cole, Academic Press, New York, USA.

Lamke, P.A. and Nash, C.H. 1974, Fungal Viruses, *Bact. Rev.* 38 : 29-56.

Linderberge, G.D. 1959, A transmissible disease of *Helminthosporium Victoriae*. *Phytopath.* 49 : 29-32.

Littlefield, I.J. and Heath, MC 1979, *Ultrastructure of Rust fungi*, Academic Press New York, USA, pp.270.

Macus, I., Harash, I., Sneh, B., Koltin, Y. and Finkler, A. 1986, Purification and characterization of pectolytic enzymes produced by virulent and hypovirulent isolates of *Rdizoctonia solani* Kuhn, *Mol. Plant Pathol.* 29 : 325-336.

Maresi, G., Giovannetti, L., Ventura, S. and Turchetti, T. 1995, Transmission of hypovirulence agents among some *Cryphonectria parasitica* strains from Italy, *Europ. J. For. Pathol.* 25 : 191–196.

Matthews, R. P.F. 1991, *Plant Virology*, Academic Press, New York, USA, pp. 835.

Mac Donald, W.L. and Fulbright, D.W. 1991, Biological control of chestnut blight : Use and limitation of transmission hypovirulence, *Plant Dis.* 75 : 656.

Mc Keen, C.D. 1995, Chestnut blight in *ontario* : Past and present status, *Can Jr. Pl. Pathol.* 17. 295 – 304.

Melzer, M.S. and Boland, G.J. 1996, Transmissible hypovirulence in *Sclerotinia minor*, *Can J. Pl. Path.* 18 : 119-128.

Mukhopadhyay, S. and Chowdhury, A.K. 1984, Application of viruses for the control of pest and diseases, In : *Interaction of Plant Pathogens in the Host*, eds. N. Samajpati and S.B. Chattopadhyay, Oxford IBH, USA pp. 192-206.

Naiki, T. and Cook, R.J. 1983, Factors in loss of pathogenecity in *Gaeumannomyces graminis* var. *tritici*, *Phytopath.* 73 : 1652-1656.

Nash, C.H., Dauthart, R.J., Ellis, L.F., Van Frank, R.M., Burnett. I.P. and Lemke, P.A. 1973, On the mycophage of *Penicillium chrysogenum*, *Can J. Microbiol.* 19 : 97-103.

Nigam, N. and Mukerji, K.G. 1991, A new mycovirus, In : *Recent Advances in Plant Biology*, eds. C.P. Malik and Y.P. Abrol, Narendra Publishing House, Delhi, pp. 335-338.

Newton, A.C., Caten, C.E. and Johnson, R. 1985, Variation for isozymes and double stranded RNA among isolates of *Puccinia striiformis* and two other cereals rusts, *Pl. Path.* 34 : 235-247.

Nowgawa, M., Kageyama, T., Nakatani, A., Taguchi, G., Shimosaka, M. and Okazaki, M. 1996, Cloning and characterization of mycovirus double-stranded RNA from the plant pathogenic fungus, *Fusarium solani* f. sp. *robiniae*, *Bioscie., Biotech. Biochem.* 60 : 784-788.

Nuss, D.L. 1992, Biological control of chestnut blight : an example of virus mediated attenuation of fungal pathogen, *Microb. Rev.* 56 : 561-576.

Nuss, D.L. and Koltin, Y. 1990, Significance of ds RNA genetic elements in plant pathogenic fungi, *Ann. Rev. Phytopath.* 28 : 37-58.

Nuss, D.L.1996, Using hypovirulence to probe and signal transduction processes underlying fungal pathogenes, *Plant Cell* 8 : 1845 – 1853.

Ou, C.S. and Hillman, B.I. 1995, Genome organization of partivirus from filamentous ascomycetes, *Alkinsonella hypoxylon*, *J. Gen. Virol.* 76 : 1461 – 1470.

Paul, C.P. and Fulbright, D.W. 1988, Double stranded RNA molecules from Muhigan hypovirulent isolates of *Endothia parasitica* vary in size and sequence homology, *Phytopath.* 78 : 751-755

Rasmussen, C.R., Mitchell, R.E. and Slack, C.I. 1968, Investigation into the infectivity of X-disease in mushroom culture, *Mush. Sci.* 7 : 385-398.

Rawlinson, C.J. and Maclean, D.J. 1973, Virus like particles in axenic cultures of *Puccinia graminis tritici*, *Tans. Br. Mycol. Soc.* 61 : 590-592.

Rigling, D. 1995, Isolation and cheracterization of *Cryphonecteria parasitica* mutant that mimic a specific effect of hypovirulence – associated ds RNA on Laccase activity, *Can. J. Bot.* 10 : 1655-1661.

Rogers, H.J., Buck, K.W. and Brasier, C.M. 1987, A mitocondrial target for double-stranded RNA in diseased isolates of the fungus that causes Dutch elm disease, *Nature* 329 : 558-560.

Rogers, H.J., Buck, K.W. and Brasier, C.M. 1988, Double stranded RNA in diseased subgroup of the Dutch elm fungus *Ophiostoma ulmi*, In : *Viruses of Fungi and Simple Eukaryotes*, eds. Y. Koltin and Leibonntz, Marcel Dekkar Inc., USA, pp. 327-351.

Saksena, K.N and Lemke, P.A. 1978, Viruses in fungi, *Comp. Virol.* 12 : 103-143.

Sanderlin, R.S. and Ghabrial, S.A. 1978, Physiochemical properties of two distinct type of virus like particles from *Helminthosporium victoriae*, *Virol.* 87 : 142 – 151.

Sen, C., Sengupta, J. and Chowdhury, A.K. 1986, Potential strategies for management of plant viruses diseases : Harvesting the viral genone, In : *IPM System in Agriculture*, Vol. II, eds., R.K. Upadhyay, K.G. Mukerji, and R.L. Rajak, Aditya Books Pvt. Ltd., New Delhi, India, pp. 527-621.

Seroussi, E., Perry, T., Ginzberg, I. and Koltin, Y. 1989, Detection of killer independent ds RNA plasmid in *Ustilago maydis* by a simple and rapid method of extraction of ds RNA, *Plasmid* 21 : 216-225.

Shapira, R., Choi., G.H. and Nuss, D.L. 1991a, Virus like genetic organization and expression strategy for a double stranded RNA genetic element associated with biological control of chestnut blight, *EMBO. J.* 10-731.

Shapira, R., Choi, G.H., Hillman, B.I. and Nuss, D. 1991b, The contribution of defective RNA to the compatibility of viral encoded double stranded RNA population present in hypovirulent strains of the chestnut blight fungus *Cryphonecteria parasitica, EMBO. J.* 10 : 741-746.

Smart, D.S. and Fullbright, D.W. 1995, Characterization of a strain of *Cryphonecteria parasitica* doubly infected with hypovirulence associated ds RNA virus, *Phytopath.* 85: 491-495.

Smart, D.S. and Fulbrught, D. 1996, Molecular biology of fungal diseases, In : *Molecular Biology of the Biological Control of Pest and Diseases of Plants*, CRC Press USA, pp. 57-65.

Stanway, C.A. 1985, Double stranded RNA viruses and pathogens is of the wheet take all fungus, *Gaumanomyces graminis* var. *trtici*, Ph. D. Thesis, University of London, England.

Styer, E.L. 1978, Electrone microscopy of internuclear virus like particles in *Phytophthora infestans*, Ph. D. Thesis Univ. Meryl and College Park, MD, USA.

Stakman, E.C. 1967, The role of plant pathology in the scientific and social development of the world, In : *Plant Pathology : Problem and Progress* 1908-1958, eds. C.S. Hotton, G.W. Fischer and R.W. Futton, Hart. Helen and S.E.A. Mc. Callan Central Book Depot, New Delhi, Dubi, India pp. 3-13.

Thiecle, D.J., Hanning, E.M. and Leibowitz, M.J. 1984, Genome structure and expression of a defective interfering mutant of the killer virus yeast, *Virol.* 137 : 20-31.

Tooley, P.W. and Hewings, A.D. 1989, Detection of double stranded RNA in *Phytophthora infestans, Phytopath.* 79 : 470-474.

Van Driesche, R.G. and Bellows, T.S. 1996, *Biological Control,* Chapman & Hall, New York, USA.

Wickner, R.B. 1976, Killer of *Saccharomyces cerevisiae* : A double stranded ribonuclic acid plasmid, *Bact. Rev.* 40 : 757-773.

Zhang, R., Pryor, A.M; Qui, B.S. and Tien, B.S. 1994, Presence of double stranded RNA in the virus like particles of *Uromyces vignale* (Barel), *Acta. Mycologia Sincia*, 13 : 48-51.

Zhang, R., Dickinson and Pryor, A. 1994, Double stranded RNAs in the rust fungi, *Ann Rev. Phytopath.* 32 : 115-133.

Zhou- Ting, Boland, G.J. and Zhou, T. 1997, Hypovirulence and double stranded RNA in *Sclerotinia homoeicarpa, Phytopath.* 87 : 147-153.

EXPLOITATION OF MICROORGANISMS AND VIRUSES AS BIO-CONTROL AGENTS FOR CROP DISEASE MANAGEMENT

R. Jeyarajan and S. Nakkeeran

Department of Plant Pathology
Tamil Nadu Agricultural University
Coimbatore- 641 003, Tamil Nadu, INDIA

1. INTRODUCTION

A large number of eco-friendly and potential fungi, bacteria and actinomycetes for biocontrol of plant diseases have been identified (Table 1). Several commercial formulations of fungal and bacterial biocontrol agents are now produced (Table 2).

In order to make biocontrol more effective, it is necessary to enhance the efficicasy of the biocontrol agents in several aspects.

2. ENHANCING EFFICACY

The present biocontrol products can be further improved to obtain greater levels of disease reductions. The various methods of enhancing the efficacy of bio-control agents under field conditions are discussed below. This will contribute towards their ultimate acceptance in disease management programmes in field by farmers. This can be achieved through the selection of crop/soil specific isolates, development of mixed inoculum and organic food base.

2.1. Crop Specific Isolates

Though one species of a biocontrol fungus like *Trichoderma* may be effective against particular species of plant pathogen, its efficacy in field is influenced by crop and soil factors. For effective biocontrol of soil-borne plant pathogen like *Macrophomina phaseolina*, it is necessary that biocontrol fungus multiplies in rhizosphere of the crop and maintains high population throughout crop growth period. This requires that root exudates of particular crop variety should serve as nutrients for the growth of biocontrol fungus. The increased activity of antagonists in the plant rhizosphere reduced the inoculum potential of a pathogen by decreasing the number of its propagules or by immobilizing the

Table 1. Biological agents effective against plant pathogens

Bio-agent	Crop	Pathogen
Fungi		
Trichoderma viride	Chilly	*Pythium ultimum*
T. harzianum	Tomato	*P. aphanidermatum*
T. virens	Gingelly	*Macrophomina phaseolina*
Laetisaria arvalis	Redgram	*Fusarium udum*
	Chickpea	*F. oxysproum* f.sp. *ciceri*
	Gram	*M. phaseolina*
	Redgram,	*M. phaseolina*
	Blackgram	
	Soybean,	*M. phaseolina*
	Greengram	
Chaetomium globosum	Apple	*Venturia inaequalis*
Bacteria		
Pseudomonas fluorescens	Rice	*Pyricularia oryzae*
	Pulses	*M. phaseolina*
	Oilseeds	
	Banana	*F. oxysporum* f.sp. *cubense*
Agrobacterium radiobactor	Crucifers	*A. tumefaciens*
Streptomyces		
Streptomyces griseoviridis	Crucifers	*P. ultimum*
		P. aphanidermatum

Table 2. Commercial formulations of fungal biocontrol agents

S.N.	Product	Biocontrol fungi	Target pathogen	Manufacturer
1.	BINAB-T	*Trichoderma* sp.	*Chondrostereum purpureum*	Bio-Innovation AB, Sweden
2.	F-Stop	*T. harzianum*	*Pythium* sp.	Eastman Kodak Co., USA.
3.	Trichodermin	*Trichoderma* sp.	*Botrytis, Pythium, Sclerotinia* and *Verticillium* sp.	Bulgarian and Soviet Governments
4.	Trichodex	*T. harzianum* (T39)	*Botrytis cinerea*	Makhteshim, Israel
5.	Tricho dowels, Tricho mini dowels, Trichojet, Trichopel, Trichoseal	*T. harzianum* *T. viride*	*Rhizoctonia solani* *Pythium* sp.	Agrimm Technologies Ltd., New Zealand
6.	Antagon - TV	*T. viride*	*R. solani* *Macrophomina phaseolina*	Green Tech Agro Products, Coimbatore, India
7.	Antagon-Combi	*T. viride* + *P. fluorescens*	*M. phaseolina*	Green Tech Agro Products, Coimbatore, India
8.	Sun-Derma	*T. viride*	*R. solani* *M. phaseolina*	Sun Agro Chemicals, Chennai
9.	Ecofit	*Trichoderma* spp.	*R. solani* *M. Phaseolina*	AgrEVo, Mumbai, India
10.	Gliogard	*T. virens* (*Gliocladium virens*)	*Pythium ultimum* *R. solani*	W.R. Grace & Co., USA.
11.	Soligard 12G	*T. virens* (*G. virens*)	*Pythium* spp.	ThermoTrilogy Corp., USA.
12.	-	*Ampelomyces quisqualis*	*Sphaerotherca fuliginea*	Ecogen, Israel.
13.	-	*Pichia guiliermondii*	*Penicillium* sp.	Ecogen, Israel.

available nutrients and reduce its vigour to establish infection. In wheat variety Apex, resistant to root rot caused by *Cochliobolus sativus* the number of fungi in the rhizosphere were 343 x 10^3 /g as against 124 x 10^3 /g in susceptible variety S165. Further 20% of bacterial population in rhizosphere of Apex were antagonistic to the pathogen, but none in S165 (Neal *et al.*, 1970). Therefore it is necessary to select suitable isolates of *Trichoderma* for each crop analogous to the selection of rhizobial strains.

2.2. Soil Specific Isolates

The survival of the biocontrol agent in soil is an important factor in its success. Studies on the biology of biocontrol fungi and bacteria have clearly established that soil moisture and pH play a decissive role not only in the growth and survival of biocontrol organisms but also in the quantity of antibiotics produced by them. *Trichoderma* is known to produce maximum quantity of antibiotics at pH 5. Therefore it is necessary to select isolates of *Trichoderma* suitable for the prevailing soil pH. Acidic pH level enhanced growth of *T. harzianum* (Chet and Baker, 1980).

Jeyarajan *et al.* (1994) reported that, *T. viride* isolates (1,2,3,6) and *T. harzianum* isolates (5,17) were able to grow at pH range of 5 to 9. Mutants of *T. viride* induced through N-methyl-N-nitro-N-nitrosoguanidine was able to grow even upto pH 9. It exhibited high antagonistic potential against *Macrophomina phaseolina* (Nakkeeran *et al.*, 1996). Pre and post emergence damping off of tomato caused by *Pythium ultimum* was effectively controlled by *Trichoderma harzianum* than cucumber seeds since tomato seeds were treated with HCl during extraction. The pH of seed leachate from tomato was 2.8 while cucumber seed leachate was 7. On the other hand the bacterial antogonist *Pseudomonas fluorescens* performs well at pH 7.0 to 8. (Muthamilan and Vidyasekaran, 1994).

Soil moisture also influences the efficacy of biocontrol agents. The antagonists *T. viride*, *T. harzianum* and *Laetisaria arvalis* showed maximum growth at low moisture stress of -4.54 bars. Sporulation of *T. viride* and *T. harzianum* at low moisture stress condition is significant in contributing to the antagonism against *R. solani* (Sangeetha and Jeyarajan, 1991). Jeyarajan *et al.* (1994) observed that *T. viride* and *T. harzianum* survived well at 40% and 60% moisture holding capacity. On the other hand *P. fluorescens* was able to perform well under high soil moisture condition (Rabindran, 1994).

2.3. Mixed Inoculum

As a single biocontrol agent may not perform well under varying soil conditions, identification of compatible combinations of biocontrol agents is desirable. Kloepper (1983) reported that mixture of two or more rhizobacteria resulted in significant reduction in tuber infestation by *Erwinia caratovora*. Four isolates of *T. viride* (1,2,3,6) recorded maximum growth at pH 5 to 8; *T. harzianum* made better growth at pH 8 and 9. This indicate that a composite culture of *T. viride* and *T. harzianum* could be developed, to perform better in soils of wide pH ranges (Jeyarajan *et al.*, 1994). *T. virens* (*Gliocladium virens*) strain 'p' produced antibiotic gliovirin and heptelidic acid. Strain 'Q' produced gliotoxin and dimethylgliotoxin. Gliovirin was inhibitory to *Pythium ultimum* and was very effective when delivered through spermosphere. Gliotoxin was more effective against *R. solani*. So the combination of both strains will broaden the disease control spectrum (Howell *et al.*, 1993). Nakkeeran *et al.* (1996) reported that combined application of *T. viride* and *P. fluorescens* recorded least incidence of panama wilt of banana incited by *Fusarium oxysporum* f.sp. *cubense*. The integration of *T. harzianum* with *Rhizobium* remarkably reduced the root rot of groundnut caused by *Sclerotium rolfsii* (Muthamilan and Jeyarajan, 1996).

2.4. Food Base

One of the main bottlenecks in the initial establishment of the biocontrol agent in the infection court is the lack of food for the antagonist, especially when applied to leaves. By adding a suitable food base to the product the growth of the antagonist after application to the crop can be boosted. Several cheap and readily available forms of organic matter have been tested as food bases for easy and quick multiplication of bio-control agents. Wheat bran-saw dust (3:1) formulation of *Trichoderma* reduced damping-off of tomato and chillies caused by *P. aphanidermatum* to 20% from 80% in control (Sivan *et al.*, 1984).

2.5. Induction of Mutants

Ahmad and Baker (1987) obtained benomyl tolerant mutants of *T. harzianum* by exposing conidia to nitrosoguanidine.The mutants T-95 and T-128, produced more cellulase than the wild strains. The mutants used cellulase as a sole carbon source to colonize plant rhizosphere, whereas the wild strain did not. Mutants of *T. virens* were developed by exposing conidia to both uv radiation and ethyl methyl sulphonate. (Papavizas *et al.*, 1990).

2.6. Genetic Engineering

Genetic engineering is considered as one of the three major scientific revolutions of the century. A successful biocontrol agent requires many characteristics. All of them may not be present in one strain. Since many biocontrol agents have infrequent / unknown sexual stages, crossing and progeny testing by conventional breeding is not possible. Genetic engineering is an useful tool to :
  (i)   transfer ability to produce antibiotics from one species to another,
  (ii)   produce location specific strains,
  (iii) improve rhizosphere competence of the antagonist and
  (iv)   improve its biocontrol potential.
By this technique important attributes like tolerance to pesticides, adverse environmental conditions like extremes of moisture, temperature and pH, vigorous growth, sporulation in deep tank fermentatom and improved shelf life can be incorporated in one organism. The strain k-1026 of *Agrobacterium radiobacter* is the first genetically manipulated antagonist produced commercially. This genetic manipulation prevent breakdown of agrocin producing avirulent strain (Cook,1993).

2.6.1. Methods of induction of new biotypes

New biotypes of biocontrol organisms can be developed by (i) exposure to fungicides, (ii) exposure to chemical and physical mutagens, (iii) protoplast fusion and (iv) protoplast transformation.

2.6.2. Exposure to fungicides

By exposing *T. harzianum* to chlorothalonil several isolates tolerant to it were developed. They differed in morphological characters, growth habit and sporulation from the wild type. Some showed enhanced biocontrol activity. This may be due to mutation, chromosomal aberration or mitotic recombination (Papavizas, 1987).

2.6.3. Exposure to mutagens

Mutants of *T. harzianum* induced through UV irradiation were found to tolerate high concentration (100-500 mg/ml) of the fungicide benomyl. The biotypes differed from the wild strain in appearance, growth habit, survival ability in soil, fungitoxic, metabolite production and in the ability to suppress damping off induced by *P. ultimum* of peas and damping off of cotton (*R. solani*). Ten mutants of *T. viride* produced from wild strain-T through UV-radiation tolerated upto 100mg of benomyl, thiobendazole and thiophanate methyl. *T. viride* (T-1-R4), *T. harzianum* (WT-6-24) and *T. virens* mutants were more effective in suppressing competitive saprophytic ability of *R. solani* in soil and controlling damping off of peas (*P. ultimum*), cotton (*R. solani*), and bean (*S. rolfsii*) (Papavizas and Lewis, 1983). UVmutant of *T. harzianum* produced one heat labile and one heat stable metabolite whereas the wild type produced only heat stable metabolite. The heat labile metabolite was identified as gliotoxin. The mutant was most effective against white rot of onion incited by *Sclerotium cepivorum*. Advantages of the mutangens inc :
(i) Development of such fungicide resistant/tolerant stains would induce the industry to take up pilot testing and commercial production.
(ii) Could be combined with minimal amount of fungicide for disease control.
(iii) Such mutants can also be used in genetic recombination programe, by introduction of specific gene encoding for increased production of antibiotics and enzymes into a recipient biocontrol fungus that possess other desirable traits.

2.6.4. Protoplast Fusion

Protoplast fusion is an excellent approach to the construction of fungal strains with novel gene combinations (Papavizas, 1987). It is possible to combine desirable traits from various parental strains to produce superior biocontrol strains. Interspecific crosses have been accomplished in *T. viride* (Stasz *et al.*, 1989).

2.6.5. Protoplast Transformation

It is the process by which naked DNA is introduced into a cell resulting in a heritage change. It involves two basic steps.
(i) Partial or complete elimination of cell barriers to allow uptake of DNA into the cell.
(ii) Stable integration, maintenance and self replication of DNA into cell. Isolated protoplasts are excellent materials for the introduction of foreign DNA. This can be done in three ways.
(a) Fusion of whole protoplasts from two strains, species or genera.
(b) Transfer of mitochondrial plasmids- plasmid mediated transformation. Hybrids contain genetic material only from the naturally inbreeding microbes from which crossed strains are derived.
(c) Transfer of DNA segments (autonomously replicating and integrated vectors) combinant DNA technique.
DNA segments are separated from donor and inserted or attached to other cells. Eukaryotic genes cloned into *E. coli*, sequence are rearranged and introduced into the new organism (Papavizas, 1987). Resistance to benomyl was constructed by transformation (Ossanna and Mischke, 1990). This was accomplished by transferring to GI 21, a wild type strain of *T. virens*, very sensitive to benomyl. The B tubulin gene from *Neurospora*

crassa that codes for resistance to benomyl was successfully expressed in *T. virens* strain GT-21. It suggest that there are tremendous possibilities for desirable transformations in biocontrol fungi, provided that the genes coding for certain desirable traits have been identified.

3. BIOCHEMISTRY OF METABOLITES

Antagonists secreted several metabolites like volatile compounds, antibiotics and enzymes.

3.1. Volatile Compounds

Strains of *T. harzianum* were found to produce two volatile metabolites like 6-n-pentenyl-2H-pyran-2 one and 6-n-pentenyl-2H-pyran-2-one (Claydon *et al.*, 1987). *T. viride* produced lactone, alcohols and terpene derivatives of pyran (Zeppa *et al.*, 1990). Claydon *et al.* (1991) found that *T. harzianum* secreted harzianolide [3-(2-hydroxyl-propyl)-4(hexa-2"-dienyl-2(5H) furanone]. Volatile substances from *T. viride*, *T. harzianum* and *T. longibrachyatum* inhibited the mycelial growth of *M. phaseolina* by 61, 51 and 22 per cent respectively (Angappan, 1992).

3.2. Antibiotics

T. viride produced larger quantities of antibiotics like trichodermin (Gotfredson and Vangedal, 1965), dermadin (Pyke and Dietz, 1966), trichoviridin (Yamano *et al.*, 1970) and sesquiterpene heptalic acid (Itoh *et al.*, 1980). The UV mutants of *T. harzianum* produced trichorzianines which are hydrophobic peptides, interact with phospholipid membranes and induced membrane permeability. *T. hamatum* produced trichoviridin 3-(3-isocyanocyclopent - 2- enzylidene) propionic acid and 3-(3-isocyano-6-oxabicyclo (3,10) hex-2-eh-5-yl) acrylic acid (Baldwin *et al.*, 1981). In addition, it produced three isonitriles among which isonitrin A was broad sprectrum in activity (Edenborough and Hebert, 1988). *T. longibrachyatum* produced trichodermin (Watts *et al.*, 1988) and trichobrachin (Brucker *et al.*, 1990).

3.3. Enzymes

The extracellular enzymes produced by *Trichoderma* spp. and *T. virens* like chitinase and glucanase are important in degradation of cell wall of pathogens. *T. viride* produced chitinase and β-1,3-glucanase (Elad *et al.*, 1982) These enzymes reduced the viability of *S. rolfsii* propagules by 95 per cent. Arlorio *et al.* (1992) reported that these enzymes affected the hyphal tip of the pathogen resulting in thinning of cell wall, leading to imbalance of turgor pressure and wall tension which caused the tip to swell and burst. Chitinase and chitobiase were produced by *T. harzianum* (Ulhaa and Peberdy, 1993). Chitin -1-4-β-chitobiosidase, n-acetyl β-D glucosaminase and endochitinase produced by *T. harzianum* had the molecular mass of 40, 35 and 41 -Kda respectively (Harman *et al.*, 1993).

4. PLANT PATHOGEN - ANTAGONIST INTERACTION

The root exudates of several crop plants are known to excert a chemotactic effect in attracting the germ tubes of plant pathogens. Abdou *et al.* (1980) reported that root exudates from *Sesamum* induced the germination of sclerotia of *M. phaseolina* in soil and attracted growth towards root region. Therefore a high population of antagonist in the rhizosphere effectively prevents infection by pathogen. Angappan (1992) found that root exudates from chickpea cultivar Gl-769 resistant to *M. phaseolina* inhibited the growth of pathogen *in vitro* (50%) and germination of sclerotia (50%), compared to root exudates of susceptible cultiar Shoba.

5. BIOCONTROL AS A SYSTEM APPROACH

Biocontrol agents bring about higher disease reduction when used as a component of integrated pest management like addition of organic matter, growing resistant variety and application of fungicides not harmful to antagonists.

5.1. Addition of Organic Matter

Application of organic matter to soil has beneficial effect on several biocontrol agents by serving as food base. The age old practise of applying organic matter to soil has been instrumental in reducing incidence of soil borne diseases. Neem cake and farm yard manure are reported to serve as food base to serve beneficial fungi in soil including *Trichoderma* (Nakkeeran *et al.*, 1993; Ushamalini *et al.*, 1997).

5.2. Addition of Antagonist

Seed treatment with *Trichoderma* gave two benefits to the crop by germinating along with seed and reduced pre-emergence damping off. The subsequent growth and colonisation in the rhizosphere afforded protection against soil borne inoculum of *M. phaseolina* in later stages of crop growth.

5.3. Addition of Sublethal Dosage of Fungicides

Application of soil fumigants like methyl bromide, indirectly helps in biological control of *Armillaria mellea* in citrus and several root diseases of tea like brown root rot, black rot and charcoal stump rot. Methyl bromide weakens the growth of these pathogens which are subsequently destroyed by *Trichoderma* which is not sensitive to such low doses of fumigant (Martosupono and Prayudi,1979). Use of fungicide metalaxyl plus *Trichoderma* has been found effective against quick wilt of pepper (*P.capsci*), since metalaxyl does not inhibit *Trichoderma*. Application of *T. harzianum* after soil fumigation with methyl bromide improved control of *S. rolfsii* and *R. solani* in peanut field. Soil fumigation alone was followed by rapid reinfestation by *S. rolfsii* and *R. solani*. The biological control agent prevented reinfestation of the fumigated soil. The combined treatment of fumigation and *T.harzianum* application, caused almost total mortality of sclerotia in soil. Transplanting tomato plants treated with *T. harzianum* into soil fumigated with methyl bromide reduced disease incidence by 93% and increased yield by 160% (Elad *et al.*, 1982).

6. FORMULATIONS AND SUBSTRATES

The Tamil Nadu Agricultural University, Coimbatore, has developed the technology of mass production of talc based formulation of *Trichoderma viride*, and *Pseudomonas fluorescens* for seed treatment. Annually it produces this formulation to cover 20,000 ha. Several private industries in Coimbatore, Bangalore, Chethali, Delhi and Chennai produce the same in large quanitites. However, the demand exceeds the present supply. The annual requirement of *Trichoderma* has been estimated as 5,000 tonnes to cover 50 per cent area in India. Indirectly it also creates self employment opportunities to the unemployed youths. The cost-benefit ratio in different crops are given in Table 3. The first commercial seed treatment formulation of biocontrol agent in India was developed by Jeyarajan *et al.*, (1994). *T. viride* was grown in molasses yeast medium for 10 days in flasks. Then 2:51 of the culture was inoculated to 50 l of sterilized molasses yeast medium in fermentor. It was incubated for 10 days with 4-8 hr aeration /day. The fungal biomass and the broth were mixed with 100kg talc (super white) powder and 500g of carboxy methyl cellulose as a sticker. It was dried in shade for 72 hr and packed in alkathene bags. The initial populatin of *Trichoderma* in the produce was 300×10^6 cfu/g. The product should contain a minimum of 20×10^6 cfu/g at the time of use. This is applied at the rate of 4 g per kg of seed. The shelf life was 4 months. Angappan (1992) found that in chickpea treated with this product the rhizosphere population was maintained at 11 to 13×10^3 cfu per gram throughout crop growth stage.

Table 3. Cost benefit ratio of *Trichoderma* seed treatment formulation.

Crop	Cost/Ha	Cost benefit ratio
Groundnut	60	1:540
Cotton	8	1:360
Gram	24	1:191
Gingelly	3	1:425
Sunflower	12	1:275
Mung/Urd bean	10	1:250

6.1. Gypsum

Renganathan *et al.* (1995) found that gypsum was a good substitute for talc but was much cheaper than talc.

6.2. Industrial Wastes

Nakkeeran and Jeyarajan (1996) tested two industrial wastes namely precipitated silica and calcium silicate in the place of talc and found them to give a population of 99, 104×10^6 cfu/g respectively compared to 143×10^6 cfu/g in talc substrate after 4 months of storage. These two substrates were also much cheaper than talc.

6.3. Diatomaceous Earth Granules

Diatomaceous earth granules were added into a broth consisting of 100 ml black strap molasses, 900 ml of water, 3 g each of KNO_3 and KH_2PO_4 till the level of saturation. It was autoclaved at 121°C for 15 minutes and spread in shallow pans to a height of 3-5 cm and autocalved again. *T. harzianum* (3 day old culture) was homogenized in a blender for 3 seconds and mixed with sterlized granules in shallow pans and incubated at 25°C for 4-7 days. Clumps were broken up and granules air dried with frequent stirring and used as an inoculum. The inoculum was mixed at 1.1(v/v) with sterilized diatomaceous earth granules impregnated with 10% molasses solution. It was applied to peanut at 140 kg/ha on 70 and 100 days after sowing to control *Sclerotium rolfsii*. The disease was reduced by 42% over control and yield increased by 13.5% (Backman and Rodriguez Kabana, 1975).

6.4. Wheat Bran : Saw Dust Formulation

Wheat bran : saw dust : tap water mixture (3:1:4 v/v) is taken in polypropylene bags and autocalved for 1 hour at 121°C for two successive days. The bags were inoculated with *Trichoderma harzianum* and incubated in illuminated chambers for 14 days at 30°C. It was applied at the time of sowing and mixed with the soil to a depth of 7-10 cm with a rotary hoe. It increased yield of beans (1500 kg/ha), tomato (300 kg/ha), cotton (500 kg/ha) and potato (400-600 kg/ha) and controlled *Sclerotium rolfsii* and *Rhizoctonia solani* (Elad *et al.*, 1986).

6.5. Wheat Bran: Peat

Wheat bran : Peat mixture (1:1 v/v) was autoclaved for 1 hour. Substrate moisture was adjusted to 50% (W/W) with sterile water, medium was inoculated with 0.1 ml of a conidial suspension containing 2×10^4 conidia/ml and incubated for seven days at 30°C, it was mixed with rooting mixture for tomato at 10% v/v to control crown rot *(Fusarium oxysporum* f.sp.*radicis-lycopersici).* The disease was reduced by 29.7% over control and yield increased by 7% (Sivan *et al.*, 1984).

6.6. Vermiculite - Wheat Bran

Trichoderma was multiplied in molasses - yeast medium for 10 days. Vermiculite (Grade 4) and milled wheat bran (250 mesh) were heated in hot air oven at 70°C for three days using metal pans. Vermiculite (100g), wheat bran (3.3 g), liquid culture (14 ml) and 0.05 N HCl (17.5 ml) were mixed and packed. This was immediately used for soil application (Lewis *et al.*, 1991). Vidya (1995) applied this formulation at 250 kg/ha to mung bean and found 41 % reduction in root rot (*M. phaseolina*) and 91% increase in yield.

6.7. Alginate Pellets

Sodium alginate (20g) was dissolved in 750 ml water at 40°C on a stirring hot plate. Wheat bran ground to pass through a 0.425 mm mesh screen was placed in a glass blender container with distilled water (50g/250ml). Kaolin can also be used in place of bran. It was autocalved for 30 min and cooled. Fermentor biomass (16-21g/l) was added to provide 7×10^6 conidia and chlamydospores. The mixture containing fungus, alginate and bran or kaolin was added dropwise into 500 ml gellant solution (0.25 M $CaCl_2$, pH 5.4). As it entered, each droplet gelled and a distinct spherical bead formed. After 20 min in the gellant, beads were separated from the solution by gentle filtration, washed and dried for 24 hr in a stream of air at 25°C. (Lewis and Papavizas, 1986).

6.8. Other Substrates

The following substrates were also found to support the growth of *Trichoderma viride* and *T. harzianum* (Kousalya and Jeyarajan, 1988) : Farm yard manure, gobar gas slurry, press mud, paddy chaff, rice bran and groundnut shell.

7. IMPROVEMENT OF FORMULATIONS

Already several commercial products of biocontrol agents are in the market. The efficacy of these products can be further improved with respect of shelf life and population.

7.1. Longer Shelf Life

Shelf life of a biocontrol agent play a crucial role in storing a formulation. It varies depending upon the nature of the food base. In general the antagonist multiplied in an organic food base has greater shelf life than that on inert of inorganic food bases. Jeyarajan et al. (1994) developed talc, peat, lignite and kaolin based formulations of T. viride, which had a shelf life of 4 months. Ranganathan et al. (1995) found that the shelf life of T. viride in gypsum based formulation was also four months. Studies on storage temperatures revealed that 20-30° was optimum to store vermiculite fermentor biomass of Trichoderma upto 75 days without loosing its viability (Nakkeeran et al., 1997).

Among the different carriers tested the shelf life of Bacillus subtilis in soybean flour was increased upto three months. Storage of 5°C increased the shelf life of T. virens and T. hamatum in granular formulations of pregelatinizing starch flour upto 6 months (Lewis et al., 1995).

7.2. Higher Population

Inorganic carriers like talc, peat, lignite and kaolin were used to prepare Trichoderma formulations. The initial population in the product of different carriers varied between 220 to 280 x 10^6 cfu/g. The population after four months of storage came down to 80, 60, 70, 60 x 10^6 cfu/g in talc, peat, lignite and kaolin respectively. Even though there was a decline in the population, it is still much above the minimum of 20 x 10^6 cfu/g of the product of various carriers (Jeyarajan et al., 1994).

8. RHIZOSPHERE COMPETENCE

Pseudomonos fluorescens has been reported to survive in the rhizosphere of rice, cotton, tomato and cucumber (Misaghi, 1990). Rabindran (1994) found that in rice rhizoshpere population of P. fluorescens was 4.3 x 10^4 /g. Ushamalini et al. (1997) found that the rhizoshpere population of Trichoderma in cowpea was 29.3 x 10^3 /g due to seed treatment as against 16.4 x 10^3 /g in the soil giving R:S ratio of 1.79.

9. APPLICATION

9.1. Delivery System

Formulated products must be delivered effectively into appropriate agricultural system. It should suit the available equipments like seed drill, seed treatment drum, etc., should be easy to perform and should be placed in the site of action (Soil, seed, wound, foliage, harvested produce).

9.2. Methods of Application

Biocontrol agents can be applied to various plant parts like seed, root, foliage, inflorescens and also the soil. For this purpose several methods have been developed.

104

9.2.1. Dry seed treatment

Delivery of antagonist to the spermosphere is the most effective and economical method. Wells *et al.* (1972) first demonstrated the efficacy of *T. harzianum* against *Rhizoctonia* spp. in the field. Seed pelleting with *T. harzianum* using 5×10^9 cfu/ml reduced the infection by *M. phaseolina* in bean and chickpea (Jeyarajan and Ramakrishnan, 1991). It reduced the root rot incidence by 90 per cent in urdbean (Jeyarajan *et al.*, 1991).

It has the following advantages over fungicides:
(i)   Not harmful to human beings and animals
(ii)   Cheaper than pesticides by 50%
(iii) Effective throughout crop growth
(iv)   Easy to apply by farmers
(v)   Improves plant growth which is spectacular
(vi)   Increases yield
(vii) High cost benefit ratio (19 to 360)
(viii) Do not affect environment
(ix)   No residue in food and ground water
(x)   No risk of the pathogen developing resistance
(xi)   Compatible with biofertilizers-*Rhizobium* and *Azospirillum*
(xii) Solubilise 'P' from soil (Anusuya and Jeyarajan, 1998).

A talc based formulation of *T. viride* was developed for seed treatment by Jeyarajan *et al.* (1994). It reduced the root rot incidence in urdbean by 60%, in chick pea by 50%, in peanut by 77% and in sesamum by 67% and increased the yield by 20,13,12 and 12% respectively. Seed treatment with antagonist *T. harzianum* and *T. viride* significantly reduced root rot incidence to 10.1% and 12.8% respectively, compared to 60% incidence in the control (Sankar and Jeyarajan, 1996). Seed treatment of pigeon pea with *Trichoderma* spp. and *Bacillus subtilis*, effectively controlled root rot and wilt of pigeon pea and enhanced the yield considerably. (Nakkeeran and Renukadevi, 1997; Nakkeeran *et al.*, 1995).

9.2.2. Solid matrix priming

It is a process in which moistened seeds are mixed with an organic carrier and the moisture content of the mixture is brought to a level just below that required for seed sprouting. The efficacy of antagonist in treated seeds can be further enhanced by solid matrix priming. *Trichoderma* grow on seed surface during priming process and increase in numbers and other microbes may not easily dislodge it. Slurry of *Trichoderma* is first added to seeds followed by addition of lignite and water. *Trichoderma* requires less moisture for growth. The primed seeds were incubated for 4 days at 80-100% RH. During this period *Trichoderma* colonises seed and sporulation was evident. In *Pythium ultimum* infested soil the stand was increased to 70-80% as against 10% in control. In tomato seed the population of *Trichoderma* increased by ten fold due to solid matrix priming (Harman *et al.*, 1989).

9.2.3. Soil

For effective management of soil-borne diseases, a high population of the antagonist in the soil is necessary. Adams (1990) defined effeciency of biocontrol agents as the ratio of number of propagules of mycoparasites required to obtain disease control to the typical inoculum density of a plant pathogen. Therefore attempts were made to develop suitable delivery system to soil. For controlling *R. solani*, *Trichoderma* population of 5 x10[6] was required for each propagule of *R. solani*. Addition of wheat bran based inoculum to soil gave 80 per cent reduction of root rot over control in chickpea and bean (Elad *et al.*, 1986). Incidence of urdbean root rot was reduced by 91.3 per cent by adding *T. viride* + *T. harzianum* to soil (Kousalya and Jeyarajan, 1988). Delivering *T. harzianum* through soil, during sowing increased the percentage of survival of peanut (90%), while in control none of the plants survived (Muthamilan and Jeyarajan, 1996).

9.2.4. Foliar spray

The biocontrol organisms can be introduced only once (Inoculative) or repeatedly applied (Inundative). Four applications of a conidial suspension of *Ampelomyces quisqualis* (12 x 10[4] conidia/ml) at weekly intervals to cucumber leaves suppressed conidial production and viability of powdery mildew. After the application of biotic agent as foliar spray, mildew colonies turned dull black with abundant pycnidia of *A. quisqualis*.

The efficacy of bio-control agents for foliar diseases is greatly affected by fluctuation of micro climate. Phyllosphere is subjected to diurnal and nocturnal, cyclic and non-cyclic variations in temperature, relative humidity, dew, rain, wind and radiation. Hence water potential of phylloplane microbes will be varying constantly. It will also vary between leaves at the periphery of the canopy and on sheltered leaves. Higher relative humidity could be observed in the shaded, dense region of the plant than that of peripheral leaves. The dew formation is greater in centre and periphery.

The concentration of nutrients like amino acid, organic acids and sugars exuded through stomata, lenticels, hydathodes and wounds varies highly. It affects the efficacy and survival of antagonist in phylloplane (Andrews, 1992). When *Pseudomonas* is applied to beet leaves, it actively competed for amino acids on the leaf surface and inhibited spore germination of *Botrytis cinerea*, *Cladosporium herbarum* and *Phoma betae* (Blakeman and Brodie, 1977).

Application of *B. subtilis* to bean leaves decreased incidence of bean rust (*Uromyces phaseoli*) by 75% equivalent to weekly treatments with the fungicide mancozeb (Baker *et al.*, 1983). A peptide containing metabolite produced by the bacterium reduced uredospore germination, prevented normal germ tube development and caused cytoplasmic abnormalities in the uredospore. The yeast *Tilletiopis minor*, *Stephanoascus flocculosus* and *S. rugulosus*, have been found to colonize and inactivate cucumber powdery mildew, *Sphaerotheca fuliginea* (Jarvis *et al.*,1989).

9.2.5. Twine coating

Gadoury *et al.* (1991) developed a novel method of applying pycnidia of the mycoparasite *Ampelomyces quisqualis* to grapes to control *Uncinula necator*. Coton twine was saturated with malt extract agar and dipped in pycnidial suspension. The twine was suspended in the trellis above grapevines, when shoots were 15 cm long. Rain splashed the conidia onto leaves, upto 3 months.

9.2.6. Cut stump application

Ricard (1981) developed a special shears to apply *T. viride* simultaneously with prunning to protect fruit trees against silver leaf disease caused by *Chondrostereum purpureum*. The shears have a reservoir for suspension of *T. viride*. Each prunning cut was covered by spore suspension.

9.2.7. Hive insert

An innovative method of application of biocontrol agent right in the infection court at the exact time of susceptibility was developed by Thomson *et al.* (1992). *Erwinia amylovora* causing fire blight of apple infects through flower and develops extensively on stigma. Colonisation by antagonist at the critical juncture is necessary to prevent flower infection. Since flowers do not open simultaneously the biocontrol agent *P. fluorescens* has to be applied to flowers repeatedly to protect the stigma. Nectar seeking insects like *Apis mellifera* can be used to deliver *P. fluorescens* to stigma. Bees deposit the bacteria on the flowers soon after opening due to their foraging habits. Thus repeated application could be avoided.

Botrytis cinerea (fruit rot of straw berry) was effectively controlled by *T. virens* through honey bees which acted as vector. Peng *et al.* (1992) developed a novel and more efficient method of applying inoculum of *T. virens* on talc and corn meal (5 x 10^8 cfu/g) which was placed inside the bee hive in a dispensor. Bees acquired the conidia on their legs and bodies as they crawled. Each bee had a spore load of 88 to 1800 x 10^3. Each flower got 1.6 to 27 x 10^3 conidia of antagonist. It suppressed *B. cinerea* on stamens (54%) and petals (47%) and controlled fruit rot.

9.2.8. Fruit spray

Pseudomonas syringae (10% wettable powder) in the modified packing line was sprayed at the rate of 10 g/l over apple fruit to control blue and grey mold of apple. The population of antagonist increased in the wounds more than 10 fold during 3 months in storage (Janisieqicy and Jeffer, 1997).

9.2.9. Fruit dip

Dipping of grapes in *Pichia guilliermondii* and *Hanseniaspora uvarum* (yeasts) cell suspension three days before harvest markedly suppressed *Rhizopus* rot for one to two weeks (Sutton and Peng, 1993).

10. POST HARVEST DISEASES

National Academy of Sciences, U.S.A has reported that nine fungicide used to control post harvest diseases are oncogenic. In U.S.A. post harvest loss in tomato amounts to 30 -50% which can feed 200 - 300 x 10^6 people per annum (Kelman, 1989). In the underdeveloped countries losses are greater due to lack of refrigeration and proper sanitation. Biological control offers great scope for their management due to the following reasons.

(i) Many fruits and vegetables are stored under controlled environment.
(ii) Since area of application for fruits and vegetables is less than in field crop it is easier to apply to target site
(iii) High value of fruits makes biocontrol agent more cost effective.
(iv) Biocontrol agents are not destroyed by U.V rays.

Their public acceptance possess no problem because people are already familier with microbial fermented food like bread and dariy products (Wilson and Pusey, 1985; Wilson *et al.*, 1991). The biocontrol agents for important post-harvest diseases of fruits and their mode of action are given in Table 4.

The biocontrol agents to manage post harvest diseases should posses the following traits.

(i) Not fastidious in nutrient requirements.
(ii) Should survive under storge conditions.

Table 4: Different modes of action of bio-control agents

Mode of action	Crop	Disease	Biocontrol Agent
Antibiosis	Apple	Blue Mould (*Penicillium italicum*) Mucor rot	*Pseudomonas cepacia* (Pyrrolnitrin) *B. subtilis* (Iturin)
	Citrus	Green Mould (*Penicillium digitatum*)	*Trichoderma* sp.
Competition	Grapes	Grey mould (*Botrytis cinerea*) *Rhizopus* rot	*Debaryomyces hansenii*
	Tomato	Grey mould (*Botrytis cinerea*) *Rhizopus* rot	*Debaryomyces hansenii*
Induced Resistance	Apple	Blue mould Grey mould	*P. syringae* *Acremonium brevae*
Enzymes	Pome fruit	*B. cinerea*	*Pichia guilliermondi* (Glucanases chitinases)

(iii) Should not produce secondary metabolites toxic to human beings
(iv) Should be compatible with other physical and chemical treatments of the commodity.
(v) Should have broad spectrum against multiple diseases of fruits and vegetables.
(vi) Should have patent potential.

During storage periods yeast play a crucial role in the management of post harvest diseases. The advantages of yeasts are :
(i) Colonize surface for long period under dry conditions
(ii) Produce extracellular polysaccharides that enhance their survival and restrict flow of nutrients to pathogen propagules
(iii) Rapidly use available nutrients and proliferate
(iv) Least affected by pesticides
(v) Competitors and don't produce toxic metabolites

In grapes two sprays with *T. harzianum* (late flowering and three weeks before harvest) reduced grey mould incidence by 50%. The first application suppressed *B. cinerea* in flowers, since pathogen first infected flowers before invading berries. The second spray limited its spread in ripening grapes and prevented entry of germ tubes through wounds and microfissures in skin (Sutton and Peng, 1993). In strawberry Peng *et al.* (1992) found that spraying a conidial suspension of *T. viride* (10^7/ml) from early flowering at 14 days interval reduced fruit rot during storage (*Botrytis cinerea* and *Mucor mucedo*) equal to the fungicide dichlorofluanid. In apple, fruit dip with a suspension of *Cryptococcus laurentii* (10^7 spores or cell/ml) gune[109] effective control of *B. cinerea* during cold storgne. Potato soft rot was reduced by 15% when seed pieces were treated with a suspension of *Pseudomonas putida*. The following biocontrol agents have been patented in U.S.A for managing post harvest diseases.

Bacillus subtilis strain B-3 (Brown rot of stone fruits)
Pseudomonas cepacea (*B. cinerea* on pome fruits)
Acremonium brevae (*Penicillium* rot of pome fruits)
Pichia guilliermondii strain US-7 and *H. uvarum* (Rots of citrus, pome fruits, stone fruits and tomato).

11. COMMERCIAL PRODUCTS

11.1. Stages in Development

Following steps in the commercial development of a biopesticide (Whipps, 1992).

i1.1.1 Isolation of antagonist

This is done from pathogen suppressive soils either by dilution plate technique or by baiting the soil with fungal structures like sclerotia of pathogen.

11.1.2. Laboratory screening

This is done *in vitro*. It is quite inexpensive and indicates the mode of action, ease to culture and environmental extremes within which the antagonist will act. A large number of isolates can be tested in a short time to select the most suitable ones.

11.1.3. Pot tests

The plant, pathogen and antagonists are provided controlled conditions. They indicate field efficacy and provide ecological data on the antagonists.

11.1.4. Field trial

Promising antagonists from pot tests are tested in field and their efficacy compared with recommended fungicides. Since variations in weather influence disease incidence, trials should be conducted for at least 15-20 sites - years (sites x years). This is costly and labour intensive.

11.1.5. Toxicological data

They include information on antagonists, safety to men, plants, animals and survival of antagonists in field.

11.1.6. Commercial data

They include cost of production, ease of use and relative efficacy with fungicides or cultural methods. If they are eco-friendly, slightly less efficacy or increased cost may be acceptable. It must fit in with integrated control.

11.1.7. Scale up

This involves methods of inoculum production, formulation, quality standards, shelf life and development of a suitable delivery system to treat soil, seed or plant parts during growth. The cropping system should be taken into account. Other disease control tactics like soil or seed treatments, foliar fungicidal spray which may interfere with the activity of the biocontrol agent should be considered. Strains suitable for each soil condition should be selected. The effect of control of target pathogen on other pathogens should be studied.

11.1.8 Fermentation methods

There are three methods of fermentation (Lewis and Papavizas, 1991).
11.1.8.1. Liquid fermentation
The first major concern in commercial production systems is achieving adequate growth of the biocontrol agent. In many cases biomass production by the bio-control agent is difficult because of the specific nutritional and environmental conditions required

for growth of the organism. Both liquid and semisolid fermentation are used for this purpose. This technology has been adopted recently to the production of bacterial and fungal biomass. A suitable medium should consist of inexpensive, readily available agricultural byproducts with appropriate nutrient balance. Acceptable materials include molasses, brewers yeast, corn steep liquor, sulphate waste liquor, cotton seed and soy flours (Lisansky, 1985). Preparation of *Trichoderma* containing chlamydospores more effectively prevented the diseases than preparations containing conidia only (Lewis *et al.*, 1990; Papavizas and Lewis, 1989). Small scale fermentation in molasses-brewers yeast medium resulted in abundent chlamydospore production (Papavizas *et al.*, 1984).

11.1.8.2. Solid Fermentation

Solid substrates for the production of inoculum of various biocontrol fungi include straws, wheat bran, saw dust, bagasse moistened with water or nutrient solutions (Papavizas, 1985).

11.1.8.3. Semi solid fermentation

It is used for the fungi which do not sporulate in liquid culture, eg., Diatomaceous earth granules impregnated with molasses (Backman and Rodriguez Kabana, 1975) wheat bran and vermiculite-wheat bran (Lewis *et al.*, 1989). This method requires more area, is labour intensive and requires low pressure air blower, and little power. The chances of contamination are high when compared to liquid fermentation.

11.2. Registration

In U.S.A registration of "Microbial pesticide" requires toxicological tests for oral, dermal, eye and other health hazards using test animals or fish. If these tests show no adverse effects and the biocontrol agent is not a pathogen, it is registered and can be sold. The cost required for research and development for biopesticide is only $ 0.8 to 1.6 x 10^6 as against $ 20 x 10^6 for chemical pesticides. The toxicological tests for a bio-control agent cost $ 0.5 x 10^6 as against 10 x 10^6 for chemical pesticides. The number of candidates to be tested to develop one bio-control product will be in 100s as against 20,000 for a chemical pesticide. It was estimated that the market size required for profit for a biocontrol agent is $ 1.6 x 10^6 per year as against $ 40 x 10^6 per year for a chemical pesticide (Cook, 1993).

11.3. Quality Control

This is very much necessary to retain the confidence of farmers on the efficacy of biocontrol formulation. Being living agents their population in a product influences the shelf life. The other contaminating microorganisms in the product should also be within permissible limits.

12. VIRUS DISEASE MANAGEMENT BY MILD STRAINS

Realizing the difficulty in managing virus diseases either directly by use of safe viricidal chemicals or indirectly by vector control, plant virologists the world over started looking for alternate methods. Vegetative propagation of fruit trees like citrus offered the advantage of cross protection as a cheap and practical method of managing the disease with a mild strain. This phenomenon has been recently reviewed by Yeh and Gonsalvas (1994) and Jeyarajan (1996).

12.1. Method of Obtaining Mild Strains

Mild strains can be obtained by selection from natural viral populations; induced mutation of natural population followed by selection; and exposing inoculated plants to high or low temperature.

12.2. Successful Cases

12.2.1. Papays ringspot virus (PRV)

Mild strains were produced by treating crude sap of plant infected by a severe Hawaiian strain (HA) with nitrous acid and the sap was inoculated to *Chenopodium quinoa* which is a local lesion host. Two mild mutants PRV HA 5-1 and 6-1 were selected. These mutants produced no conspicuous symptoms in papaya in the green house (Yeh and Gonsalvas, 1984) and provided a high degree of protection against the severe wilting and mosaic strains prevalent in Taiwan. The protected trees yielded 32% more with better fruit quality resulting in 111% increased income (Wang *et al.*, 1987). The mild strain was propagated in *Cucumis metuliferus*. The leaves were ground in 0.01 M cold potassium phosphate buffer pH 7.0 using 50 ml/g tissue. After straining the extract through cheese cloth, carborundum powder was added at 40g/l and taken in a metal tank connected to spray gun attached to an air compressor. The seedlings were inoculated by pressure sprayer (8 kg/sq cm). One person can inoculate 10,000 seedlings in 2h. After four weeks, all the inoculated plants showed positive results by ELISA. It has now become a routine practise in Taiwan. Upto 1993, papaya orchards in 1722 ha (3,444,000 plants) have been planted with protected plants. The cost worked out to just 28 $/ha. In Hawai, the protection was of very high degree (90-100%) and gave economic production for at least 24 months (Mau *et al.*, 1990). No breakdown was observed even after 30 months.

12.2.2. Tomato mosaic virus

It is a major problem especially in glass house. Cross protection has solved this problem (Oshima, 1979). Rast (1972) produced a symptomless mutant (MII-16) by nitrous acid treatment of the virus. It is commercially produced and widely used in the Netherlands and U.K. It increased yield by 6-15%. Crude sap of heat (34°C) attenuated mild strain diluted to 1:10 or commercial preparation diluted to 1:100 so as to contain 28-1200 μg of virus/ml was effective. The inoculum was prepared by mixing 100 ml of diluted virus suspension with 1g carborundum powder and taken in a commercial paint sprayer. The seedlings in nursery were inoculated with a pressure of 1.28 kg/cm² by keeping the spray nozzle 10 cm from the plants and moving it at 1m/sec (Fletcher and Rowe, 1975). Yield of protected plants was doubled. In Japan, Oshima (1975) produced an attenuated mutant (L11A) by exposing severe strain infected plants to high temperature (34°C). It is successfully used on a large scale and the method is called vaccination. It gave complete control against Ohio IV strain. In china, Tien and Chang (1983) produced a symptomless mutant by nitrous acid treatment. The seedlings were infected by simply immersing the root in virus suspension which is an easy method. Under field condition it increased the yield by 60%.

12.2.3. Citrus tristeza virus

This is a serious disease in Argentina, Brazil, U.S.A., Spain, South Africa, India and Isreal where it destroyed over 50 x 10⁶ trees. In Brazil 8 x 10⁶ sweet orange trees were protected by inoculation with a mild strain (Cost and Muller, 1980). Since citrus is a perennial fruit plant propagated by budding, it is a one time operation and hence easy to adopt.

In Tamil Nadu, the tristeza problem in acid lime has been solved by a mild strain which is able to give effective cross protection againts severe strain in field. From 1976, over one lakh seedlings have been cross protected without any breakdown. Balaraman and Ramakrishnan(1978) found that two mild strains M1 and M2 gave effective protection in acid lime in Karnataka and increased true vigour and yield. However, recently breakdown of protection has been reported. In Maharastra, Sawant and Khade (1995) reported the efficacy of this strain and there was no breakdown of cross protection.

13. CONCLUSION

While reviewing a century of root disease investigations (Garret,1955) stated that marriage between plant pathology and soil microbiology gave birth to a progeny biocontrol. This child was burried under a tombstone inscribed biological control by inochlation of soil with antagonistic microooryunisms. However, interest in biological control of plant diseases was revived in 1965 when the First International Conference of Plant Pathology was held. Consequently the scepticism of plant pathologists and farmers on the feasibility of biocontrol of plant diseases was dropped. Today several commercial products of biocontrol agents are already in the market. This products has paved the way for further research. So that in near future for formulation of biocontrol agent against biocontrol of foliar disease will also be developed. Recent advances in biotechnology have thrown a wide range of scope for exploiting microorganisms for plant disease biocontrol. Virus disease which has defied mens attempt to manage by direct methods can also be effectively tackled by use of mild strains. Therefore by the turn of the century, farmers can expect biocontrol as a panacea for solving bring some of the most difficult plant diseases.

REFERENCES

Abd-el moity, T.H., Papavizas, G.C. and Lewis,J.A. 1982, Induction of new isolates of *Trichoderma harzianum* tolerant to fungicides and their experimental use for control of white rot on onion, *Phytopath.* 72 : 396 - 400.

Abdou,T.A., Elhassan, S.A and Abbas.H,K. 1980, Seed transmission and pycnidial formulation in sesame wilt disease caused by *Macrophomina phaseolina* (Maubl) Ashby, *Agric. Res. Rev.* 57 : 63-69.

Adams, P.B. 1990, The potential of microparasites for biological control of plant diseases, *Ann. Rev. Phytopath.* 28 : 59-72.

Ahmad, T.S. and Baker, R. 1987, Competitive saprophytic ability and cellulolytic activity of rhizosphere competent mutants of *Trichoderma harzianum*, *Phytopath.* 77 : 358 -362.

Andrews, J.H 1992, Biological control in the phyllosphere, *Ann. Rev. Phytopah.* 30 : 603-635.

Angappan, K. 1992, Biological control of chickpea root rot caused by *Macrophomina phaseolina* (Tassi) Goid, M.Sc(Ag) thesis, TNAU, Coimbatore, pp. 114.

Anusuya, D. and Jayarajan, R. 1998, Solubilization of phosphorus by *Trichoderma viride*, *Curr. Sci.* 74 : 464 - 466.

Arlorio, M., Ludwig, A., Beller, J and Bonfante, P. 1992, Inhibition of fungal growth by plant chitinase and B-1, 3-glucanase, *Protoplasma* 171 : 34 - 43.

Backman, P.A. and Rodrigue-Kabana, R. 1975, A system for the growth and delivery of biological control agents to the soil, *Phytopath.* 65 : 819-821.

Baker, S.C., Stavely, J.R., Thomas, C.A., Sasser, M. and Mac Fall, S.J. 1983, Inhibitory effect of *Bacillus subtilis* on *Uromyces phaseali* and on development of rust pustules on bean leaves, *Phytopath.* 73 : 1148 - 1152.

112

Balaraman, K. and Ramakrishnan, K. 1978, Cross protection for control of citrus tristeza virus, *Indian Hort.* 23 : 22 -23.

Baldwin, J.E., Derome, A.E., Field,L., Gallagher, P.T., Taha, A.A., Thaller, V., Brewer, D. and Taylor, A. 1981, Biosynthesis of a cyclopentyl dienyl isonitrile acid in cultures of fungus *Trichoderma hamatum*, *J. Chem. Soc. Chem. Commn.* 1227-1229.

Blakeman, J.P. and Brodie, I.D.S. 1977, Competition for nutrients between epiphytic microorganisms and germination of spores of plant pathogens on beet root leaves, *Physol. Pl. Path.* 10 : 29-42.

Bruckner, H., Reinecke, C., Kripp, T. and Kieb, M. 1990, Screening, isolation and sequence determination of a unique group of polypeptide antibiotics from filamentous fungi, *Proc. Int. Mycol. Cong.* Regensburg, Federation Republic Germany, 224 p.

Chalutz, E. and Wilson, C.L. 1990, Post harvest biocontrol of green and blue mold and sour rot of citrus fruit by *Debaryomyces hansenii*, *Plant Dis.* 74 ; 134-137.

Chet, I. and Baker, R. 1980, Induction of suppressiveness to *Rhizoctonia solani* in soil., *Phytopath.* 70 : 994 - 998.

Claydon, N., Allan, M., Hanson, J.R. and Avent, A. 1987, Antifungal alkyl pyrones of *Trichoderma harzianum*, *Trans. Br. Mycol. Soc.* 88 : 503 - 513.

Claydon, N., Hanson, J.R., Avent, A.G. and Trunch, A. 1991, Harzianolide, a butenolide metabolite from cultures of *Trichoderma harzianum*, *Phytochem.* 30 : 3802 - 3803.

Cook, R.J. 1993, Making greater use of introduced microorganisms for biological control of plant pathogens, *Ann. Rev. Phytopath.* 31 : 53 - 80.

Costa, A.S. and Muller, G.W. 1980, Tristeza control by cross protection : A.U.S. Brazil cooperative sucess, *Plant Dis.* 64 : 538 - 841.

Edenborough, M.S. and Hebert, R.B. 1988, Naturally occuring isocyanides, *Nat. Pro. Reptr.* 5 : 229-245.

Elad, Y., Chet, I., Boyle, P. and Henis, Y. 1982, Parasitism of *Trichoderma* sp. on *Rhizoctonia solani* and *Sclerotium rolfsii* - scanning electron microscopy and fluorescent microcopy, *Phytopath.* 72 : 85-88.

Elad, Y., Zuiel, Y. and Chet, I. 1986, Biological control of *Macrophomina phaseolina* (Tassi), Gold, by *Trichoderma harzianum*, *Crop Prot.* 5 : 288-292.

Fletcher, J. T. and Row, J.M. 1975, Observations and experiments on the use of an avirulent mutant strain of tobacco mosaic virus as a means of controlling tomato mosaic, *Ann. Appl. Biol.*, 81 : 171 -179.

Gadoury, D.M., Pearson, R.C. and Seem, R.C. 1991, Reduction of the incidence and severity of grape powdery mildew by *Ampelemyces quisqualis*, *Phytopath.* 81 : 122.

Garrett, S.D. 1955, A century of root disease investigation, *Ann. Appl. Biol.*, 42 : 211 - 19.

Gotfredson, W.O. and Vangedal, S. 1965, Trichodermin, a new sesquiterpened antibiotic, *Acta Chem. Scand.* 19 : 1088 - 1102.

Harman, G.E. 1991, Seed treatments for biological control of plant disease, *Crop Prot.* 10 : In press.

Harman, G.E., Hayes, C.K., Lorito, M., Broadway, R.M., Pietro, A., Peterbauer, C. and Transmo, A. 1993, Chitinolytic enzymes of *Trichoderma harzianum* : purification of chitobiosidase and endo chitinase, *Phytopath.* 83 : 313 - 318.

Harman, G.E., Taylor, A.G. and Stasz, T.E 1989, Combining effective strains of *Trichoderma harzianum* and solid matrix priming to improve biological seed treatment, *Plant Dis.* 7 3 : 631-637.

Howell, C.R., Stipanovic, R.D. and Lumsden, R.D. 1993, Antibiotic production by strains of *Gliocladium virens* and its relation to the biocontrol of cotton seedling diseases, *Biocont. Sci. Technol.* 3 : 435 - 441.

Itoh, Y., Kodama, K., Furuya, K., Takahash, S., Heneishi, J., Taki-guchi, Y. and Arai, M. 1980, A new sesquiterpene antibiotic heptelidic acid producing organisms, Fermentation, Isolation and Characterisation, *J. Antibio.* 33 : 468 - 473.

Janisiewicz, W.J. and Jeffers, S.N. 1997, Efficacy of commercial formulation of two biofungicides for control of blue mold and gray mold of apple in cold storage, *Crop Prot.* 16 : 629-633.

Jarvis, W.R., Shaw, L.A. and Traquiair, J.A. 1969, Factors affecting antagonism of cucumber powdery mildew by *Stephanoascus flocculosus and S.rugulosus*, *Mycolo. Res.* 92 : 162-165.

Jeyarajan, R. and Ramakrishnan, G. 1991, Efficacy of *Trichoderma* formulation against root rot disease of grain legumes, *Petria* 1 : 137-138.

Jeyarajan, R., Ramakrishnan, G. Dinakaran, D. and Sridar, R. 1994, Development of products of *Trichoderma viride* and *Bacillus subtilis* for biocontrol of root rot diseases, In : *Biotechnology in India*, ed.B.K. Dwivedi, Bioved Research Society, Allahabad, India, pp. 25.

Jeyarajan, R. 1996, Virus disease management by cross protection, *Indian J. Mycal. Pl. Path.* 26 : 147 - 152.

Jeyarajan, R. and Nakkeeran, S. 1996, Exploitation of biocontrol potential of *Trichoderma* for field use, In : *Current Trend in Life Sciences*, Vol. XXI, eds. K. Manibhushan Rao and A. Mahadevan, Today and Tomorrow Printers and Publishers, New Delhi, India, pp 61-66.

113

Kelman, A. 1989, Introduction : The importance of research on the control of post harvest disease of perishable food crops, *Phytopath.* 79 : 1374.

Kloepper, J.W. 1983, Effect of seed piece inoculaion with plant growth promoting rhizobacteria on poulations of *Erwinia carotovora* on potato roots and daughter tubers, Phytopath. 73 : 217 - 219.

Kousalya, G. and Jeyarajan, R. 1988, Techniques for mass multiplication of *Trichoderma viride* Pers. fr and *T. harzianum* Rifai, National seminar on management of crop diseases with plant products/ biological agents, Agricultural college and Research Institute, Madurai, pp. 32-33.

Lewis, J.A. and Papavizas, G.C. 1986, Reduced incidence of *Rhizoctonia* damping off of cotton seedlings with preparations of biocontrol fungi, *Biol.Cult.Tests.* 1 : 48.

Lewis, J.A., Fravel, R.D., Lumsden, R.D. and Shasha, B.S. 1995, Application of biocontrol fungi in granular formulations of pregelatinized starch- flour to control damping off diseases caused by *Rhizoctonia solani*, *Biolo. Cont.* 5 : 397 - 404.

Lewis, J.A., Papavizas, G.C. and Lumsden, R.D. 1989, A new biocontrol formulation for *Trichoderma* and *Gliocladium*, *Phytopath.*, 79 : 1160.

Lewis, J.A., Papavizas, G.C. and Lumsden, R.D. 1991, A new formulation system for the application of biocontrol fungi to soil, *Biocont. Sci. Technol.* 1 : 59-69.

Lewis, J.S., Barksdale, T.H and Papavizas, G.C. 1990, Green house and field studies on the biological control of tomato rot caused by *Rhizoctonia solani*, *Crop Prot.* 9 : 8 - 14.

Lisansky, S.G. 1985, Production and commercialization of pathogens, In: *Biological Pest Control*, eds. N.W. Hussey and N.Scopes, Blandford Press, Poole, UK, pp. 210-218.

Martosupon,M. and Prayudi, B. 1979, Root disease control on tea by fungicides, *Proc. Indon. Phytopath.*, Congr.V, Malang, Indonesia.

Mau, R.F.L., Gonsalves, D. and Bautista, R. 1990, Use of cross protection to control papaya ring spot virus at waiantae, *Proc. 25th Annual Papaya Industry Association Conference*, 1989, pp. 77 - 84.

Misaghi, I.J. 1990, Screening bacteria for root colonizing ability by a rapid method, *Soil Biol. Biochem.* 22 : 1085-1088.

Muthamilan, M. and Jeyarajan, R. 1996, Integrated management of *Sclerotium* root rot of groundnut involving *Trichoderma harzianum*, *Rhizobium* and *Carbendaziam*, *Indian J. Mycol. Pl. Pathol.* 26 : 204 - 209.

Muthamilan, M. and Vidhyasekaran, P. 1994, Development of *Pseudomonas fluorescens* as a commercial biopesticide product, Paper presented in National Symposium at Tamilology Agriculture University, Coimbator.

Nakkeeran, S . and Jeyarajan, R. 1996, Exploitation of antagonistic potential of *Trichoderma* for field use. Paper presented in National Symposium on disease of plantation crops and their management at Institute of Agriculture, Sriniketan.

Nakkeeran, S. and Renukadevi, P. 1997, Seed borne microflora of pigeon pea and their management, *Pl. Dis. Res.* 12 : 103 - 107.

Nakkeeran, S., Gangadharan, K. and Renukadevi, P. 1995, Efficacy of organic amendments and antagonists on root rot and wilt of pigeon pea, In : *Proc. National Symposium on Organic Farming*, T.N.A.U., Madurai, pp. 141 - 142.

Nakkeeran, S., Jeyarajan. R, Samyappan, R., Thayamanavan, B. and Sankar, P. 1996, Induction and biological characterization of mutants of *Trichoderma viride*, *Indian J. Mycol. Pl. Pathol.* 26 : 121.

Nakkeeran, S., Subramanian, S. Ramamurthy, R., and Sankar, P. 1996, Various approach in the management of panama wilt of banana, *Indian J. Mycol. Pl. Pathol.* 26 : 121.

Nakkeeran, S., Sankar, P. and Jeyarajan, R. 1997, Standardization of storage conditions to increase the shelf life of *Trichoderma* formulations, *J. Mycol. Pl. Pathol.* 27 : 60-63.

Nakkeeran, S. and Kousalya, G. 1993, Role of neem cake in the management of root rot and wilt of pigeon pea, Paper presented in World Neem Conference (Abs.), Bangalore, India, p.41.

Oshima, N. 1979, Theory and practice of tomato vaccination in Japan, *Pac. Sci. Cong.*, ISSR Khabarousk 14 : 32.

Oshima, N., 1975, The control of tomato mosaic disease with attenuated virus of tomato strain of TMV, *Rev. Pl. Prot. Res.* 8 : 126 - 135.

Ossamna,N. and Mischke, S. 1990, Transformation of the biocontrol fungus *Gliocladium virens* to benomyl resistance, *Appl. Envir. Microbiol.* 56 : 3052-3056.

Papaviza, G.C. and Lewis, J.A. 1989, Effect of *Gliocladium* and *Trichoderma* on damping off and blight of snap bean caused by *Sclerlotium rolfsii*, *Plant Pathol.* 38 : 277-286.

Papavizas, G.C. and Lewis, J.A. 1983, Physiological and biocontrol characteristics of stable mutants of *Trichoderma viride* resistant to MBC fungicides, *Phytopath.* 73 : 407 - 411.

Papavizas, G.C. 1985, *Trichoderma* and *Gliocladium* : Biology, ecology and potential for biocontrol, *Ann. Rev. Phytopath.* 23 : 23 - 54.

114

Papavizas, G.C. 1987, Genetic manipulation to improve the effectiveness of biocontrol fungi for plant disease control, In : *Innovative Approaches to Plant Disease Control*, ed. I.Chet, John Wiley & Sons, New York, USA, pp. 193 - 212.

Papavizas, G.C., Roberts, D.P. and Kim, K.K. 1990, Development of mutants of *Gliocladium virens* resitatnt to benomyl, *Can. J. Microbiol.* 16 : 121 - 129.

Papavizas, G.C. and Lumsden, R.D. 1980, Biological control of soil borne fungal propogules, *Ann. Rev. Phytopath.* 18 : 389-418.

Papavizas,G.C. 1985, *Trichoderma* and *Gliocladium* : Biology, ecology and potential for biocontrol, Ann. *Rev. Phytopath.* 23 : 23-54.

Peng, G., Sutton, J.C. and Kevan, P.G. 1992, Effectiveness of honey bees of applying the biocontrol agent *Gliocladium roseum* to strawberry flowers to suppress *Botrtyis cinerea, Can. J. Plant Pathol.* 14 : 117-129.

Pusey, P.L., Wilson, C.L., Hotchkiss, M.W. and Franklin, J.D. 1986, Compatibility of *Bacillus subtilis* for post harvest control of peach brown rot with commercial fruit waxess, dicloran and cold storage conditions, *Plant Dis.* 70 : 587-590.

Pyke, T.H. and Dietz, A. 1966, U-21, 9263, a new antibiotic, discovery and biological activiry, *Appl. Microbiol.* 14 : 506 - 510.

Rabindran, R. 1994, Biological control of rice sheath blight caused by *Rhinzoctonia solani* and blast caused by *Pyricularia oryzae* using *Pseudomonas fluroscens*, Ph.D thesis, Dept of Plant pathology, T.N.A.U, Coimbatore.

Rast, A.T.B. 1972, MII-16 an artificial symptomless mutant of tobacco mosaic for seedling inoculation of tomato crops, *Neth. J. Plant Pathol.* 78 : 110 -112.

Renganathan, K., Sridar, R. and Jeyarajan, R. 1995, Evaluation of gypsum as a carrier in the formulation of *Trichoderma viride, J. Biol. Cont.* 9 : 61 - 62.

Ricard, J.L. 1981, Commercialization of a *Trichoderma* based myco-fungicide, Some problems and solutions, *Biocont. News Infor.* 2 : 95-98.

Sangeetha, P. and Jeyarajan, R. 1991, Effect of osmotic water potential on growth and sporulation of *Trichoderma* spp. and *Rhizoctonia solani* Kuhn, *Indian J. Mycol. Pl. Pthol.* 21 : 254 - 256.

Sankar, P. and Jeyarajan, R. 1996, Biological control of sesamum root rot by seed treatment with *Trichoderma* spp. and *Bacillus subtilis, Indian J. Mycol. Pl. Pathol.* 26 : 147 - 153.

Sawant, D.M. and Khode, V.M. 1995, Management of citrus tristeza virus in khagzi lime by cross protection with mild strain, *Indian J. Mycol. Pl. Pathol.* 25 : 103.

Sivan, A., Elad, Y. and Chet, I. 1984, Biological control, Effect of new isolates of *Trichoderma harzianum* on *Pythium apnanidermatum, Phytopat.* 74 : 498 - 501.

Stasz, T.E., Harman, G.E. and Gullion, L. 1989, Limited vegatative compatibility following intra and interspecific protoplast fusion in *Trichoderma, Experim Mycol.* 13 : 364 - 371.

Sutton, J.C. and Peng, G. 1993, Manipulation of vectoring of biocontrol organisms to manage foliage and fruit diseases in cropping systmes, *Ann. Rev. Phytopath.* 31 : 473-493.

Thomson, S.V., Hansen, D.R., Flint, K. and Vandenberg, S.D. 1992, Dissemination of bacteria antagonistic to *Erwinia amylovara* by honey bees, *Plant Dis.* 76 : 1052-1056.

Tien, P. and Chang, V.H. 1983, Control of two seed borne virus disease in china by the use of protective inoculation, *Seed Sci. Technol.* 11 : 969 - 972.

Ulhoa, C.J. and Peberdy, J.F. 1993, Effect of carbon sources on chitobiose production by *Trichoderma harzianum, Mycol. Res.* 97 : 45 - 48.

Ushamalini, C., Rajappan. K. and Kousalya, G. 1997, Control of cowpea wilt by non chemical means, *Pl. Dis. Res.* 12 : 122-129.

Vidhya, R. 1995, Studies on biological control of mungbean root rot [*Macrophomina phaseolina* (Tessi.) Gold.] by *Trichoderma viride* Pers, M.Sc.(Ag.) thesis, TNAU, Coimbatore, India, 143 p.

Wang, H.L., Yeh, S.D., Chiu, R.J. and Gonsalves, D. 1987, Effectiveness of cross protection by mild mutants of papaya ring spot virus for control of ringspot disease of papaya in Taiwan, *Plant Dis.* 71 : 491 - 497.

Watts, R., Dahiya, J., Chaudary, K. and Tauro, P. 1988, Isolation and characterization of a new antifungal metabolite of *Trichoderma reesii, Plant Soil* 107 : 81-84.

Wells, H.D., Bell, D.K. and Jawarsi, C.C 1972, Efficacy of *Trichoderma harzianum* as a biocontrol agent for *Sclerotium rolfsii, Phytopath.* 62 : 442 - 447.

Whipps, J.M. 1992, Concepts in mycoparasitism and biological control of plant diseases, In : *New Approaches in Biological Control of Soil Borne Diseases*, ed. D.F. Jenson, International Union of Biological Science Copenhagen, IOBC/WPRS Bulletin 1992/XV/1 : 54 - 59.

Wilson , C.L. and Pusey, P.L. 1985, Potential for biological control of post harvest diseases, *Plant Dis.* 69 : 375-378.

Wilson , C.L., Wisniewski, M.E., Biles, C.L., Mclaughlin, R., Chaultz, E. and Droby, S. 1991, Biological control of post harvest diseases of fruits and vegetables, *Crop Prot.* 10: 172-177.

Yamano, T., Hemmi, S., Yamamoto, I. and Trubaki, K. 1970, Tichoviridin new antibiotic. Japanese kokai, 70 : 5435, Chemical abstracts 73 : 65093.

Yamaya, J., Yoshioka, M., Meshi, T., Okada, Y. and Obno, T. 1988, Expression of tobacco mosaic virus RNA in trangenic plants, *Mol. Gen. Genet.* 211 : 520- 525.

Yeh, S.D. and Gonsalves, D. 1994, Practices and perspective of control of papaya ringspot virus by cross protection, *Adv. Dis. Vector Res.* 10 : 237-257.

Yeh, S.D. and Gonsalves, D. 1984, Evaluation of induced mutants of papaya ringspot virus for control by cross protection, *Phytopath.* 74 : 1086-1091.

Zeepa, G., Allegrone, G., Barbeni, M. and Guarda, P.A. 1990, Variability in the production of volatile metabolities by *Trichoderma viride*, *Annale di microbiologia Entomologia* 90 : 171-176.

SUSTAINABLE MANAGEMENT OF ARBUSCULAR MYCORRHIZAL FUNGI IN THE BIOCONTROL OF SOIL-BORNE PLANT DISEASES

M. P. Sharma and A. Adholeya

Microbial Biotechnology
Tata Energy Research Institute, Darbari Seth Block,
Habital Place, Lodhi Road, New Delhi - 110 003, INDIA

1. INTRODUCTION

The management of crop diseases caused by root pathogens has become one of the challenging research areas in plant pathology. Increasing knowledge and concern about the environmental consequences of pesticide applications have aroused interest in alternative methods of plant protection. The roots of most plant species live in symbiosis with certain soil fungi by establishing what are known as mycorrhizae. The fungus biotrophically colonizes the root cortex and develops an extramatrical mycelium that helps plant to acquire mineral nutrients from soil (Harley and Smith, 1983). It has been recognised that mycorrhizal symbiosis plays a key role in nutrient cycling in the ecosystem and also protects plants against environmental and cultural stress (Barea and Jeffries, 1995). Most of the major plant families are able to form mycorrhiza naturally, the arbuscular mycorrhizal (AM) associations being the most common mycorrhizal type involved in agricultural systems (Barea *et al.,* 1993; Mukerji, 1999; Singh *et al.*, 2000).

Generally, AM fungi cause few changes in root morphology, but the physiology of the host plant changes significantly. The improved potential for mineral uptake from the soil accounts for changes in the nutritional status of the host tissues, which, in turn, changes the structural and biochemical aspects of root cells. This can alter membrane permeability and thus the quality and quantity of root exudation. Altered exudation induces changes in the composition of microorganisms in the rhizospheric soil, which has been termed as 'mycorrhizosphere' (Bansal and Mukerji, 1994). The net effect of these alterations is a tolerant plant better able to withstand environmental stress and plant diseases (Linderman, 1988). To evaluate the influence of AM fungi on diseases incidence and development, the variable factors,namely plant pathogens, the symbiotic fungus, and environmental conditions have been considered. Mostly, the interactions between pathogen and the symbiont are mediated by the host. The characterization of these interactions

117

should therefore include information on the mechanisms involved, because interactions vary with specific host-symbiont- pathogen combination, it is difficult to make generalizations on the effect of AM fungi on diseases.

The purpose of this review is to attempt to (1) discuss the potentiality of AM fungi and mechanisms involved in the biocontrol of plant diseases, (2) to understand the behaviour and interaction of mycorrhizospheric organisms with AM fungi, and (3) exploit the possibilities of mass production of AM fungi.

2. ARBUSCULAR MYCORRHIZAE IN SUPRESSION OF SOIL-BORNE FUNGAL-PATHOGENS

Many studies have demonstrated that AM fungi can reduce plant diseases (Caron, 1989; Graham, 1986; Perrin, 1990). However, large-scale studies have been restricted only to the interaction. In the past 20 years, correlations between endomycorrhizal and enhanced resistance to soil-borne pathogens have been established in a number of host-pathogen interactions, which have been reviewed by Caron (1989), Dehne (1982) and Mukerji, (1999). Although much attention has been paid to such correlations, the exact mechanism by which the AM fungi protect their hosts against pathogens infection still remains to be understood and, in many cases, is controversial (Jalali and Jalali, 1991; Lieberei and Feldmann, 1989). Several hypotheses have been put forward to explain the effect of AM fungi, which play a crucial role in plant protection against root pathogens, but very few of them have explained on the basis of biochemical and cytological investigations.

Since arbuscular mycorrhizae are established in the roots of host plants, research on the correlation between mycorrhizae and disease incidence has been concentrated on

Table 1. Influence of arbuscular mycorrhizal (AM) fungi on soil-borne pathogenic fungi

Fungal pathogen	Host plant	Disease incidence	Reference
Fusarium oxysporum f. sp. *radicis lycoperisici*	Tomato	-	Caron *et al.,* 1986
F. oxysporum lycopersici	Tomato	-	Dehne and Schoenbeck, 1979
Pythium aphanidermatum	Tomato	. . .	Hedge and Rai, 1984
Fusarium avenacium	Clover	-	Dehne, 1982
F. oxysporum cepa	Onion	-	Dehne, 1982
F. solani	Soybean	. . .	Zambolin and Schenck, 1983
F. oxysporum cucumerinum	Cucumber	-	Dehne, 1977
Sclerotium rolfsii	Peanut		Krishna and Bagyaraj, 1983
Pythium ultimum	Soybean	. . .	Chou and Schmitthenner, 1974
Pythium ultimum	Poinsettia	-	Stewart and Pfleger, 1977
Phytophthora cinnamoni	Pine	-	Baertschi *et al.,* 1982
P. megasperma	Alfalfa	+	Davis *et al.,* 1978
P. palmmivora	Papaya	. . .	Ramirez, 1974
Macrophomina phaseolina	Soybean	. . .	Krishna and Bagyaraj, 1983
Thielaviopsis basicola	Tomato	. . .	Giovanetti *et al.,* 1991
Thielaviopsis basicola	Tobacco	-	Baltruschat and Schonbeck, 1975
Thielaviopsis basicola	Citrus	. . .	Davis, 1980
Thielaviopsis basicola	Cotton	-	Schonbeck and Dehne, 1977
Verticillum dahliae	Cotton	+	Davis *et al.,* 1979
Pyrenochaeta terrestris	Onion	-	Safir, 1968

- Decrease   + Increase

diseases caused by soil-borne pathogens. A summary of the results is given in Table 1. In general, mycorrhizal plants suffer less damage and the incidence of the disease is decreased or pathogen development inhibited (Dehne, 1982).

The interactions of AM fungi with pathogenic microorganisms in the rhizosphere have received considerable attention and a variety of responses can be summarized as follows (Sharma *et al.*, 1992).

(i) AM infections, in general, protect plants from soil-borne diseases.
(ii) Higher nutrient concentrations in mycorrhizal plants make such plants more susceptible to foliar pathogens.
(iii) No definite relationship appears to exist between bacterial infection and mycorrhization.
(iv) Pre-mycorrhizal infection of transplanted crops protects the plants from nematode infections.

Most of the studies related to the effect of AM fungi on pathogen suppression are based on improved nutrition (Smith, 1988). Increased phosphorus (P) assimilation has been proposed to explain both a higher susceptibility of mycorrhizal cotton plants to *Verticillium* wilt (Davis *et al.*, 1979) and a higher resistance to of mycorrhizal citrus to *Phytophthora* root rot (Davis and Menge, 1980). However, other authors observed no effect of P fertilization on pathogen population (Caron *et al.*, 1986b). It is apparent frommany reports that AM fungi can lead to lower disease incidence on mycorrhizal plants (Caron, 1989; Perrin, 1990; St-Arnand *et al.*, 1994).

Caron *et al.* (1986) showed that root colonization by *Glomus intraradices* significantly reduced the population of *Fusarium* root-rot in tomato caused by *Fusarium oxysporum* f. sp. *radices-lycopersici* causing crown and root-rot in tomato. Establishment of *G. intraradices* in the soil was sufficient to decrease the intensity of root rot and the population of *F. oxysporum*. Further experiments conducted on tomato showed that an increase in available phosphorus in roots and leaves had no effect on the population of *F. oxysporum* f. sp. *radicis-lycopersici* and on root necrosis but decreased the extent of root colonization by *G. intraradices* (Caron *et al.*, 1986b). Thus it seems that the presence of *G. intraradices* by itself suppresses the populations of *F. oxysporum* and root necrosis but has no significant effect on growth parameters irrespective of phosphorus levels. St-Arnand *et al.* (1994) observed that inoculations of substrate with *G. intraradices* reduced the populations of *Pythium ultimum* on *Tagetes patula*. They showed that the extent of colonization bearing arbuscules or vesicles of AM was unrelated to P nutrition and to the observed reduction of *P. ultimum* in the roots or in the substrate. This finding contradicts statements that high level of mycorrhization are necessary to induce protection against pathogens (Graham and Menge, 1982; Smith, 1988; Smith *et al.*, 1986) and agrees with the results of Caron *et al.* (1986a), who could protect tomato plants against *F. oxysporum* even at very low colonization levels.

The reduction in the population of *F. oxysporum* on tomato was not affected by the sequence of inoculation : whether tomato plants were inoculated with both the microorganisms simultaneously or the inoculation with *G. intraradices* occurred four weeks before or after that with *F .oxysporum*, the results were similar. However, root colonization of *G. intraradices* was significantly increased when *F. oxysporum* was inoculated simultaneously with or four weeks before the VAM fungus (Caron *et al.*, 1986).

The influence exerted by AM fungi by stimulating plant defense reactions was investigated using an *in vitro* system in which RiT-DNA-transformed carrot roots were infected with *Fusarium oxysporum* f.sp. *chrysanthemi*. In the non-inoculated roots, the pathogen multiplied abundantly through much of the tissue, including the vascular stele, whereas in mycorrhizal plants its growth was restricted to the epidermis and the outer cortex (Benhamou *et al.*, 1994).

The reduction in the infection by *F. oxysporum* due to inoculations with *G. mosseae* in tomato and *Capsicum* roots was also reported by Al-Momany and Al-Radded (1988).

In the tomato plants inoculated with *G. mosseae,* the incidence of *Fusarium* wilt was only 11%, as against 45% in non-mycorrhizal plants (Ramraj *et al.,* 1988). Rosandahl and Rosandahl (1990) showed that inoculation of cucumber seedling with *G. etunicatum* and *Glomus* sp. before or simultaneously with *Pythium ultimum* increased the survival of the seedlings and saved the plants from damping off . Giovannetti *et al.* (1991) showed that tobacco plants inoculated with mycorrhizal fungus *Glomus monosporum* showed increased tolerance to *Thielaviopsis basicola,* higher root and leaf dry weight, lower root infection and fewer - chlamydospores than non-inoculated infected plants. Inoculations of *G. fasciculatum* reduced the number of sclerotia produced by *Sclerotium rolfsii* in peanuts (Krishna and Bagyaraj, 1983).

3. ARBUSCULAR MYCORRHIZAE IN SUPPRESSION OF SOIL-BORNE NEMATODES

Since mycorrhizae affect a broad range of soil-borne fungal pathogens, it may be concluded that their beneficial influence is to some extent independent of pathogens. The parasitization of plants by nematodes (mainly endoparasites) can be influenced by the establishment of arbuscular mycorrhiza (Table 2). Symptoms of nematode infection are generally reduced and often, but not always, nematode populations are also reduced (as indicated by number of galls, juveniles or eggs per unit root length) (Hussey and Roncadori, 1978). Sitaramaiah and Sikora (1982) showed that inoculations of tomato transplants with *Glomus fasciculatum* significantly reduced root penetration by juveniles and the development of *Rotylenchulus reniformis* compared to controls. Development of the gelatinous matrix was delayed and fewer eggs per egg mass were produced on incoulated plants. The number of galls formed by *Meloidogyne incognita* on tomato was significantly lower in AM-inoculated tomato plants. However, the AM colonization did not prevent the penetration by the larvae (Suresh *et al.,* 1985). Baghel *et al.* (1990) showed that inoculation of *G. mosseae* stimualted the growth of *Citrus jambhiri* seedlings. Simultaneous inoculations of AM fungus and the citrus nematode, *Tylenchulus semipenetrans* partly neutralized the adverse effect of the nematode. Pre-inoculations of *Piper nigrum* cv. *panniyar* colonizing with *G. fasciculatum* or *G. etunicatum* significantly reduced the root-knot index (*Meloidogyne incognita*) by 32.4 and 36.0% respectively and also the growth of piper plants improved (Mukerji, 1999; Shivaprasad *et al.,* 1990).

Krishna Prasad (1991) reported that the percentage of root-knot nematode (*M. incognita*) infestations in tobacco (*Nicotiana tabaccum*) seedlings was 67.5% at 50 days and 95% at 75 days after sowing in non-mycorrhizal plants and 48% to 52% at 50 days and 73% at 75 days after sowing in mycorrhiza (*G. fasciculatum*) inoculated plots. The number of galls, egg masses per plant, and eggs per egg mass of infested plants was reduced by 61-89% as a result of AM inoculation.Transplanting of mycorrhizal tobacco seedlings in to root-knot nematode infested soil showed improved growth and yield compared to non-mycorrhizal plants.

Al-Radded (1995) showed that preinoculations of *G. mosseae* significantly reduced root knot infections and reproduction of root-knot nematode on tomato. The ability of mycorrhizal plants to grow well despite infection by nematodes is generally considered to be the principal affect of mycorrhizal fungi or the interaction of host plants with parasitic nematodes (Hussey and Roncadori, 1982).

Carling *et al.* (1996) showed that AM fungi made peanut plant more tolerant to the nematode and offset the reduction in growth caused by *M. arenaria* at the two lower P levels. However, AM fungi and added P increased the number of galls and egg production in *M. arenaria,* thereby increasing peanut susceptibility to nematode attack

120

Table 2. Influence of arbuscular mycorrhizae on nematodes

Nematode	Plant	Incidence	Reference
Meloidogyne arenaria	Grape	+	Atilano *et al.*, 1981
M. arenaria	Peanut	+	Hussey and Roncadori, 1982
M. hapla	Carrot	. . .	Sikora and Schonbeck, 1975
	Tomato	-	Cooper and Grandison, 1986
	Alfalfa	-	Grandison and Cooper, 1986
M. incognita	Tobacco	. . .	Fox and Spasott, 1972
	Tomato	. . .	Sikora and Schonbeck, 1975
	Soybean	-	Kellam and Schonbeck, 1980
	Cucumber	-	Priestel, 1980
	Cotton	-	Roncadori and Hussey, 1977
	Tomato	-	Suresh and Bagyaraj, 1985
	Tomato	-	Sharma *et al.*, 1994
	Cotton	-	Saleh and Sikora, 1984
	Cotton	-	Smith *et al.*, 1986
M. javanica	Chickpea	-	Diederichs, 1986
	Tomato	-	Singh *et al.*, 1990
Radopholus similis	Citrus	-	O'bannon and Nemec, 1979
R. citrophilus	Rough lemon	-	Smith and Kaplan, 1988
Rotylenchus reniformis	Cotton	-	Sitaramaiah and Sikora, 1982
	Bean	-	Sitaramaiah and Sikora, 1982
	Cucumber	-	Sitaramaiah and Sikora, 1982
	Tomato	-	Sitaramaiah and Sikora, 1982
Pratylenchus brachyurus	Cotton	-	Hussey and Roncadori, 1977
P. penetrans	Cucumber	. . .	Priestel, 1980
	Pigeonpea	+	Elliott *et al.*, 1984

- Decrease   + Increase

and had only a minimal effect on AM colonization. Furthermore, a comparison of mycorrhizal and P-fertilized non-mycorrhizal plants of similar size showed that the latter are more susceptible to nematode attack indicating the likely involvement of factors other than P nutrition in the interactions (Smith. 1987, 1988). On the other hand, mycorrhizal inoculations of tamarillo (*Cyphomandra betacea*) against root-knot nematode, *M. incognita* could not be duplicated by adding P and fertilizer was not therefore due merely to improved P nutritions of the host (Cooper and Grandison, 1987). Heald *et al.* (1989) showed that *M. incognita* suppressed the growth of non-mycorrhizal *Cucumis melo* plants by 84% as compared to 21% in AM inoculated plants at 50 mg/g P. A similar trend was observed in soil with 100 mg/g phosphorus. Preinoculations of *G. fasciculatum* in tomato var. *pusa ruby* significantly reduced the number of galls/plant, egg masses per plant and eggs per egg mass due to *M. incognita*. They also observed that the offset of AM fungi on reducing root-knot nematode was not the same in pots and in the field.

Nematode populations was reduced up to 58% in the pots and only up to 36% in the field (Sharma and Bhargava, 1993). The time of AM inoculations, whether before, simultaneously or after inoculations with the pathogens or nematode, greatly affects the efficacy of AM fungi in controlling nematodes. Studies conducted by serval workers demonstrate that preinoculations of AM fungus suppressed nematode reproduction and development in roots to a greater degree when compared to plants inoculated simultaniously with both the organisms (Jain and Sethi, 1987; Suresh and Bagyaraj, 1984; Taha and Abdel-Kader, 1990).

4. MECHANISMS INVOLVED IN PATHOGEN SUPPRESSION BY AM FUNGI

The hypotheses proposed to explain AM fungal effects on soil-borne plant pathogens generally have been considered to have either a physical or physiological basis (Hussey and Roncadori, 1982; Linderman, 1985). The role played by AM fungi in biological control of plant diseases has been the subject of several reviews (Bagyaraj, 1984; Caron, 1989; Jalali and Jalali, 1991), but mixed responses and intrepretations have precluded any clear conclusion that AM fungi always suppress plant diseases . Such inconsistancies should be expected, however, considering the diverse experimental approaches and the use of different AM fungi on different hosts in different soils (Schenck, 1987). In reviews of biological control of plant dieseases, mycorrhizae are thought to be biological controlling agent against plant diseases primarily by means of stress reduction (Baker, 1986; Cook and Baker, 1982). The biological control of plant diseases may be strongly affected by AM fungi by one or more mechanisms involving.

4.1. Physical Mechanisms

4.1.1. Morphological changes in roots and root tissues

Localized morphological influences of AM fungi have been shown in mycorrhizal roots. Lignification is reported to prevent penetration of mycorrhizal plants by pathogens. Dehne and Schonbeck (1978) showed increased lignification of cells in the endodermis of mycorrhizal tomato and cucumber plants, and speculated that such responses accounted for reduced incidence of *Fusarium* wilt (*F. oxysporum* f. sp. *lycopersici*). Baker (1976) reported a similar effect on pink root of onion (*Pyrenochaeta terrestris*). Increased wound-barrier formation inhibited *Thielaviopsis* black root rot (*T. basicola*) of mycorrhizal holly (*Ilex crenata*) plants.

A stronger vascular system in mycorrhizal plants increases flow of nutrients, imparts greater mechanical strength and reduces the effect of vascular pathogens (Schonbeck, 1979). Smaller syncytia with fewer cells increase the resistance of the host to nematodes (Fassuliotis, 1970).

4.2. Physiological Mechanisms

4.2.1. Nutritional changes

AM fungi have been shown to affect root growth, root exudations, nutrient absorptions and physiological responses of the host to environmental stresses. Many of these factors are related to P physiology and nutrition (Graham *et al.*, 1981; Hayman, 1982; Pacovsky *et al.*, 1986). AM fungi enhance root growth, expand absorptive capacity and affect cellular processes in roots (Hussey and Roncadori, 1982; Smith and Gianinazzi-Pearson, 1988). Davis (1980) showed that the effect of AM fungi on *Thieloviopsis* root rot of citrus was larger than non-inoculated plants unless the latter were supplied with additional P. These mycorrhizal-induced compensatory processes may explain the increased tolerance of mycorrhizal and P- fertilized plants because many plants can compensate for loss of root mass or function caused by pathogens (Wallace, 1973). Greater tolerance is also attributable to increased root growth and phosphate status of the plant (Cameron, 1986).

In addition to P, AM fungi can enhance the uptake of Ca, Cu, Mn, S and Zn (Pacovsky *et al.*, 1986; Smith and Gianinazzi-Pearson, 1988). Host susceptibility to pathogens and tolerance of the disease process can be influenced by the nutritional status

122

of the host and fertility status of the soil (Cook and Baker, 1982) . For example, nematode-damaged plants frequently show deficiencies of B, N, Fe, Mg and Zn (Good, 1968). In the absence of AM, phosphate can combine with minor elements to create deficiencies which would predispose plants to root-knot nematodes (Smith et al., 1986). Thus AM fungi may increase tolerance to pathogens by increasing the uptake of essential nutrients other than P that would be deficient in a non-mycorrhizal plant. Caron et al. (1986a,b,c)compared responses between AM and non-AM tomato plants with a relatively low P threshold requirement to root and crown rot disease caused by Fusarium oxysporum f. sp. radicis-lycopersici. Added P did not reduce disease incidence severity and pathogen populations in rhizosphere of non-AM plants but did so with AM plants, even though plant growth and tissue P were not different in the two treatments.

Phosphate and root isoflavonoid accumulations by AM fungi appear to be distinct responses. The nutritional effect of AM does not seem to be directly implicated in increased isoflavonoid content of mycorrhizal roots (Morandi et al., 1984).

4.2.2. Competition for host photosynthates and infection site

AM fungi are almost totally dependant on soluble carbohydrates produced by the host for their carbon source (Harley and Smith, 1983; Reid, 1984). However, there is little evidence of a competition between the symbiont and pathogens of host resources warrents investigation, particularly in interactions with sedentary endoparasitic nematodes, because of the obligate requirement of both groups of organisms for host-derived compounds. Mechanisms affecting nematode activities in a mycorrhizal root systems that are not related to improved P nutrition or increased photosynthesis may involve utilization of photosynthates by the AM fungus at the expense of nematode reproduction or the conversion of carbohydrates - received from the host into forms not usable by the nematode. Similarly, P transport by AM fungi to the casparian strip in roots is via fungal structures (Reid, 1984). This contrasts with mass flow and diffusion of P through soil and root symplast on a non-mycorrhizal root (Bieliski, 1973). Dehne (1982) indicated that fungal root pathogens could occupy root cortical cells adjacent to those colonized by AM fungi, indicating a lack of competitions. It has been suggested that nematode pathogens, on the otherhand, require host nutrients for reproduction and development and direct competition with AM fungi has been hypothesized as a mechanism of their inhibiters (Dehne, 1982; Smith, 1988).

Since AM fungi, soil-borne fungal pathogens, and plant parasitic nematode occupy similar root tissues, direct competition for space has been postulated as a mechanism of pathogen inhibition by AM fungi (Davis and Menge, 1980; Hussey and Roncadori, 1982; Linderman, 1985). Competition betweetn an AM fungus and P. parasitica was proposed in citrus (Davis and Menge, 1980). On spilt root systems, the amount of mycorrhized root tissue was reduced only when the AM fungus and P. parasitica were in direct association. However, this hypothesis has been contradicted since many root pathogens infect the root tip where AM fungal structures do not occur (Harley and Smith, 1983; Linderman, 1985).

4.2.3. Biochemical changes or changes in chemical constituents of plant tissues

Available evidence demonstrates an intimate relationship between root exudation and disease initiation (Cook and Baker, 1982). Influence of AM fungi and P nutrition on root exudation as a regulatory process in root disease has been investigated (Graham and Menge, 1982). The influence of P nutrition on membrane permeability in root cells, concentration, exudation of amino acids and reducing sugars has been proposed as a mechanism regulating mycorrhizal root penetration and colonization (Ratnayake *et al.*,1978). Root colonization may effect qualitative and quantitative changes in root exudates to alter rhizosphere or rhizoplane microorganisms (Linderman, 1985). Tomato roots inoculated with *Glomus fasciculatum* had increased concentrations of phenylalanine and sorine; these two amino acids being inhibory to root-knot nematode development (Parvatha Reddy, 1974). Higher amounts of catechols, which inhibit *Sclerotium rolfsii* growth *in vitro,* have been reported in mycorrhizal roots (Krishna and Bagyraj, 1986). Behamou *et al.* (1994) reported that production of unusal material in plants infected with *Fusarium oxysporum* f. sp. *chrysanthemi* whether colonized, with AM or not and coating of most of the intercellular spaces with similar substances was the most typical host reactions. Such deposits according to the authors, may be infused with phenolics as the deposited material often interacted physically with the walls of involving hyphae exhibiting morphological and cytological changes. These deposits, in addtion to acting as a barrier to fungal spread, also display fungitoxic activity. A few electron-opaque structures resembling the deposits were found in some cells and intercellular spaces of non-infected mycorrhizal carrot roots but were absent in infected, non-mycorrhizal carrot roots. Restriction of pathogen growth together with an increase in hyphal alteration and accumulation of new plant products in mycorrhizal roots and the absence of such phenomena in non-mycorrhizal roots show that mycorrhizal infection is responsibile at least in part for active of plant defence systems.

Dehne *et al.* (1978) demonstrated increased concentration of antifungal chitinase in AM roots and they also suggested that increased arginine accumulations in AM roots suppresses sporulation in *Thielaviopsis*, a mechanism previously suggested by Baltruschat and Schonbeck (1975). Morandi *et al.* (1984) found increased concentrations of phytoalexin-like isoflavonoid compounds in AM roots compared to those in non-AM soybean. They postulated that such materials could account for the increased resistance to fungal and nematode root pathogens of AM plants compared to non-AM plants.

Enhanced accumulation of coumestrol in soybean helps explain how mycorrhizal infection decreases the development of pathogenic nematodes more consistantly than that of fungi (Hussey and Roncadori, 1982; Kellam and Schenck, 1980). Suresh and Bagyraj (1984) reported that AM inoculations reduced root-knot infestations and such plants have increased quantities of sugars and amino acids, which play a role in suppressing nematode reproduction. Pre-inoculation of tomato roots with *G. fasciculatum* coupled with such biochemical changes as increased amount of lignins and phenols made tomato resistant to the root-knot nematode *Meloidogyne incognita* (Singh *et al.*, 1990).

Cordier *et al.* (1998) showed that the induction of resistance to *Phytophthora parasitica* in mycorrhizal tomato plants is governed by the cellular and molecular phenomena underlying bio-protections, using a spilt root experimental systems, they showed that the control of *P. parasitica* in mycorrhizal tomato root systems involves induction of localized resistance in arbuscules containing cells and systemic resistance in non-mycorrhizal tissues. Ultrastructural investigations coupled with histochemical immunocytochemical analyses have provided evidence that decreased pathogen development in both mycorrhizal and non-mycorrhizal parts of mycorrhizal root systems is associated with modifications in host cells, together with the accumulation of defence-related molecules. Further investigations were aimed at characterizing plant genes expressed during the bioprotection of mycorrhizal tomato infected with *P. parasitica*. The induction of defence-related enzymes in mycorrhizal tomato roots against *Phytophthora parasitica* was also reporeted by Pozo Maria-Jose *et al.* (1998) and observed hydrolytic enzymes chitinase chitosanase and β-1,3 - glucanase.

124

4.3. Biological Mechanisms

4.3.1. Microbial changes in the mycorrhizosphere

The concept of 'mycorrhizosphere' implies that mycorrhizae significantly influence the microflora of the rhizosphere by altering root physiology and exudation. In addition, extra-radical hyphae of AM fungi provide a physical or nutritional substrate for bacteria. Meyer and Linderman (1986b) and Secilia and Bagyaraj (1987) showed qualitative and quantitative changes in the mycorrhizosphere soil of AM plants compared to the rhizosphere of non-AM plants. These microbial shifts were clearly time-dependent and dynamic. Meyer and Linderman (1986a) showed that the production of sporangium and zoospere by the root pathogen *Phytophthora cinnamomi* was reduced in the presence of rhizosphere leachates from sweet corn and chrysanthemum. Similarly, populations of bacteria and actinomycets in pot cultures of different AM fungi were quantitatively and qualitatively analysed by Secilia and Bagyaraj (1987). They showed that pot cultures of *G. fasciculatum* harboured more actinomycetes antagonistic to *Fusarium solani* and *Pseudomonas solanacearum* than those of non-mycorrhized plants or plants colonized by other mycorrhizal species tested in the study.

Other studies have indicated that pathogen suppression by AM fungi involved changes in mycorrhizosphere microbial populations. Caron *et al.* (1985, 1986a,b) showed a reduction in *Fusarium* population in the mycorrhizosphere soil of tomatoes and a corresponding reduction in root rot in AM plants when compared to non-AM plants, probably due to the increased antagonism in the AM mycorrhizosphere. They have also concluded that the reduction in disease was not due to P nutrition.

Bartschi *et al.* (1981) reported that protection of hosts roots against *P. cinnamomi* root rot is due to pre-inoculation with mixed AM cultures taken from pot cultures. Further, Secilia and Bagyaraj (1987) demonstrated that the effect is due to a build up of antagonists in the pot cultures. Changes in mycorrhizsphere populations of species antagonistic to pathogens seem a likely explanation for many of the reported effects of AM on diseases.

5. STRATEGIES FOR THE MANAGEMENT OF ARBUSCULAR MYCORRHIZAE IN DISEASE CONTROL

5.1. Integrated Approach

If VAM contribute to disease suppression, then agricultural practices that reduce populations of AM fungi and associated antagonists could result in an increased need for fungicides and nematicides. To ensure that compatible combinations of AM and antagonists occur in the same soil, growers could inoculate the seeds with appropriate cultures and then use less than the recommended levels pesticides. Saleh and Sikora (1989) showed that the fungicides benomyl and carbendazim significantly reduced the antagonistic activity of *G. fasciculatum* that affects reproduction in *M. incognita* on cotton when applied as a drench at planting time. The number of nematodes increased significantly on mycorrhizal plants when benomyl was applied at planting time but no significant change was noticed with similar applications of carbendazim, which is less toxic to *G. fasciculatum* than benomyl. However, when the fungicides were applied 25 days after planting, there were fewer nematodes in mycorrhizal plants. Neither of the fungicides altered nematode populations on non-mycorrhizal plants. Similarly, *G. fasciculatum* did not affect nematode populations density when applied 30 days after or

125

simultaneous inoculations at planting time but when it was applied 50 days later, it reduced the nematode population significantly. Co-inoculation efficacy of mycorrhizae with other biocontrol agents need to be optimized for both pot culture and field conditions.

5.2. Host-symbiont-soil Fertility

Research is needed to assess how AM fungi affect the host-pathogen relationship differently from P. Growth responses to AM vary depending on the mycorrhizal dependency of the host, the endophyte strain, and the soil fertility level (Harley and Smith 1983). Preliminary research to determine whether a relationship is dependent on an AM fungal species or mixture of species that are indigenous to an area and stimulate plant growth at moderately fertile soils. This is necessary assessing, to a closer approximation, the response of mycorrhizal plants to pathogen infection for the soil type and fertility level occuring in the field. A better approach would be to use varying soil P levels that encompass field recommendations to produce non-AM plants of equivalent size and similar P status as mycorrhizal plants. These conditions are necessary to find out the real mechanism of the tolerance to pathogen attack shown by AM plants.

5.3. Statistical Models for Mycorrhizae - Pathogen Interactions

Factorial regression analyses that evaluate the response of qualitative treatments (mycorrhizal versus non-mycorrhizal plants receiving a particular P regime) over varying quantitative treatments (Pathogen Pi) are essential for valid statistical interpretation of data. Also, frequency values of nematode cycles over physiological time have been used to compare the effects of AM fungi and P nutrition on nematode parasitism (Smith *et al.*, 1986). Another quantative approach adaptable to mycorrhiza - nematode interactions is the Seinhorst damage functions, $Y = m + (1-m)Z^{P-T}$, to determine if the model parameters T (tolerance limit), m (minimum yield), or Z (nematode virulence) are affected differently in mycorrhizal and non-mycorrhizal P-fertilized plants (Wallace, 1983).

5.4. Initial Inoculum Density and Inoculations Sequence of Pathogen and Symbiont

A high pathogen inoculum density can overwhelm biocontrol agents (Cook and Baker 1983) and this has been shown in AM studies as well (Schenck, 1987). It is difficult to draw firm conclusions about the potential for biocontrol unless a range of pathogen inoculum densities is used. While varying the inoculum densities of pathogen to produce different levels of disease incidence or yield reduction, however, one should avoid the use of a Pi (inoculum density) that so severally stunts or kills plants that AM fungi are not given an opportunity to colonize roots and stimulate growth (MacGuidwin *et al.*, 1985).

Because root infections by fungal and nematode pathogens precede mycorrhizal root colonization (Linderman, 1985) the sequence in which plants are inoculated may affect the nature of the interaction (Smith, 1987). Thus, an inoculation method used in many studies has been to inoculate plants with the AM fungus 2 - 4 weeks before inoculating them with the pathogen (Cooper and Grandison, 1986; Smith *et al.*, 1986). This technique allows AM fungi enough time to colonize roots before they are challenged by the pathogen, however, it is restricted to containerized or tansplanted hosts (Hussey and Roncadori, 1982). Since, AM fungi are generally considered slow colonizers it seems logical that a minimum level of AM fungal root colonization is required for the fungus, to have any effect on the affect pathogen. AM fungal inoculum dose of 0.5 - 5.0 spores per

126

gram of soil has been used to produce optimal growth responses and maximum root colonization levels. Although such thresholds of the levels of colonization required to affect nematode activites have been reported (Grandison and Cooper, 1986; Smith *et al.*, 1986), high levels of AM fungal root colonization have not reduced the degree of root infections by fungal pathogens (Davis and Menge, 1980; Graham and Menge, 1982).

6. INOCULUM PRODUCTION OF AM FUNGI

Since AM fungi are obligate symbionts, they must be produced on living roots. This is problematic when mass production is needed, and also runs the risk of contamination with plant pathogens. AM fungi depend for reproduction on photosynthates of the plant (Harley and Smith, 1983) or on root components when the root is manipulated to grow meristematically on artificial growth media under axenic conditions (Mugnier and Mosse, 1987).

The sources of AM inocula are defined by the biology of AM fungi. All infective structures of the fungi can be used including chlamydospores or pieces of the mycelium produced inside or outside the host roots. Infected roots and the infected substrates (carriers containing infected roots or/and AM mycelium and spores) are common sources of AM inoculum. Different methods of inoculum production, type of inoculant, and the practical applications for each are schematically presented in Fig. 1.

6.1. Multiplication of Starter Inoculum

6.1.1. Conventional method or soil-based inoculum production

Culturing AM fungi on plants growing in disinfected soil has been the most frequently used technique for increasing the number of propagules (Menge, 1984).

A highly susceptible trap plant should be used to ensure that maximum inoculum levels are achieved (Bagyaraj and Manjunath, 1980). After isolating fungal propagules or acquiring a starter culture, the particular isolate must be multiplied and stock cultures in the form of colonized roots must be maintained. These roots or spores are used to produce large amounts of inoculum on soil-based media in pots (Sieverding and Barea, 1991). Such soil-based inocula are quite easy to produce and can be obtained in 6-12 months however, this type of inoculum is too heavy for extensive use in agriculture but may be appropriate for plant production systems which involve a transplanting stage and for raising healthy seedlings in forestry sector, and is common in horticulture (Gianinazzi *et al.*,1990a). Frequent applications of high N fertilizer increase spore production in several isolates of AM fungi, but sporulation of an isolate of *G. intraradices* was severally reduced by this treatment (Douds and Schenck, 1990a). Sporulation is often depressed by high P concentration (Same *et al.*, 1983). Soil inoculum from pot cultures of many *Glomus* species may be stored for several months or years. However, for more reliable storage of a wider range of AM fungi, the fungi can be dried *in situ* with the host and their frozen at -70°C (Douds and Schenck, 1990b).

6.1.2. Procedure for on-farm inoculum production of AM

This is done in five stages :

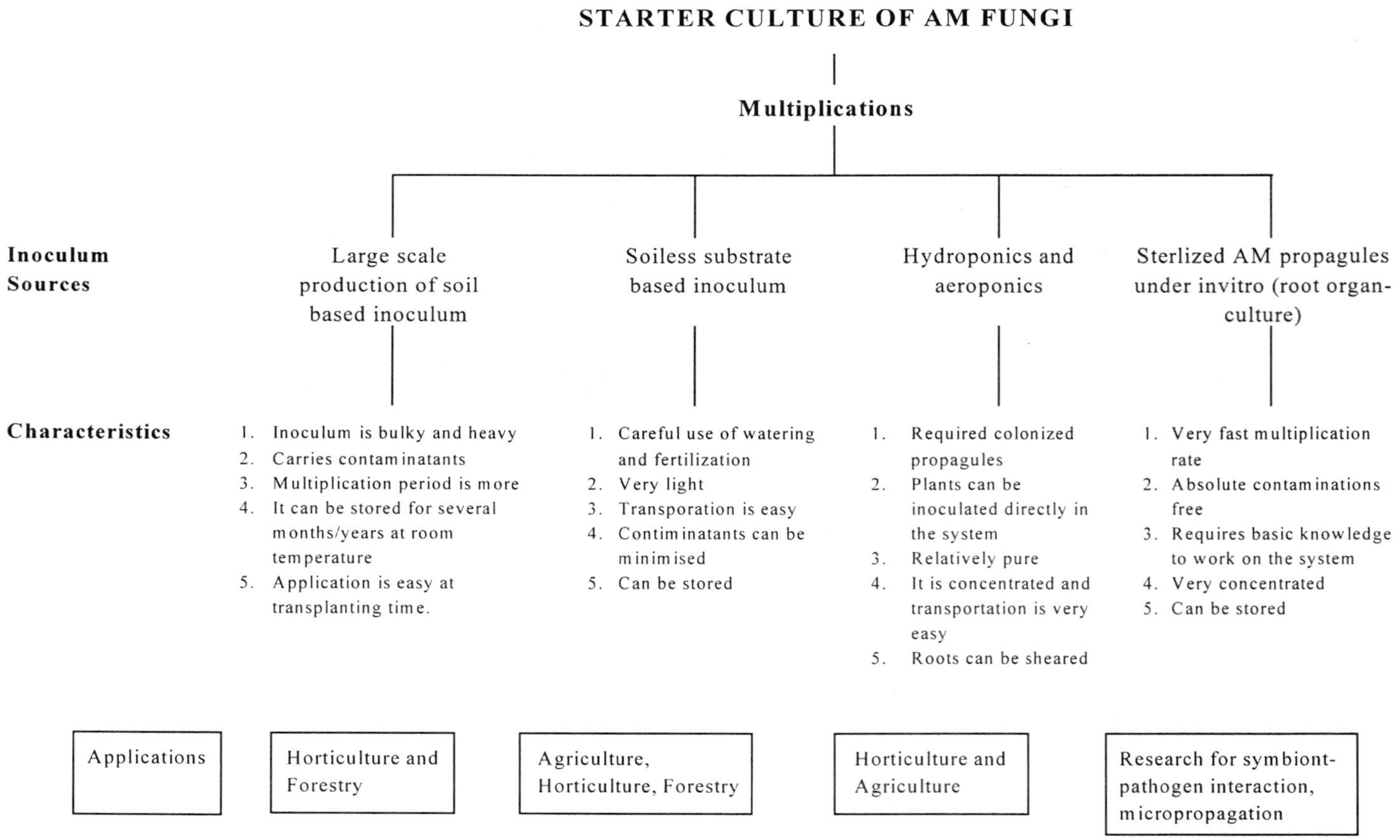

STARTER CULTURE OF AM FUNGI

Multiplications

	Large scale production of soil based inoculum	Soiless substrate based inoculum	Hydroponics and aeroponics	Sterlized AM propagules under invitro (root organ-culture)
Inoculum Sources				
Characteristics	1. Inoculum is bulky and heavy 2. Carries contaminatants 3. Multiplication period is more 4. It can be stored for several months/years at room temperature 5. Application is easy at transplanting time.	1. Careful use of watering and fertilization 2. Very light 3. Transporation is easy 4. Contiminatants can be minimised 5. Can be stored	1. Required colonized propagules 2. Plants can be inoculated directly in the system 3. Relatively pure 4. It is concentrated and transportation is very easy 5. Roots can be sheared	1. Very fast multiplication rate 2. Absolute contaminations free 3. Requires basic knowledge to work on the system 4. Very concentrated 5. Can be stored
Applications	Horticulture and Forestry	Agriculture, Horticulture, Forestry	Horticulture and Agriculture	Research for symbiont-pathogen interaction, micropropagation

Figure 1. Different methods of inoculum production, tyre of inoculant and pratical application of AM fungi.

6.1.2.1. Starter culture

Large scale production of AM fungi begins with a starter culture (TERI). The starter culture can be procured either by isolating or by ordering it from various laboratories that maintain pure cultures of specific interest.

6.1.2.2. Disinfection of soil

Soil in nursery beds should be sterilized either with methyl bromide or formalin by drenching it up to at least 18 inches depth with either of the solutions. In one of the study conducted at Forest Research Institute, Dehradun, it was found that sunlight is equally effective for soil sterilization. After treating with chemicals the soil should be covered for 3 days and then kept open for at least 8 days before commencing any operation. Sun light for sterilization, involves covering the soil with transparent polythene sheets for a minimum of 20 days.

6.1.2.3. Preparation of nursery bed

The soil of nursery should be raised up to 30 cm. This can be done making surrounding furrows of the similar depth. Soil should be thoroughly mixed and preferably sieved. If the soil is compact, sand may be mixed for good mycorrhizal development in a ratio of 2:1.

6.1.2.4. Sowing and inoculation

Furrows of 6 cm depth can be made in the nursery beds and AM propagules, mixed with any suitable carrier placed in the furrows. The inoculum should contain at least 30-40 spores per gram of substrate. The inoculum should be covered with a thin layer of soil on which host seeds preferably monocots should be sown.

6.1.2.5. Maintenance and monitoring

The beds should be watered when required and should be kept free from weeds. After 3 months , the extent of colonization and spore production could be assessd.

6.1.3. Soilless substrate based system

Many fungi have been cultured in soilless media (Table 3). The ideal soilless mixture should hold sufficient water for plant growth but also allow good aeration. Bark, calcined clay, expanded clay, pertile, soilrite and vermiculite have been used as inert substrates. Frequent additons of nutrient solutions (Douds and Schenck, 1990a), incorporation of slow-release fertilizers (Coltmon et al., 1988) or the use of less available forms of P (Thompson, 1986) are strategies for nutrient management in soilless media.

6.1.4. Hydroponics and aeroponics

Colonized roots, spores and hyphae of endophyte could produced in hydroponics or aeroponics systems. Pre-colonized plants on sterile substrate are needed for these systems. However, plants can be infected directly in the aeroponic system (Hung et al., 1991).

Inoculum can be propagated using the nutrient film technique (NFT) by growing precolonized plants in a defined nutrient solution which flows over the host roots (Mosse and Thompson, 1984). In an aeroponic system, (a fine mist of defined nutrient solutions) suspended in air, roots of the host are bathed in (Zobel et al., 1976). This method of AM multiplication is free from any substrate and can be sheared which resulted in to very high propagule number (Jarstfer and Sylvia, 1995; Sylvia and Jarstfer, 1992).

6.1.5. Root-organ culture or axenic culture of AM fungi

Ri-plasmid transformed root culture offers an efficient method of growing infected roots as no plant growth regulators are required for sustained growth (Mugnier and

Table 3. Soilless media used in the culture of arbuscular fungi. (Jarstfer and Sylvia, 1992)

Material	AM Fungi
Bark-shreeded douglas fir	*Glomus fasciculatum*
Calcined Montmorillonite Clay	*Glomus macrocarpum, G.monosporum, G.versiforme, Scutellispora calospora*
Expanded Clay Aggregates	*Acaulospora laevis, Glomus constrictum, G. etunicatum, G.fasciculatum, G.intraradix, G.macrocarpum, G.mossae, Scutellispora pellucida*
Peat	
hypnum	*Glomus fasciculatum*
sphagnum	*A.caulospora spinosa, Glomus mosseae*
+ perlite and vermiculite	*G.intraradices, G.fasciculatum*
+ pumice (2:1)	*Gigaspora margarita, Glomus fasciculatum, G. tenue*
Pertile	
+ soilrite[a] (1:1)	*Glomus fasciculatum*
	G.fasciculatum
Sand	
basaltic	*Glomus aggregatum*
river	*Acaulospora spinosa, Glomus fasciculatum, G. mosseae*
silica	*G.fasciculatum, G.mosseae,*
+ vermiculite (3:1)	*G.mosseae, G.etunicatum, G.fasciculatum*
+ grit (2:1)	*G.mossea, G.fasciculatum*
Vermiculite	*Acaulospora spinosa, Glomus fasciculatum, G.mosseae*

[a] (perlite, vermiculite and peat, 1:1:1)

Mosse, 1987). However, N and sucrose supply (Mugnier and Mosse, 1987), P concentration (Becard and Fortin, 1988) and pH (Mosse and Hepper, 1975) have been shown to influence the spread of AM fungi in cultured roots.

6.1.5.1. Production of AM

The root organ culture method using Ri T-DNA-transformed roots of *Daucus carota* is used for the production of AM fungal inocula. At present, three AM fungi are being produced, namely *Gigaspora margarita, G. gigantea* and *Glomus intraradices* at The Tata Energy Research Institute, New Delhi. A three stage production system has been standardized, consisting of (TERI):

(i) Nunc plate,

(ii) Petriplate, and

(iii) Jar bottles.

The Nunc plate (plug) was introduced for producing actively growing mycelial system as it has some unique advantages. The smaller area for the growth of the actively germinating spore shows high amount of incidence due to limited area for the growth of the transformed carrot roots.The result of which is the mycelia is able to colonize the root with high amount of incidence to the branching root. The plug containing actively growing mycelia would be used for the initiation of bulking up process. Another feature of the plug-based mode of production is the limited amount of medium with the consequent faster depletion of nutrients which enables the fungi to complete the life cycle in a shorter time (6 - 8 weeks) as compared to 12 to 14 weeks required in conventional plate-based system. As there is a subtle balance between the depletion of nutrients in the medium and the growth of the of the mycelia to a critical level at which sporulation can begin, the plug-based system offers an attractive method of lagre-scale production of inocula scale up procedure.

Jars containing 100 ml of M medium were used for the production of the spores as a long-term form of cultivation of the inocula .The seed inoculum for bulking up was either the plug or the spores or colonized bits of host roots produced from cultivation in petri plates.

6.1.5.2. Harvesting of inocula

The ROC (root organ culture) plate containing the dual system was kept for 2-3 weeks after the formation of spores. This was to allow hardening of the daughter spores, which takes the form of secondary spore thickening. Premature harvesting of the spores resulted in lower percentage of germination during subsequent cycles. The inocula were harvested after solubilization of the medium from the ROC plate, in 10 MM citrate buffer (pH= 6.0). After the spores were picked up, they were washed in sterile distilled water and spread on the M medium for germination to reinitiate the dual culture. The colonized roots obtained from ROC were also tested for their infectivity potential.

6.1.5.3. Estimation of infectivity potential

Four inocula namely, ERRC4 (*Glomus intraradices*), ERRC-6 (*Glomus mosseae*), TR2 (indigenous consortium) and *Gigaspora margarita* produced *in vitro* were subjected to an Infectivity Potential (IP) test. Plastic pots of 100 ml capacity (2.5 × 6 cm) were filled with the inoculum. Eight seedlings of jowar or sudan grass *(Sorghum vulgare)* were planted in each pot and maintained in a greenhouse at 35±2°C, with 60% relative humidity . The pots were watered at regular intervals and on the 12th day, the plants were harvested . The roots permeating the soil were cut at the collar and were cleared by dipping in 5% Calgon and washing gently in fresh water. The roots were stained as described by Philip and Hayman (1970).

To analyse the random distribution of the primary entry points along the root length of the root , complete root systems were spread out in a 2 × 2 cm grided petri plate. The number of intersects across the grid lines were counted and root length was calculated accordingly to the method of Tennat (1975). The roots were then chopped into 1cm pieces and 50 root bits per replicate were selected and randomly mounted on glass slides in lactoglycerol. The total number of Primary Entry Points (PEP) was counted and the number of entry points formed per cm root length was assessed.The value thus obtained were multiplied with the root length to attain a final count of entry points on the whole root system by each inoculum.

Based on the quantification, the amount of ERRC-4, ERRC-6 and TR2 inocula required to produce 1000 infective units was assessed. Plastic tubes of 200 ml capacity (2.5cm × 12cm) were filled with sand sieved with 30 BSS mesh size and a dose of 1000 propagules of the inoculum was administered to each tube. Eight seedlings of sudan grass/ sorghum were planted in each tube. After the 60th day, the tubes were allowed to desiccate for one month. The infected roots thus produced in each tube were subjected to an IP test (as mentioned above) to check the effect of reinitiation and another parallel IP test was run for spore-based cultures sieved out of the tube after homogenization.

The set up was harvested, stained and analysed as discussed earlier. The total number of infectious propagules produced per gram of inoculum was multiplied with the total root biomass produced in each tube to calculate the total number of infective propagule produced for the tube root-based culture. For spore-based cultures, the number of infective propagules produced per gram was multiplied with the amount of total substrate excluding root biomass to calculate the total IP produced per tube. The results were compared with those of *in vitro* production considering the initial dose of 1000 propagule per tube.

The results (Table 4) clearly indicate that the *in vitro* system is more superior to the conventional system (pot-based production) for mass production of mycorrhizal inoculum.

Statistical analysis shows a significant variation in the total number of infective propagules contributed by roots and spores in different systems. The pot-based production give a 2-4 fold increase in one cycle (90 days) whereas one cycle of *in vitro* system results in more than 2000 fold increase.

Table 4. Comparison of inoculum production by *in vitro* and conventional method

Isolate	Initial viable propagule	Secondary spore production (viable count)	No. of propagule in root bits	Total no. of infective propagules	Propagule production/ Initial propagule
In vitro production					
Gigaspora margarita	1	257d	2449b	2706	2706.00
Pot based production					
Glomus intraradics	1000	500c	3959a	4459	4.00
Glomus mosseae	1000	1940a	1563c	3503	3.00
Indigenous consortium	1000	1721b	747d	2468	2.00

The columns sharing the same letters are not significantly different at P>0.05 level.

6.2. Known Sources of Commercial AM Inoculum

This information is provided for the convenience of the reader and does not constitute an endorsement of any product by Sylvia (University of Florida). Any claims should be independently validated.

6.2.1. North America

Bio-Enhancement Technologies (Camarillo, CA), Phone: 805-383-0910, Fax: 805-398-3773
BioScientific, Inc. (Avondale, AZ) 85232, Phone: 602-93294588
Bio-Terra Technologies, Inc. (Las Vegas, NV), Phone: 702-256-6404, FAX: 702-255-2266
First Fruits (Triadelphia, WV), Phone: 888-489-0162
Horticultural Alliance, Inc. (Sarasota, FL), Phone: 800-628-6373, 941-358-0164, FAX: 941-359-9005
Mycorrhizal Applications (Grants Pass, OR), Phone 541-476 3985 and Fax 541 476 1581
Mikko-Tek Labs (Timmons, Ontario), Phone: 705-268 3536
Plants Health Care (Pittsburgh, PA) 1-800-421-9051
Premier Enterprises (Red Hill, PA), Phone: 800-424-2554, FAX: 215-679-4119
Premier Tech (Quebec, Canada), Phone: 418-867-8883 ext. # 392 or 1-800-606-6926
Reforestation Technologies (Monterey, CA), Phone: 800-784-4769
Roots Inc. (Indepence MO), Phone 800-342-6173
Tree of Life Nursery (San Juan Capistrano, CA), Phone: 714-788-0685, Fax: 714-728-0509
Tree Pro (West Lafayette, IN), Phone: 800-875-8071

6.2.2. Other locations

Biological Crop Protection Ltd.(Wye, Ashford, Kent, UK), Phone: 0233 813240 FAX: 0233 813383
Bio-Organics (Medillin, Columbia)
Biorize (Dijon, France), Phone: 33 1 42 33 97 99, Fax: 33 1 42 24 70 15
Central Glass Col., Chemicals Section (Tokyo, Japan)
Global Horticare (Lelystad, The Netherlands), Phone: 31 320 232167, FAX: +31 320 281175
IJemitsu Kosan Co. (Sodegaura, Chile)
MicroBio, Ltd. (Royston, Herts, UK), contact Ingrid Arias at iarias@bbsrc,ac.uk

7. CONCLUSIONS

In view of the increased concern for environmental quality, sustainable technologies need to be incorporated into agricultural systems. Management of AM fungi is an important aspect of such an approach (Linderman and Bethlenfalvay, 1992). However, the obligate symbiotic nature of AM fungi and limited inoculum supplies continue to impede research aimed at managing these beneficial fungi. The application of selected AM fungi will not only benefit plant growth and development, but it offers the possibility of increasing the resistance to soil-borne plant pathogens as well. Under natural conditions this can be regarded as a positive side effect of AM but not as an evidence of its potential in biological control. In most instances, AM significantly change the physiology and chemical constituents of the host, the pattern of root exudation , and the microbial composition of the rhizosphere. Experimental studies differed in fungal symbiont , inoculum doses, and pathogen types . The combination of all factors is responsible for successful disease suppression. A low level of fungal aggresiveness and a weak plant reaction are two factors that permit establishment of a symbiotic relationship (Gianinazzi, 1984). A weak but permanent activity may explain the enhanced resistance of AM plants to certain plant pathogens (Bagyaraj, 1984). In this respect , the mechanism involved although the tendency is to implicate only one. Since AM effects on plant nutrition, especially P uptake, are often so striking, reports implicate P nutrition as a mechanism of disease control . However, many reports mention mechanisms excluding P, such as biochemical changes in the tissues/ cells, as a physiological mechanism, and antagonistic microorganisms of the mycorrhizosphere. Most of the studies do not cover the mechanisms based on morphological changes and changes in the mycorrhizosphere , probably are consolidated and multiple. Since microbial composion in soil reaches a new equilibrium as a result of the selective pressure induced by mycorrhizae, it should be considered a cornerstone foundation component in rhizosphere . In managing rhizosphere populations for biological control of plant diseases, compatible endophyte and effective antagonists should be moved to the production system to ensure that their potential in the manegement of plant diseases is fully used.

Greater attention needs to be placed on the evaluation of the level, vigour and production required for combating plant diseases. Species or isolate-specific screening of AM and of AM -plant combinations needs to be carried out for specific target pathogens.

REFERENCES

Al-Momany, A. and Al-Radded, A. 1988, Effect of vesicular-arbuscular mycorrhizae on *Fusarium* wilt of tomato and pepper, *Alexandria J. Agril. Res.* 33(1) :249-261.

Al-Raddad, A. and Adhmad, M. 1995, Interaction of *Glomus mosseae* and *Paecilomyces lilacinus* on *Meloidogyne javanica* of tomato, *Mycorrhiza* 5(3): 233-236.

Atilano, R. A., Menge J .A . and Van Gundy, S. D. 1981, Interaction between *Meloidogyne arenaria* and *Glomus fasciulatum* in grape, *J. Nematol.* 13: 52-57.

Baghel, P. P. S., Bhatti, D. S. and Jalali, B. L. 1990. Interaction of VA mycorrhizal fungus and *Tylenchulus semipenetrans* on citrus, In : *Current Trends in Mycorrhizal Reserach* eds. B.L. Julali and H. Chand, *Proc. National Conference on Mycorrhiza*, Haryana Agricultural University, Hisar, India, pp. 118-119.

Bagyaraj, D.J. 1984, Biological interactions with VA mycorrhizal fungi. In: *VA Mycorrhiza*, eds C.L. Powel, and D.J. Bagyaraj, CRC Press, Boca Raton, Florida, USA, pp.131-153.

Bagyaraj, D.J. and Manjunath, A. 1980, Selection of suitable host for mass production of vesicular-arbuscular mycorrhizal inoculum, *Plant Soil* 55 : 495-498.

Baker, R. 1986, Biological control: an overview, *Can. J. Pl. Pathol.* 8: 218-221

Baltruschat, H. and Schoenbeck, F. 1975, The influence of endotrophic mycorrhiza on the infestation of tobacco by *Thielaviopsis basicola*, *Phytopathologische Zetschrift* 84: 172-188

Bansal, M. and Mukerji, K.G. 1994, Positive corelation between VAM induced changes in root exudation and mycorhizosphere mycoflora, *Mycorrhiza* 5: 39-44.

Barea, J.M. and Jeffries, P. 1995, Arbuscular mycorrhizas in sustainable soil plant systems. In: *Mycorrhiza : Structure, Function , Molecular Biology and Biotechnology*, eds. A. Varma and B. Hock, Springer-Verlag, Heidelberg, Germany, pp. 521-560.

Barea, J.M., Azcon, R. and Azcon-Aguilar, C. 1993, Mycorrhiza and crops, In: *Advances in Plant Pathology*, Vol. 9, *Mycorrhiza: A Synthesis*, ed. I. Tommerup, Academic Press, London, pp. 167-189.

Bartschi, H., Gianinazzi-Pearson, V. and Vegh, I. 1981, Vesicular-arbuscular mycorrhiza formation and root rot disease (*Phytophthora cinnamomi*) development in *Chamaecyparis lawsoniana*, *Phytopath. Z.* 102: 213-218.

Becard, G. and Fortin, J.A. 1988, Early events of vesicular arbuscular mycorrhiza formation on Ri T-DNA transformed roots, *New Phytol.* 108: 211-218.

Becker, W. N. 1976, Quantification of onion vesicular-arbuscular mycorrhizae and their resistance to *Pytenochaeta terrestine* dissertation, Univ. Illionois, Urbana, 72 pp.

Benhamou, N., Fortin, J.A., Hamel, C., St-Arnaud, M. and Shatilla, A. 1994, Resistance responses of mycorrhizal RiT-DNA- transformed carrot roots to infection by *Fusarium oxysporum* f. sp.*chrysanthemi*, *Phytopath.* 84 : 958-968.

Bieliski, R.L. 1973, Phosphate pools, phosphate transport and phosphate availability, *Annu. Rev. Plant Physiol.* 24: 225-252

Cameron, G.C. 1986, Interactions between two vesicular -arbuscular mycorrhizal fungi , the soybean cyst nematode, and phosphorus fertility on two soybean cultivars, M.S.thesis, Univ. Georgia, Athens. 38 pp.

Carling, D.E., Roncadori, R.W. and Hussey, R.S. 1996, Interactions of arbuscular mycorrhizae, *Meloidogyne arenaria*, and phosphorus fertilization on peanut, *Mycorrhiza* 6: 9-13.

Caron, M. 1989, Potential use of mycorrhizae in control of soilborne disease, *Can. J. Plant Pathol.* 11: 177-179.

Caron, M., Fortin, J. A. and Richar, C. 1986a, Effect of phosphorus concentration and *Glomus intraradices* on *Fusarium* crown and root rot of tomatoes, *Phytopath.* 76: 942-946.

Caron, M., Fortin, J.A. and Richard, C. 1986b, Effect of *Glomus intraradices* on infection by *Fusarium oxysporum* f. sp. *radicis-lycopersici* in tomatoes over a 12-week period, *Can. J. Bot.* 64: 552-556.

Caron, M., Richard, C. and Fortin, J. A. 1986c, Effect of preinfestion of the soil by a vesicular-arbuscular mycorrhizal fungus, *Glomus intraradices* on *Fusarium* crown and root rot of tomatoes, *Phytoprotec.* 67: 15-19.

Caron, M., Fortin, J. A. and Richar, C. 1985, Influence of substrate on the interaction of *Glomus intraradices* and *Fusarium oxysporum* f. sp. *radicis-lycopersici* on tomatoes, *Plant Soil* 87: 233-239.

Chou, I. G. and Schmithenner, A. F. 1974, Effect of *Rhizobia japonicum* and *Endogone mosseae* on soybean root rot caused *Pythium ultimum* and *Phytophthora megasperma*, *Plant Dis.Rep.* 58: 221-225.

Coltman, R.R., Waterer, D.R. and Huang, R.S. 1988, A simple method for production of *Glomus aggregatum* inoculum using controlled -release fertilizer, *Hort. Scie.* 23: 213-215.

Cook, R.J. and Baker, K.F. 1982, *The Nature and Practice of Biological Control of Plant Pathogens*, APS Press, St. Paul, USA.

Cooper, K. and Grandison, G.S. 1986, Interaction of vesicular-arbuscular mycorrhizal fungi and root knot nematode as cultivars of tomato and white clover susceptible to *Meloidogyne hapla*, *Ann. Appl. Biol.* 108: 555-565.

Cooper, K.M. and Grandison, G.S. 1987, Effects of vesicular-arbuscular mycorrhizal fungi on infection of tamarillo (*Cyphomandra betacea*) by *Meloidogyne incognita* in fumigated soil, *Plant Dis.* 71: 1101-1106.

Cordier, C., Nuria Ferrol., Barea, J.M., Gianinazzi, S. and Gianinazzi- Pearson, V. 1998, Cellular and molecular events underlying the induction of *Phytophthora parasitica* in mycorrhizal tomato plants, In : *Proc. International Conference on Mycorrhizae* (Abs.) (IICOM2), Swedish Univ. of Agril. Sciences, Uppsala, Sweden .

Davis, R.M., Menge, J.A. and Zentmyer, G.A. 1978, Influence of vesicular-arbuscular mycorrhizae on *Phytophthora* root rot of alfalfa crop polants, *Phytophath.* 68: 1614-1617.

Davis, R. M. and Menge J. A. 1980, Influence of *Glomus fasciculatum* and soil phosphorus on *Phytophthora* root rot of citrus, *Phytopath.* 70: 447-452.

Davis, R. M., Menge, J. A. and Erwin, D. C. 1979, Influence of *Glomus fasciculatum* and soil phosphorus on *Verticillium* wilt of cotton, *Phytopath.* 69: 453-456.

Davis, R. M. 1980, Influence of *Glomus fasciculatum* on *Thielaviopsis basicola* root rot of citrus, *Plant Dis.* 64: 839-840.

Dehne, H. W. 1982, Interaction between vesicular-arbuscular mycorrhizal fungi and plant pathogens, *Phytopath.* 72: 1114-1119.

Dehne, H. W. and Schonbeck, F. 1979, Untersuchungen zum Einfluss der endotrophen Mykorrhiza auf Pflanzenkrankheiten, II, Phenolstoffwechsel und Lignifizierung, (The influence of endotrophic mycorrhiza on plant diseases, II, Phenolmetabolism and lignification), *Phytopathol. Z.* 95: 210-216

134

Dehne, H. W. and Schoenbeck, F. 1978, Investigation on the influence of endotrophic mycorrhiza on plant diseases, 3, Chitinase activity and ornithine cycle, *Zeitschrift fur Pflanzenkrankheiten und Pflanzenschuty* 85: 666-678.

Dehne, H. W., Schonbeck, F. and Baltruschat, H. 1978, Untersuchungen zum einfluss der endotrophen Mycorrhiza auf Pflanzenkheiten, 3, Chitinase-aktivitat and und Ornithinzyklus, (The influence of endotrophic mycorrhiza on plant diseases, 3, Chitniase activity and ornithinecycle), *Z. Pflkrankh.* 85: 666-678.

Douds, D.D. and Schenck, N.C. 1990a, Increased sporulation of vesicular-arbuscular mycorrhizal fungi by manipulation of nutrient regimes, *Appl. Environ. Microbiol.* 56 : 413-418.

Douds, D.D. and Schenck, N.C. 1990b, Cryopreservation of spores of vesicular-arbuscular mycorrhizal fungi, *New Phytol.* 115: 667-674.

Elliott, A. P., Bird, G .W. and Safir, G. R. 1984, Joint influence of *Pratylenchus penetrans* (Nematoda) and *Glomus fasciculatum* (Phycomyceta) on the ontogeny and *Phaseolus vulgaris*, *Nematropica* 14: 111-119.

Fassuliotis, G. 1970, Resistance in *Cucumis* sp. to the root knot nematode *Meloidgyne incognita acrita*, *J. Nematol.* 2: 174-178.

Fox, J. A. and Spasoff, L. 1972, Interaction of *Hterods solanacearum* and *Endogone gigantea* on tobacco, (Abstr.), *J. Nematol.* 4: 224-225.

Gianinazzi, S. 1984, Genetic and molecular aspects of resistance induced by infections or chemicals, In: *Plant Microbe Interactions, Molecular and Genetic Perspectives*, Vol.1, eds. T. Nester and E.W. Kosuge, Macmillan, New York, USA, pp. 321-342.

Gianinazzi, S., Gianinazzi-Person, V. and Trouvelof, A. 1990a, Potentials and procedures for the use of endomycorrhizas with special emphasis on high value crops, In: *Biotechnology of Fungi for Improving Plant Growth* eds.J.M. Whipps and B. Lumsden, Cambridge University Press, Cambridge, pp. 41-54.

Gianinazzi, S., Trouvelot, A. and Gianinazzi-Pearson, V. 1990b, Role and use of mycorrhizas in horticultural crop production, In : *Proc. XXIII International Horticulture Congress*, Florence, pp.25-30.

Giovannetti, M., Tosi, L., Delatoree, G. and Zazzerini, A. 1991, Histological, physiological and biochemical interactions between vesicular-arbuscular mycorrhizae and *Thielaviopsis basicola* in tobacco plants, *Phytopathol. Z.* 131: 265-274.

Good, J.M. 1968, Relation of plant parasitic nematodes to soil management practices, In: *Tropical Nematology*, eds. G. C. Smart and V. G. Perry, University of Florida, Gainsville, pp. 113-138.

Graham, J.H. 1986, Citrus mycorrhizae : potential benefits and interactions with pathogens, *Hort. Scie.* 21: 1302- 1306.

Graham, J. H., Leonard, R. T. and Menge, J. A. 1981, Membrane mediated decreases in root exudation responsible for phosphorus inhibition of vesicular-arbuscular mycorrhiza formation, *Plant Physiol.* 68:548-552.

Graham, J. H. and Menge, J. A. 1982, Influence of vesicular-arbuscular mycorrhizae and soil phosphorus on take all diesease of wheat, *Phytopath.* 72: 95-98.

Grandison, G. S. and Cooper, K. M. 1986, Interaction of vesicular-arbuscular mycorrhizae and cultivars *Meloidgyne hapla,*. *J. Nematol.* 18: 141-149.

Guillemin, J.P., Gianinazzi, S., Gianinazzi-Pearson, V. and Marchal, J. 1994, Contribution of arbuscular mycorrhizas to biological protection of micropropagated pineapple [*Ananas comosus* (L) Merr.] against *Phytophthora cinnamomi* Rands, *Agric. Sci. Finl.* 3: 241-251.

Harley, J.L. and Smith, S.E. 1983, *Mycorrhizal Symbiosis*, Academic Press, New York, USA, 483 pp.

Hayman, D S. 1982, The physiology of vesicular-arbuscular endomycorrhizal symbiosis, *Can. J. Bot.* 6: 944-963.

Heald, C. M., Bruton, B. D. and Davis, R. M. 1989, Influence of *Glomus intraradices* and soil phosphorus on *Meloidogyne incognita* infecting *Cucumis melo*, *J. Nematol.* 21(1): 69-73.

Hedge, S .V. and Rai, P. V. 1984, Influence of *Glomus fasciculatum* on damping-off of tomato, *Curr. Sci.* 53: 588-589.

Hung, L.L. and Sylvia, D.M. 1988, Production of vesicular -arbuscular mycorrhizal fungus inoculum in aeroponic culture, *Appl. Environ. Microbiol.* 54: 3.

Hung, L.L., O'Keefe, D.M. and Slyvia, D.M. 1991, Use of a hydrogel as a sticking agent and carriers of vesicular-arbuscular mycorrhizal fungi, *Mycol. Res.* 95: 427-429.

Hussey, R. S. and Roncadori, R. W. 1977, Interaction of *Pratylenchus brachyurus* and an endomycorrhizal fungus on cotton, *J. Nematol.* 9: 270-271.

Hussey, R. S. and Roncadori, R. W. 1978, Interaction of *Pratylenchus brachyurus* and *Gigaspora margarita* on cotton, *J. Nematol.* 10: 18-20.

Hussey, R. S. and Roncondori, R. W. 1982, Vesicular-arbuscular mycorrhizae may limit nematode activity and improve plant growth, *Plant Dis.* 66: 9-14.

Jain, R. K. and Sethi, C. L. 1987, Pathogenicity of *Heterodera cajani* on cowpea as influenced by the presence of VAM fungi *Glomus fasciculatum* or *G. epigaeus*, *Indian J. Nematol.* 17(2): 165-170.

Jalali, B. L. and Jalali, I. 1991, Mycorrhiza in plant disease control, In: Handbook of Applied Mycology, Vol.1, *Soil and Plants*, eds. D.K. Arora, B. Rai, K.G. Mukerji and G.R. Knudsen, Marcel Dekker, Inc., New York, USA, pp. 132-154.

Jarstfer, A.G. and Sylvia, D.M. 1992, Inoculum production and inoculation strategies for vesicular-arbuscular mycorrhizal fungi, In: *Soil Microbial Ecology: Application in Agriculture and Environmental Management,* ed. B. Meting, Marcel Dekker, Inc., New York, USA, pp. 349-377.

Jarstfer, A.G. and Sylvia, D.M. 1995, Aeroponic culture of VAM fungi, In: *Mycorrhiza : Structure, Function, Molecular Biology and Biotechnology*, eds. A. Varma and B. Hock, Springer-Verlag, Heidelberg, Germany, pp. 521-559.

Jose-Maria, P., Azcon-Aguilar, C., Dumas-Gaudot, E. and Barea, J.M . 1998, Induction of defence enzymes in tomato roots by arbuscular mycorrhiza and its implication in the protective effect against *Phytophthora parasitica*, In : *Proc. International Conference on Mycorrhizae* (Abs.), (IICOM2), Swedish Univ. of Agril. Sciences, Uppsala, Sweden.

Kellam, M. K. and Schenck, N. C. 1980, Interaction between a vesicular-arbuscular mycorrhizal fungus and root knot nematode on soybean, *Phytopath.* 70: 293-296.

Krishna, K. R. and Bagyaraj, D. J. 1983, Interaction between *Glomus fasciculatum* and *Sclerotium rolfsii* in peanut, *Can. J. Bot.* 41: 2349-2351.

Krishna Prasad, K. S. 1991, Influence of a vesicular arbuscular mycorrhiza on the development and reproduction of root-knot nematode affecting flue cured tobacco, *Afro-Asian J. Nematol.* 1(2): 130-134.

Krishna, K. R. and Bagyaraj, D. J. 1986, Phenolics of mycorrhizal and uninfected groundnut var. MGS-7, *Curr. Res.* 15: 51-52.

Lieberei, R. and Feldmann 1989, Physiological changes in roots colonized by vesicular arbuscular mycorrhizal fungi - reactions in multualistic and parasitic interactions, *Agric. Ecosys. Environ.* 29: 251-255.

Linderman, R.G. 1985, Microbial interactions in the mycorrhizosphere, In: *Proc. 6th N. Am. Conf. on Mycorrhizae*, ed. R. Molina, pp. 117-120.

Linderman, R.G. 1988, Mycorrhizal interactions with the rhizosphere effect, *Phytopath.* 78: 366-371.

Linderman, R.G. and Bethlenfalvay, G.J. 1992, VA-Mycorrhiza and sustainable agriculture, Soil Sci. Soc. America, Madison, WI, USA.

MacGuidwin, A.E., Bird, G.W. and Safir, G.R. 1985, Influence of *Glomus fasciculatum* on *Meloidogyne hapla* infecting *Allium cepa.*, *J. Nematol.* 17: 389-395.

Menge, J.A. 1984, Inoculum production, In: *VA-mycorrhiza*, eds. C.L. Powel, and D.J. Bagyaraj, CRC Press Inc., Boca Raton, Florida, USA, pp. 187-203.

Meyer, J.R. and Linderman, R.G. 1986a, Response of subterranean clover to dual inoculation with vesicular-arbuscular mycorrhizal fungi and a plant growth-promoting bacterium, *Pseudomonas putida*, *Soil Biol. Biochem.* 18 : 185-190.

Meyer, J.R. and Linderman, R.G. 1986b, Selective influence on populations of rhizosphere or rhizoplane bacteria and actinomycetes by mycorrhizas formed by *Glomus fasciculatum*. *Soil Biol. Biochem.* 18, 191-196.

Morandi, D., Bailey, J. A and Gianinazzi-Pearson, V. 1984, Isoflavonoid accumulation in soybean roots infected with vesicular-arbuscular mycorrhizal fungi, *Physiol. Pl. Pathol.* 24: 356-364

Mosse, B. and Hepper, C. 1975, Vesicular-arbuscular mycorrhizal infections in root organ cultures, *Physiol. Pl. Pathol.* 5: 215-223.

Mugnier, J. and Mosse, B. 1987, Spore germination and viability of a vasicular arbuscular mycorrhizal fungus *Glomus mosseae*, *Trans. Brit. Mycol. Soc.* 88: 411-413.

Mukerji, K.G. 1999, Mycorrhizae in control of plant pathogens : Molecular Approaches, In : *Biotechnological Approaches in Biocontrol of Plant Pathogens*, eds. K.G. Mukerji, B.P. Chamola and R.K. Upadhyay, Kluwer Academic / Plenum Publishers, New York, USA, pp. 135-156.

O'Bannon, J. H. and Nemec, S. 1979, The response of citrus seedlings to a symbiont *Glomus etunicatum*, and a pathogen *Radopholus similis*, *J. Nematol.* 11: 270-275.

Pacovsky, R.S. 1986, Micronutrient uptake and distribution in mycorrhizal and phosphorus-fertilized soybeans, *Plant Soil* 95: 379-388.

Pacovsky, R.S., Bethelenfalvay, G. J. and Paul, E. A. 1986, Comparisons between P-fertilized and mycorrhizal plants, *Crop Scie.* 26: 151-156.

Parvatha Reddy, P. 1974, Studies on the action of amino acids on the root knot nematode *Meloidogyne incognita*, Ph.D. thesis, University of Agricultural Sciences, Bangalore, India.

Perrin, R. 1990, Interactions between mycorrhizae and diseases caused by soil-borne fungi, *Soil Use Manag.* 6: 189-195.

Phillips, J.M. and Hayman, D.S. 1970, Improved procedures for clearing roots and staining parasitic and vesicular-arbuscular mycorrhizal fungi for rapid assessment of infection, *Trans. Br. Mycol. Soc.* 55: 158-160.

Priestel, G. 1980, Wechseibeziehung zwischen der endotropics Mycorrhiza und dem Wurzelgallennematoden *Melotdogyne incognita*, (Kofoid and White, 1919) Chitwood, 1949 an Gurke. Dissertation, Hannover, Germany, 103 pp.

Ramirez, B. N. 1974, Influence of endomycorrhizae on the relationship of inoculum density of *Phytophthora palmivora* in soil to infected papaya roots, M. S. thesis, University of Florida, Gainesville, 45 pp.

Ramraj, B., Sahnmugam, N. and Dwarkanath Reddy, A. 1988, Biocontrol of *Macrophomina* root rot of cow-pea and *Fusarium* wilt of tomato by using VAM fungi, In : *Mycorrhizae for Green Asia: Proc. First Asian Conference on Mycorrhizae*, pp. 250-251.

Ratnayake, M., Leonard. R. T. and Menge, J. A. 1978, Root exudation in relation to supply of phosphorus and its possible relevance to mycorrhizla formation, *New Phytol.* 81: 543-552.

Reid, C. P. P. 1984, Mycorrhizae: A root-soil interface in plant nutrition, In: *Microbial-Plant Interactions*, eds. R.L. Todd and J. E. Giddens, ASA Special Pub. 47 pp.

Roncadori, R. and Hussey, R. 1977, Interaction of the endomycorrhizal fungus *Gigaspora margarita* and root-knot nematode on cotton, *Phytopath.* 67: 1507-1511.

Rosendahl, C. N. and Rosendahl, S. 1990, The role of vescular-arbuscular mycorrhiza in controlling damping-off and grwoth reduction in cucumber caused by *Pythium ultimum, Symbiosis* 9: 363-366

Safir, G. 1968, The influence of vesicular-arbuscular mycorrhiza on the disease of onion in *Pyrenocharia terrestris*, M. S. thesis, University of Urbana, Urbana, 36pp.

Saleh, H. and Sikora, R. A. 1984, Relationship between *Glomus fasciculatum* root colonization of cotton and its effect of *Meloidgyne incognita, Nematol.* 30: 230-237

Saleh, H.M. and Sikora, R.A. 1989, Effect of quintozen, benomyl and carbendazim on the interaction between the endomycorrhizal fungus *Glomus fasciculatum* and the root knot nematode *Meloidogyne incognita* on cotton, *Nematol.* 34 (4): 432-442.

Same, B.I., Robson, A.D. and Abbott, L.K. 1983, Phosphorus, soluble carbohydrates and endomycorrhizal infection, *Soil Biol. Biochem.* 15: 593-597.

Schenck, N.C. 1987, Vesicular-arbuscular mycorrhizal fungi and the control of fungal root diseases, In: *Innovative Approaches to Plant Disease Control*, ed. I. Chet, John Wiley and Sons Inc., New York, USA, pp. 179-191.

Schenck, N.C. and Kellam, M.K. 1978, The influence of vesicular-arbuscular mycorrhizae on disease development, Fla. Agric. Exp. Stn. Bull. 798.

Schoenbeck, F. 1979, Endomycorrhiza in relation to plant diseases, In : *Soil-borne Plant Pathogens,* eds. B. Schipper and W. Gams, New Yrok, USA, pp. 271-280.

Schoenbeck, F. and Dehne, H.W. 1977, Damage to mycorrhizal cotton seedling by *Thielaviopsis basicola, Plant Dis. Rep.* 61: 266-268.

Schoenbeck, F. and Dehne, H.W. 1981, Mycorrhiza and plant health, *Sesunde Pflanzen* 33: 186-190.

Secilia, J. and Bagyaraj, D.J. 1987, Bacteria and actinomycetes associated with pot cultures of vesicular-arbuscular mycorrhizas, *Can. J. Microbiol.* 33, 1069-1073.

Sharma, M. P. and Bhargava, S. 1993, Potential of VAM fungus *G. fasciculatum* against the root–knot nematode *Meloidogyne incognita* on tomato, *Mycorrhiza News* 4(4): 4–5.

Sharma, M. P., Bhargava, S., Verma, M. K. and Adholeya, A. 1994, Interaction between the endomycorrhizal fungus *Glomus fasciculatum* and the root-knot nematode *Meloidogyne incognita* on tomato, *Indian J. Nematol.* 24 (2): 34–39.

Sharma, A. K., Johri B. N. and Gianinazzi, S. 1992, Vesicular-arbuscular mycorrhizae in relation to plant diseases, *World J. Microbiol. Biotech.* 8: 559-563.

Shivaprasad, P., Jacob, A., Nair, S. K. and George, B. 1990, Influence of VA mycorrhizal colonization on root-knot nematode infestation in *Piper nigrum* L., In: *Current Trends in Mycorrhizal Research, Proc. National Conference on Mycorrhiza*, eds. B.L. Jalali, H. Chand, Haryana Agricultural University, Hisar, pp. 100-101.

Sieverding, E. and Barea, J.M. 1991, Perspectives de la inoculation de sistemas de production vegetal con hongos formadores de micorrizas VA, In: *Fijacion y Movilizacion Biologica de Nutrients, Coleccion Nuevas Tendencias*, Vol. II, eds. J. Olivares and J.M. Bbarea, C.S.I.C., Madrid, pp. 221-245.

Sikora, R.A. and Schonbeck, F. 1975, Effect of vesicular-arbuscular mycorrhiza on the population dynamics of the root-knot nematodes, VIII Int. Congr. Pl. Protect. 5: 158-166.

Singh, R., Adholeya, A. and Mukerji, K.G. 2000, Mycorrhiza in control of soil-borne pathogens. In : *Mycorrhizal Biology,* eds. K.G. Mukerji, B.P. Chamola and J. Singh, Kluwer Academic / Plenum Publishers, New York, USA, pp. 173-197.

137

Singh, Y.P., Singh, R.S. and Sitaramaiah, K. 1990, Mechanism of resistance of mycorrhizal tomato against root-knot nematode, In: *Current Trends in Mycorrhizal Research, Proc. National Conference on Mycorrhiza,* ed. B.L. Jalali and H. Chand, Haryana Agricultural University, Hisar, India, pp. 96-97.

Sitaramaiah, K. and Sikora, R. A. 1982, Effect of mycorrhizal fungus *Glomus fasciculatum* on the host-parasitic relationship of *Rotylenchus reniformis* in tomato, *Nematologica* 28: 412-419.

Smith, G. S. 1987, Interaction of nematodes with mycorrhizal fungi, In : *Society of Nematology Vistas,* eds. J. A. Veech and D. W. Dickson pp. 133-143.

Smith, G.S. 1988, The role of phosphorus nutrition in interactions of vesicular-arbuscular mycorrhizal fungi with soilborne nematodes and fungi, *Phytopath.* 78: 371-374.

Smith, G.S. and Kaplan, D.T. 1988, Influence of mycorrhizal fungus, phosphorus and burrowing nematode interactions on growth of rough lemon citurs seedlings, *J. Nematol.* 20: 539-544.

Smith, S.E. and Gianinazzi-Pearson, V. 1988, Physiological interactions between symbionts in vesicular-arbuscular mycorrhizal plants, *Ann. Rev. Pl. Phys. Mole. Biol.* 39: 221-244.

Smith, G.S., Roncadori, R.W. and Hussey, R.S. 1986a, Interaction of endomycorrhizal fungi, superphosphate and *Meloidgyne incognita* on cotton in microplot and field studies, *J. Nematol.* 18: 208-216.

Smith, G.S., Hussey, R. S. and Roncadori, R.W. 1986b, Penetration and post-infection development of *Meloidogyne incognita* as affected by *Glomus intraradices* and phosphorus, *J. Nematol.* 18: 429-435.

St-Arnaud, M., Hamel, C., Caron, M. and Fortin, J.A. 1994, Incidence of *Pythium ultimum* in roots and growth substrate of mycorrhizal *Tagetes patula* colonized with *Glomus intraradices, Pl. Path.* 16: 187-194.

Steward, E.L. and Pfleger, F.L. 1977, Development of poinsettia as influenced by endomycorrhizae, fertilizer and root rot pathogens *Pythium ultimum* and *Rhizoctonia solani, Florist's Rev.* 159, 37: 79-80.

Suresh, C.K., Bagyaraj, D.J. and Reddy, D.D.R. 1985, Effect of vesicular-arbuscular mycorrhiza for survival, penetration and development of root-knot nematode in tomato, *Plant Soil* 87: 305-308.

Suresh, C.K. and Bagyaraj, D.J. 1984, Interaction between a vesicular arbuscular mycorrhiza and a root knot nematode and its effect on growth and chemical composition of tomato, *Nematolgia Mediterranea* 12(1): 31-39.

Sylvia, D.M. and Jarstfer, G.J. 1992, Sheared-root inocula of vesicular -arbuscular mycorrhizal fungi, *Appl. Environ. Microbiol.* 58: 229-232.

Taha, A. H .Y. and Abdel-Kader, K. M. 1990, The reciprocal effects of prior invasion by root-knot nematode or by endomycorrhiza on certain morphological and chemical characteristics of Egyptian clover plants, *Ann. Agricul. Sci.* (Cairo) 35(1): 521-532.

Tennant, D. 1975, A test of modified line intersect method of estimating root length, *J. Ecol.* 63: 995-1001.

Thompson, J.P. 1986, Soilless cultures of vesicular-arbuscular mycorrhizae of cereals: Effects of nutrient concentration and nitrogen source, *Can. J. Bot.* 64: 2282-2294.

Wallace, H. R. 1973, *Nematode Ecology and Plant Disease,* Alden Press, London and Oxford, USA.

Wallace, H.R. 1983, Interactions between nematodes and other factors on plants, *J. Nematol.* 15: 221-227.

Zambolim, L. and Schenck, N.C. 1983, Reduction of the effects of pathogenic, root-infecting fungi on soybean by the mycorrhizal fungus, *Glomus mosseae, Phytopath.* 73: 1402-1405.

THE POTENTIAL OF ARBUSCULAR MYCORRHIZAL ASSOCIATIONS FOR BIOCONTROL OF SOIL-BORN DISEASES

P. J. Joseph and P. Sivaprasad

Kerala Agricultural University
College of Agriculture, Vellayani - 695 522
Thiruvananthapuram, Kerala, INDIA

1. INTRODUCTION

The increasing awareness of the possible deleterious effects of fungicides on the ecosystem and environment and the growing interest in pesticide free agricultural produce have created much enthusiasm among scientists on viable alternate methods such as biological control for the management of many soilborne diseases. The success stories of biocontrol in many plant diseases, inconsistent performance of pesticides and the prohibitive investment associated with finding new pesticides further accelerated the interest in biological control. Biological control of plant pathogens is presently accepted as a key practice in sustainable agriculture as it is based on the management of natural resources.

Arbuscular mycorrhizal (AM) associations are receiving considerable attention in recent years since the symbiosis enables better plant growth by higher uptake of nutrients (Harley and Smith, 1983; Marschner and Dell, 1994) and reduces infection by soilborne plant pathogens (Azcon - Aguilar and Barea, 1996; Dehne, 1982) apart from several other beneficial attributes. Encouraging results have been reported on the arbuscular mycorrhizal fungi (AMF) and root - pathogen interactions and the potential of AMF in suppressing many serious soilborne plant pathogens. In fact, the possibility of using mycorrhizal system to suppress soilborne pathogens and to promote plant growth is under vigorous exploration. It has been stated that the degree of protection conferred by AM associations varies with the pathogen involved and can be modified by soil and environmental conditions.

Mycorrhizal associations are so prevelent in terrestrial ecosystems that nonmycorrhizal plant is an exception rather than the rule. Of the different groups of mycorrhizal associations, AMF is often referred as the 'universal plant symbiont' and is the most common type in agricultural crops and natural ecosystem (Gianinazzi and

Schuepp, 1994). This mutualistic association is formed by a group of Zygomycetous soil fungi belonging to the order Glomales (Morton and Benny, 1990; Mukerji 1999; Rosendahl *et al.*, 1994; Singh *et al.*, 2000) which are characterized by forming intracellular arbuscules and extra - matrical mycelium.

2. THE BIOCONTROL POTENTIAL OF ARBUSCULAR MYCORRHIZAL ASSOCIATIONS

Terrestrial plants are dependent on the intense biological activity that surrounds the root system for their nutrition, growth and health. In the rhizosphere AMF occupy a unique ecological position as they are partly inside and partly outside the host (Read *et al.*, 1985). Pathogenic fungi, bacteria and nematodes could directly interact with AMF in the mycorrhizosphere. Stewart and Pfleger (1977) observed that once AMF association was established in the root system before the pathogen's invasion, it could deter the infection and reproduction of the pathogen, but when pathogen and AMF were inoculated at the same time the pathogen infected the plant faster than AMF.

Study on the role of AM symbiosis in protection against plant pathogens began in 1970's and a great deal of information has been published on the subject. An attempt is made here to review all the available work. There are excellent review articles summarizing and discussing results on the role of AMF in the biocontrol of soilborne plant diseases which include those by Schoenbeck (1979); Dehne (1982); Bagyaraj (1984); Schenck (1987); Smith (1988); Caron (1989); Jalali and Jalali (1991); Paulitz and Linderman (1991); Perrin (1991); Sharma and Mukerji (1992, 1999); Hooker *et al.* (1994); Linderman (1994); Azcon-Aguilar and Barea (1996); Mukerji *et al.* (1997) and Sharma *et al.* (1998).

One of the earlier studies concerning the interaction of AMF and soil microorganisms demonstrated that mycorrhizal roots of yellow birch (*Betula lutea*) were predominated by saprophytic fungi such as *Trichoderma*, *Penicillium* and *Paecilomyces* spp. while nonmycorrhizal roots had fungi in the pathogen related genera such as *Pythium*, *Fusarium* and *Cylindrocarpon* spp. (Katznelson *et al.*, 1962). A similar observation by Schenck (1981) also showed that the saprophytic fungi in and on mycorrhizal tomato roots were found increased and incidence of pathogenic fungi like *Pythium* and *Fusarium* were decreased.

The first reported evidence on the biocontrol efficiency of AMF in reducing soilborne diseases had been given by Safir (1968) in onion against pink rot caused by *Pyrenochaeta terrestris* which was further tested and confirmed using two AMF species, viz., *Glomus fasciculatum* and *Gigaspora margarita* by Becker (1976). In another important early investigation, Baltruschat and Schoenbeck (1972a, b) observed that inoculation with *G. mosseae* enhanced the resistance of tobacco and alfalfa to root rot incited by *Thielaviopsis basicola* and inhibited the production and germination of chlamydospores of the pathogen. However, when the pathogen inoculum density was increased, there was decrease in the inhibition of the resting spore. The role of AMF in reducing the damage caused by *Urocystis tritici* to wheat root (Khan and Khan, 1974), root rot of strawberry (Nemec, 1974) and *Cylindrocarpon destructans* in strawberry (Paget, 1975) are other examples of early reports emphasising the significance of using AMF as tool for biocontrol.

2.1. AMF and *Fusarium* Diseases

Fusarium spp. are potential pathogens found in all types of soils of the world causing considerable damage to cultivated crops in the form of vascular wilts, root rots and

140

damping off. Much of the early work on biocontrol using AMF was against *F. oxysporum* f. sp. *lycopersici*, the pathogen which causes vascular wilt in tomato. Successful biocontrol of the disease was achieved by pre-inoculation with *G. mosseae* (Dehne and Schoenbeck, 1975, 1979 a) which was further tested and proved by Schenck and Kellam (1978) using *G. fasciculatum* and *G. mosseae* and by Al-Momany and Al-Raddad (1988) with seven isolates of *Glomus* spp.

AM association significantly reduced the incidence and intensity of wilt of cucumber incited by *F. oxysporum* f. sp. *cucumerinum* (Dehne, 1977; Dehne and Schoenbeck, 1979b). Ames and Linderman (1978) obtained reduced crop damage due to *F. oxysporum* in mycorrhizal easter lilly. The incidence of wilt of chickpea incited by *F. oxysporum* f. sp. *ciceri* was significantly reduced by preinoculation with *G. fasciculatum* and *G. mosseae* (Jalali and Thareja, 1981). Caron *et al.* (1986) also obtained significant reduction in the incidence and development of crown and root rot of tomato incited by *F. oxysporum* f. sp. *radices-lycopersici* using AMF. The pathogen population and root necrosis were also significantly reduced. Other examples of biocontrol include suppression of fusarial wilt in *Cassia tora, Albizia procera* and *Dalbergia sissoo* due to pre-inoculation with *G. fasciculatum* and *G. tenue* (Chakravarthy and Misra, 1986 a, b); reduction of disease incidence and pathogen population of *F. oxysporum* f. sp. *vasinfectum* in pepper with different isolates of *Glomus* spp. (Al-Momany and Al-Raddad, 1988) and *F. oxysporum* f. sp. *medicaginis* in alfalfa using *G. fasciculatum* and *G. mosseae* (Hwang *et al.*, 1992).

Reduction of population of *Fusarium* sp. and disease suppression in legumes (Jalali, 1983); *F. solani* in groundnut (Parvathi *et al.*, 1985); *Fusarium* sp. in tomato (Caron, 1989) and *F. moniliformae* causing damping off of cardamom (Thomas *et al.*, 1994) were also reported due to AM symbiosis.

2.2. AMF and *Phytophthora* Diseases

Phytophthora spp. are considered to be one of the most serious groups of pathogens causing severe economic loss to important tropical and temperate crops. The significance of AM associations in reducing the incidence of *Phytophthora* incited diseases was initially recorded in papaya (Ramirez, 1974) and later in citrus in which root rot incidence and intensity caused by *P. parasitica* were significantly reduced (Davis and Menge, 1980; Schenck *et al.*, 1977) due to *G. fasciculatum*. Woodhead (1978) found that association of *G. calidonius* and *G. etunicatum* reduced the harmful effects of *P. megasperma* c.v. *sojae* in soybean in steamed soil.

In "lawson pine" (*Chamecyparis lawsoniana*) the infection and development of *P. cinnamomi* was greatly reduced due to AM association (Baertschi *et al.*, 1981). They further observed that mixed mycorrhizal population was considerably more effective in reducing pathogen development than a pure culture. Based on evidence from green house experiments, Bisht *et al.* (1985) proved, the potential inhibitory effect of *Gigaspora calospora* on the development of pigeonpea blight (*P. drechsleri* f. sp. *cajani*). An AMF conferred 'P' nutritional advantage in sweet orange seedlings was recognised by Graham and Egel (1988) which increased the resistance / tolerance of seedlings to *Phytophthora* root rot.

Another economically important disease which has effectively been suppressed by AMF is "foot rot" of black pepper incited by *P. capsci*. Initial report of disease suppression of 'foot rot' with *G. fasciculatum* was made by Anandaraj *et al.* (1993). Subsequent reports (Anandaraj and Sarma, 1994) further established the suppressive effect of AMF on 'foot rot' incidence and stimulation of root regeneration. A negative relationship between 'foot rot' incidence and percentage mycorrhizal root colonization and

141

spore count in black pepper was noticed by Sivaprasad *et al.* (1995 b). When AMF inoculated black pepper cuttings were planted in a diseased field, there was significant reduction of disease incidence in mycorrhizal plants (16.5 %) as compared to non mycorrhizal plants (28.5 %) (DARE, 1996). It was further suggested by Sarma *et al.* (1996) that mycorrhizal development in black pepper compensated the root damage caused by the pathogen and showed considerable reduction in 'foot rot' incidence, foliar yellowing and defoliation under natural condition. Considerable reduction in 'foot rot' incidence could be achieved with native isolates of AMF which are more adapted to local soil conditions (Robert, 1998; Sivaprasad *et al.*, 1997).

Azhukal disease of cardamom is another devastating disease incited by *P. nicotianae* which was remarkably reduced both in green house and field due to AMF inoculation at the time of planting in the secondary nursery (Sivaprasad, 1995). Cordier *et al.* (1996) studied the colonization of root tissues of mycorrhizal tomato by *P. nicotianae* and revealed that the pathogen related loss of root biomass and function was compensated by AMF.

2.3. AMF and *Pythium* Incited Disease

Pythium spp. are soil born fungi causing serious diseases like damping off and soft rot in several cultivated crops bringing about considerable economic loss. Attempts of biological control of *Pythium ultimum* using AMF was initiated by Stewart and Pfleger (1977) who observed significant reduction in damage caused by the pathogen in mycorrhizal poinsettia plants. The suppressive effect of *G. fasciculatum* and *G. mosseae* on *P. ultimum* in poinsettia was further emphasised by Schenck and Kellam (1978) and Kaye *et al.* (1984).

Biocontrol of damping off of cucumber seedlings caused by *P. splendans* was investigated by Kobayashi (1990) and observed that AMF inoculation alone induced a weak response but combined use of AMF and charcoal compost drastically reduced the disease on 2 - 3 week old seedlings while in vermiculite grown cucumber, the incidence of damping off by *Pythium* sp. was greatly reduced in the presence of *Glomus* spp. (Rosendahl *et al.*, 1992).

Effect of inoculation of *G. fasciculatum* and *G. intraradices* on growth of saifoin and incidence of damping off by *Pythium* sp. was studied by Hwang *et al.* (1993) and found that AMF reduced the severity and damage caused by the pathogen. In *Tagetes patula* also mycorrhizal association with *G. intraradices* tended to reduce the infection and population of *Pythium* sp. considerably (St. Arnaud *et al.*, 1994).

In spice crops, the biocontrol potential of AMF against *Pythium* incited diseases has been amply described. Inoculation with *G. multiculae* and *G. fasciculatum* significantly reduced rhizome rot incidence incited by *P. aphanidermatum* in ginger (Iyer and Sundararaju, 1993 ; Sivaprasad, 1993) whereas in cardamom *G. fasciculatum* effectively checked damping off by *P. vexans* (Thomas *et al.*, 1994). Joseph *et al.* (1995) observed that there is a negative relationship between natural incidence of rhizome rot and mycorrhizal colonization in ginger. It was further emphasised that native isolates of *Glomus* spp. significantly reduced the incidence and intensity of rhizome rot of ginger and improved plant growth both in green house and field conditions (Joseph, 1998; Joseph and Sivaprasad, 1997).

2.4. AMF and *Rhizoctonia* Diseases

Another major soilborne pathogen which causes severe damage such as damping off, seedling blight, leaf blight, etc. in several cultivated crops is *Rhizotonia solani*. The

142

biocontrol potential of mycorrhizal fungi against *R. solani* was initially tested and found successful in poinsettia (Schenck and Kellam, 1978; Stweart and Pfleger, 1977) in legumes (Jalali, 1983) and in groundnut (Parvathi *et al.*, 1985). The biological deterrent activity of AMF against damping off of cauliflower seedlings by *R. solani* was tested by Iqbal and Nasim (1988) in Pakistan and observed that the seedlings preinoculated with AMF survived better and resisted the pathogen attack significantly. The biocontrol potential of AMF was further highlighted by Kobayashi (1990) by observing that inoculation of AMF alone induced a weak response but combined use of AMF and charcoal compost drastically reduced damping off of cucumber. Incidence of damping off in cardamom caused by *R. solani* was also significantly reduced due to AMF inoculation (Thomas *et al.*, 1994).

2.5. AMF and *Thielaviopsis basicola*

Another soil borne pathogen which causes damping off in several crops is *T. basicola*. A detailed investigation on the interaction of AMF and *T.basicola* which causes root rot in tobacco and alfalfa revealed that *G. mosseae* inoculation increased host resistance and inhibited the production and germination of the pathogen (Baltruschat and Schoenbeck, 1972a). The protective effect of AMF against this pathogen was further tested on cotton (Schoenbeck and Dehne, 1977) and on sweet orange seedlings (Davis, 1980) and found to be very effective.

2.6. AMF and *Macrophomina* Incited Diseases

Root rot caused by *M. phaseolina* was effectively managed by mycorrhizal associations in legumes such as soybean (Stewart and Pfleger, 1977) and mungbean (Jalali *et al.*, 1990). The disease incidence was as high as 77.9 per cent in non mycorrhizal mungbean plants whereas the incidence was reduced drastically to 13.3 per cent in mycorrhizal plants.

2.7. AMF and *Verticillium* Wilt Diseases

Wilts caused by *Verticillium* spp. are recognised as major vascular diseases of many crops. The wilt of aubergine caused by *V. albo-atrum* was found to be significantly reduced by *Gigaspora* sp. (Melo *et al.*, 1985). They further observed that interaction between different AMF species and the pathogen was highly specific and the ability of disease suppression by AMF varied greatly with species. Hwang *et al.* (1992) also showed that inoculation with *G. fasciculatum* and *G. mosseae* significantly reduced the population of *V. albo-atrum* in alfalfa.

2.8. AMF and Other Soil Borne Plant Pathogens

The efficiency of AMF in reducing the incidence and severity of several other soilborne diseases have been proved, beyond doubt, in many crop plants of economic importance. Reduction of infection level of *Cylindrocladium scoparium* in mycorrhizal yellow poplar (Stewart and Pfleger, 1977); *Olpidium brassicae* in tobacco (Schoenbeck and Dehne, 1979); take all disease (*Gaeumanomyces graminis*) of wheat (Graham and Menge, 1982); *Sclerotium rolfsii* in peanuts (Krishna and Bagyaraj, 1983); root rot (*Aphanomyces euteichus*) of pea (Rosendahl, 1985) are some of examples in this regard. In most of these instances the need for pre-inoculation of the host plant with the appropriate AMF as a prerequisite for obtaining reasonable level of disease reduction was emphasised.

The influence of AMF on multiple infection of host plants by different pathogens was studied by Chabbra *et al.* (1992). They noticed that when *G. fasciculatum* inoculated maize plants were infected by maize leaf blight (*Helminthosporium maydis*) seedling blight and stalk rot (*Fusarium moniliforme*) and *Acremonium* stalk rot (*Acremonium kiliense*), the plants exhibited complete resistance to *F. moniliforme* while there was no effect on the other two diseases. Hence the resistance offered by AM associations is not general, but specific to each AMF - host - pathogen complex.

In another major study, the interaction between AMF and phytotoxic micromycetes, which are the causative agents of apple replant disease was investigated by Catska (1994). Inoculation of *G. fasciculatum* and *G. macrocarpum* suppressed the population of these microorganisms in the soil whereas enhanced that of *Azospirillum*. He concluded that use of AMF can replace chemical treatment of soil against this disease.

3. INCREASE IN DISEASE INTENSITY DUE TO AMF

Increased damage and disease incidence due to mycorrhizal associations are not uncommon. Ross (1972) observed increased damage and higher disease incidence by *Phytophthora megasperma* c.v. *sojae* in mycorrhizal soybean. Baltruschat *et al.* (1973) observed that high inoculum density of *T. basicola* in mycorrhizal cotton rhizosphere rendered the mycorrhizal fungi ineffective for biocontrol. In another study Chou and Schmitthenner (1974) showed that there was no reduction in disease due to *P. megasperma* c.v. *sojae* and *Pythium ultimum* in mycorrhizal soybean. The study indicated that the symbiosis neither dispose the host to infection nor enhance the severity of infection. Greater disease incidence and severity due to AM associations were also reported by Davis *et al.* (1978) in avocado, alfalfa and citrus by *P. cinnamomi, P. megasperma* and *P. parasitica* respectively in which the effect of AM associations was negated in the presence of the pathogen.

In cotton, *Verticillium* wilt was more severe in AMF inoculated plants (Davis *et al.*, 1979). The reasons attributed were increased avenues for penetration of the pathogen due to rapturing the cortex by abundant chlamydospores of AMF, reduced concentration of K in mycorrhizal plants and larger growth of the plants due to better nutrient absorption. Increased disease incidence was also reported in mycorrhizal tomato due to wilt caused by *F. oxysporum* f. sp. *lycopersici* (Mcgraw and Schenck, 1981) where as no effect was observed by Zambolin and Schenck (1981, 1983) due to AMF inoculation in soybean against *F. solani, R. solani* and *M. phaseolina* and against *Verticillium* wilt of tomato (Baath and Hayman, 1983).

The results of all these investigations indicated that AM associations in plants can decrease, increase or have no effect on the development and severity of plant diseases. In most instances, precolonization of the host by AMF before the pathogenic infection was found to be a prerequisite for effective protection against the pathogen. The effect of disease suppression also depends on the virulence of the pathogen. A higher inoculum level of the parasite tended to decrease the positive effects of the symbiosis (Dehne and Schoenbeck, 1979 a). Hence higher virulence of the pathogen will be detrimental to both the host as well as the endophyte. Moreover, successful suppression of a pathogen not only depends on the mode of parasitism and the virulence of the pathogen but also on the specific potential of the AMF to induce resistance in each host - endophyte - pathogen system.

4. MECHANISMS OF INTERACTIONS BETWEEN AMF AND FUNGAL ROOT PATHOGENS IN BIOLOGICAL CONTROL

The study on the possible role of AM symbiosis for protection against plant diseases began in the 1970's and a great deal of information has been generated on the subject, but still very little is known about the underlying mechanisms (Hooker *et al.*, 1994; Linderman, 1994). Various mechanisms had been proposed, but with little experimental data. It is postulated that biocontrol of plant diseases may be accomplished by AMF in one or more of the mechanisms as detailed below.

4.1. Improved Host Nutrition

The most obvious contribution of AM symbiosis in plants is the increased uptake of 'P' and other mineral nutrients resulting in more vigorous plants better able to resist / tolerate root diseases or lead to disease escape (Davis, 1980 ; Graham and Menge, 1982; Smith, 1988). However, improved nutrition of mycorrhizal plants may also lead to an increase in disease incidence (Davis and Menge, 1980; Davis *et al.*, 1978; Ross, 1972). What is good for the plant is also good for the pathogen. Experiments conducted by several workers (Caron *et al.*, 1986; Graham and Egel, 1988) showed that mechanisms of disease suppression other than enhanced 'P' uptake operated in mycorrhizal plants.

4.1.1. Competition for host photosynthates

Both AMF and root pathogens need photosynthates of the host for their growth and may compete for the carbon compounds reaching the roots (Linderman, 1994; Smith, 1988). When AMF precolonized the roots, they have primary access to these compounds which limits the carbon demand of the pathogen. However, there is little evidence to support this hypothesis.

4.1.2. Competition for infection/colonization sites

Pre-inoculation of plants with AMF reduced the development of disease in several instances (Dehne, 1982; Joseph and Sivaprasad, 1997; Schenck and Kellam, 1978). It has been stated that infection and proper development of AMF symbiosis well before the attack by the pathogen in the host is a prerequisite for increased resistance of the host. Simultaneous competition by both fungi for the root infection sites almost completely diminished the protection from disease offered by the symbiont (Baertschi *et al.*, 1981). Due to the higher degree of plant susceptibility to AMF, they may be more successful for host colonization in direct competition with pathogens (Dehne, 1982). Possible infection sites in the host root may be saturated with the endophyte inducing resistance to pathogenic infection. It has been shown that *Phytophthora* development was reduced in AMF colonized and adjacent regions of AMF root system and the pathogen did not penetrate arbuscule containing cells (Cordier *et al.*, 1996). It is a classical example for localized competition.

5. ANATOMICAL AND MORPHOLOGICAL CHANGES IN THE ROOT SYSTEM

Mycorrhizal roots are more lignified than nonmycorrhizal ones especially in the stelar tissues (Dehne, 1977). Increased lignification of AM plant cells such as in pink rot affected onion (Becker, 1976) and fusarial wilt affected tomato and cucumber root (Dehne

145

and Schoenbeck, 1979 b) were accounted for reduced incidence of disease. Wick and Moore (1984) elucidated increased wound barrier formation that inhibited *Thielaviopsis* black root rot of holly plants. It has been demonstrated that remarkable changes occur in root system morphology such as increased root branching and the meristamatic and nuclear activities of root cells due to AMF colonization which might affect infection by the pathogen and its further development (Atkinson *et al.*, 1994).

6. MICROBIAL CHANGES IN MYCORRHIZOSPHERE

It has been observed that AMF establishment can change both total population and specific functional groups of microorganisms in the rhizosphere soil (Mayer and Linderman, 1986). They further showed that propagules produced by *Phytophthora cinnamomi* were reduced by the application of extracts of rhizosphere soil of mycorrhizal plants. Secilia and Bagyaraj (1987) isolated more pathogen antagonistic actinomycetes from the rhizosphere of AM plants. There is evidence of microbial shifts in the mycorrhizosphere which in turn influence the growth and health of plants. Mycorrhizal associations alter the host physiology which play a decisive role in the root exudation pattern and the resultant microbial changes that occur in the mycorrhizosphere, both qualitatively and quantitatively (Amora-Lazcano *et al.*, 1998; Azcon - Aguilar and Bago, 1994). The altered microflora belong to a wide range of taxonomic groups and are known to stimulate the establishment and stability of mycorrhizal symbiosis (Singh, 1998). Hence the effect of altered microbiota in the rhizosphere of AM plants, though not specifically evaluated as a mechnasim of biocontrol, there are indications that such a mechanism does operate (Azcon - Aguilar and Barea, 1992; Barea *et al.*, 1996; Mukerji *et al.*, 1999).

7. STIMULATION OF PLANT DEFENCE MECHANISMS

Right from the earlier studies, it had been speculated that elicitation of specific defence mechanisms and altered host physiology are probable mechanisms of biological control achieved by AM symbiosis in plants. Baltrauschat and Schoenbeck (1975) proposed that accumulation of amino acids particularly arginine and citrulline is the mechanism of root rot suppression in AM tobacco plants caused by *T. basicola*. Further evidence on the accumulation of arginine in mycorrhizal tomato, bean and corn was provided by Dehne and Schoenbeck (1979b) and the reason was attributed to interruption of ornithine cycle in AM plants. Roots colonized by AMF exhibited high chitinolytic activity and the enzymes involved may as well be effective against fungal pathogens (Dehne *et al.*, 1978).

During the life cycle, plants develop a number of defence responses elicited by various signals. Phytoalexins, enzymes of phenyl propaniod path way, chitinases, glucanases, peroxidases, pathogenesis related proteins (PR proteins), hydroxy proline rich glycoproteins, and phenolics are the common compounds which are involved in plant defence and studied in relation to AM formation (Gianinazzi - Pearson *et al.*, 1994). It has been shown that the plants produced a number of new proteins (endomycorrhizins) in response to AM formation (Dumas-Gaudot *et al.*, 1994; Gianinazzi-Pearson and Gianninazzi, 1995). However, it is not known whether the altered pattern of protein synthesis is related to defence reactions.

Phytoalexins accumulate as a defence response in plants consequent to pathogen attack. Although not detected during initial stages of AM formation, Morandi *et al.* (1984) observed the accumulation of isoflavanoids in soybean during later stages of symbiosis
146

and suggested as the mechanism of protection against soilborne pathogens. Wyss *et al.* (1991) also showed that there was increased production of glyceollin in mycorrhizal roots of soybean in response to infection by *R. solani*. However, another phytoalexin, medicarpin, was increased transiently during early stages of AM colonization, but decreased to low levels later (Harrison and Dixon, 1993) in *Medicago* sp. Although there are indications of low activation of phenyl propanoid related enzymes like phenyl alanine ammonium lyase and chalcone isomerase during early stages of AM formation (Lambias and Mehdy, 1993 ; Volpin *et al.*, 1994), their influence on host-endophyte-pathogen systems were not studied in detail.

There is no direct evidence of increased production of chitinases and β - 1, 3 glucanases as a defence response of AM plants against pathogen attack. However, it had been shown that m RNA's encoding for chitinase (Blee and Anderson, 1996) and β - 1, 3 glucanase (Lambias and Mehdy, 1995) had been accumulated in and around arbuscule containing cells suggestive of localised induction of defence related genes. Wall bound peroxidase activity associated with epidermal and hypodermal cells increased in mycorrhizal roots which might contribute to higher resistance to root pathogens (Gianinazzi and Gianinazzi-Pearson, 1992).

Significant enhancement of phenolic content of AM sesamum plants was reported by Selvaraj and Subramonian (1990). Histochemical studies revealed accumulation of different types of lipids and catechol tanins in vesicles. Increased levels of phenolics in mycorrhizal plants and their role as a probable mechanism of disease resistance were also highlighted by other workers (Krishna and Bagyaraj, 1984; Sivaprasad *et al.*, 1995).

In short it can be concluded that only weak defence responses are produced by AM plants with regard to lignification, production of phytoalexins, peroxidases, phenolics, etc., indicating AMF do not elicit typical defence responses. Although in small quantities, these compounds could sensitize the host root to pathogen attack and enhance the defence mechanism of the host. This was amply illustrated by Benhamou *et al.* (1994) by comparing the responses of AM and non AM transformed carrot roots to infection by *F. oxysporum* f. sp. *chrysanthemi*. In mycorrhizal roots the growth of pathogen was restricted to epidermis and cortical tissues and hyphae exhibited a high level of structural disorganisation while the pathogen developed further and reached even the vascular stele in nonmycorrhizal roots and no structural disorganization of hyphae occurred. This is suggestive of activation of plant defence responses by mycorrhiza formation by promoting quicker and stronger reaction against pathogen and provides protection (Mukerji, 1999).

8. COMBINED APPLICATION OF AMF AND OTHER ANTAGONISTIC RHIZOSPHERE MICROFLORA FOR BIOCONTROL

There is increasing interest in the microbial interactions in the rhizosphere of arbuscular mycorrhizae which may play decisive role in the mycorrhizal establishment and its success as a biofertilizer cum biocontrol agent. It is mentioned earlier in this chapter that the changed microflora of the mycorrhizosphere exert tremendous influence on the growth and health of plants. It has now been increasingly emphasised that the prophylactic property of AMF could be exploited in association with other rhizosphere microorganisms known to be antagonistic to root pathogens such as *Trichoderma, Gliocladium* and fleurocent Pseudomonads which are being used as biocontrol agents (Azcon- Aguilar and Barea, 1996 ; Barea *et al.*, 1996).

In several reports the interaction between saprophytic / antagonistic fungi and AMF differed widely even when the same genus was involved. *Trichoderma* spp. had been reported to have both antagonistic (Sylvia and Schenck, 1983; Wyss *et al.*, 1992) and stimulatory effect on AMF (Calvet *et al.*, 1992, 1993).

One of the earliest studies on combined use of AMF and antagonists for the management of soil borne diseases was made by Kohl and Schlosser (1989). They observed that infection and colonization of maize roots by *G. etunicatum* was almost unaffected by strains of *T. hamatum* and *T. harzianum* and suggested the combined application of the two biocontrol agents to promote plant health. The effect of *Gliocladium virens*, another well known antagonist, on mycorrhizal colonization of cucumber was studied by Paulitz and Linderman (1991) and found that dual inoculation of the antagonist along with *G. etunicatum* and *G. mosseae* was compatible. Calvet *et al.* (1992) observed that *Trichoderma* spp. stimulated the germination of AMF spores by producing volatile compounds and could be used along with AMF to ensure better plant growth and health.

A very convincing study on the combined effect of two groups of biocontrol agents in the suppression of plant diseases was conducted by Kumar *et al.* (1993). They tested mixed biocontrol cultures of *G. epigaeus* and *T. viride* singly and in combination against three wheat root rot pathogens *viz.*, *Bipolaris sorokiniana*, *F. avanaceum* and *F. javanicum*. The mixed cultures protected wheat plants against all the three pathogens. Tripartite system of interaction of AMF - antagonist - pathogen was further tested in marigold (*Tagetes erecta*) by Calvet *et al.* (1993) and found that the combined effect of *G. mosseae* and *T. aureoviride* produced a synergistic effect on growth of marigold and reduced pathogen population considerably.

Anandaraj *et al.* (1996) obtained considerable reduction of "foot rot" incidence in field trials with black pepper cuttings inoculated with AMF at the time of nursery planting and *T. harzianum* and *G. virens* added every year in the field. In a recent study for the management of rhizome rot of ginger caused by *Pythium aphanidermatum*, it was observed that incidence and intensity of rhizome rot was significantly reduced both in green house and field trails when native isolates of AMF (*Glomus* spp.) and antagonists (*T. viride* and *Aspergillus fumigatus*) were delivered to the production system at the time of planting (Joseph, 1998). Similar synergistic effects of combined application of native isolates of AMF and antagonists were also obtained for the management of 'foot rot' of black pepper (Robert, 1998).

9. CONCLUSIONS

It becomes apparent that in sustainable management systems, the prophylactic property of AM symbiosis could well be exploited in association with other microbial antagonists of fungal pathogens, either fungi or PGPR (Plant growth promoting rhizobacteria). Several studies indicated that they do not antagonize AMF, but can improve the development of mycosymbiont and facilitate AM formation (Barea *et al.*, 1996). Development of efficient strains with better ecological adaptability for effective interaction through genetic manipulation is currently one of the thrust areas of research in integrated disease management.

REFERENCES

Al-Momany and Al-Raddad, A. M. 1988, Effect of vesicular arbuscular mycorrhizae on *Fusarium* wilt of tomato and pepper, *Alexandria J. Agric. Res.* 33 : 249-261.
Ames, R. W. and Linderman, R. G. 1978, The growth of Easter lily (*Lilium longiflorum*) as influenced by vesicular arbuscular mycorrhizal fungi, *Fusarium oxysporum* and fertility level. *Can. J. Bot.* 56 : 2773.
Amora-Lazcano, E., Vazquez, M. M. and Azcon, R. 1998, Response of nitrogen- transforming microorganisms to arbuscular mycorrhizal fungi,. *Biol. Fertil.Soils* 27 : 65-70

Anandaraj, M. and Sarma, Y. R. 1994, Biological control of black pepper diseases, *Indian Cocoa Arecanut Spices J.* 18 (1) : 22-23.

Anandaraj, M., Ramana, K. V. and Sarma, Y. R. 1993, Suppressive effects of VAM on root pathogens of black pepper, a component of Western Ghats forest ecosystem, In : *IUFRO Symposium,* Kerala Forest Research Institute, Peechi, p. 64.

Anandaraj, M., Venugopal, M. N. and Sarma, Y. R. 1996, Biological control of diseases of spice crops, Annual Report 1995-96, IISR, Calicut, Kerala, pp. 70-71.

Atkinson, D., Berta, G. and Hooker, J. E. 1994, Impact of mycorrhizal colonization on root architecture, root longevity and formation of growth regulators, In : *Impact of Arbuscular Mycorrhizas on Sustainable Agriculture and Natural Ecosystems,* eds. S. Gianinazzi, and H. Schuepp, Birkhauser, Basel, pp. 84-99.

Azcon-Aguilar, C. and Bago, B. 1994, Physiological characteristics of the host plant promoting an undisturbed functioning of the mycorrhizal symbiosis, In : *Impact of Arbuscular Mycorrhizas on Sustainable Agriculture and Natural Ecosystems,* eds. S. Ginaninazzi, and H. Schuepp, Birkhauser, Basel, pp. 47-60.

Azcon-Aguilar, C. and Barea, J. M. 1992, Interactions between mycorrhizal fungi and other rhizosphere microorganisms, In : *Mycorrhizal Functioning - An Integrative Plant Fungal Process,* ed. M.J. Allen, Chapman and Hall, New York, pp. 163-198.

Azcon-Aguilar, C. and Barea, J. M. 1996, Arbuscular mycorrhizas and biological control of soil-borne plant pathogens - an overview of the mechanism involved, *Mycorrhiza* 6 : 457-464.

Baath, E. and Hayman, D. S. 1983, Plant growth responses to vesicular- arbuscular mycorrhiza XIV, Interactions with *Verticillium* wilt on tomato plant, *New Phytol.* 95 : 419.

Baertschi, H., Gianinazzi-Pearson, V. and Vegh, I. 1981, Vesicular arbuscular mycorrhiza and root-rot disease (*Phytophthora cinnamomi* Rands.) development in *Chamaecyparis Iawsoniana* (Murr.) Parl. *Phytopath.* 102 : 213.

Bagyaraj, D. J. 1984, Biological interactions with VA mycorrhizal fungi, In : *VA Mycorrhiza,* eds. C.L. Powell, and D.J. Bagyaraj, CRC, Press, Boca Raton, Florida, USA, pp. 131-153.

Baltruschat, H. and Schoenbeck, F. 1972a, Influence of endotrophic mycorrhiza on chlamydospore production of *Thielaviopsis basicola* in tobacco roots, *Phytopath. Z.* 74 : 358.

Baltruschat, H. and Schoenbeck. F. 1972b, The influence of endotrophic mycorrhiza on the infestation of tobacco by *Thielaviopsis basicola, Phytopath. Z.* 74 : 172-188.

Baltruschat, H. and Schoenbeck, F. 1975, The influence of endotrophic mycorrhiza on the infestation of tobacco by *Thielaviopsis basicola, Phytopathol. Z.* 84 : 172-188.

Baltruschat, H., Sikora, R. A. and Schoenbeck, F. 1973, Effect of VA mycorrhiza (*Endogone mosseae*) on the establishment of *Thielaviopsis basicola* and *Meloidiogyne incognita* in tobacco, *Abstr. 2nd Intern. Cong. Plant Pathol.* pp. 661.

Barea, J. M., Azcon-Aguilar, C. and Azcon, R. 1996, Interactions between mycorrhizal fungi and rhizosphere microorganisms within the context of sustainable soil-plant systems, In : *Multitrophic Interactions in Terrestrial Systems,* eds. A.C. Gange, and V.K. Brown, Black Well, Oxford, USA.

Becker, W. N. 1976, Quantification of onion vesicular arbuscular mycorrhizae and their resistance to *Pyrenochaeta terrestris,* Ph.D. thesis, University of Illinois, Urbana, USA.

Benhamou, N., Fortin, J. A., Hamel, C., St. Arnaud, M. and Shatilla, A. 1994, Resistance response of mycorrhizal Ri T-DNA- transformed carrot roots to infection by *Fusarium oxysporum* f. sp. *chrysanthemi, Phytopath.* 84 : 958-968

Bisht, V. S., Krishna, K. R. and Nene, Y. L. 1985, Interaction between vesicular arbuscular mycorrhiza and *Phytophthora drechsleri* f sp. *cajani, Internat. Pigeon Pea News Lett.* 4 : 63-64.

Blee, K. A. and Anderson, A. J. 1996, Defense related transcript accumilation in *Phaseolus vulgaris* L. colonized by arbuscular mycorrhizal fungus *Glomus intraradices* Schenek and Smith, *Pl. Physiol.* 110 : 675-688

Calvet, C., Barea, J. M. and Pera, J. 1992, *In vitro* interactions between the vesicular arbuscular mycorrhizal fungus *Glomus mosseae* and some saprophytic fungi isolated from organic substrates, *Soil Biol. Biochem.* 24 : 775-780

Calvet, C., Pera, J. and Barea, J. M. 1993, Growth response of marigold (*Tagetes erecta* L.) to inoculation with *Glomus mosseae, Trichoderma aureoviride* and *Pythium ultimum* in a peat-perlite mixture, *Plant Soil* 148 : 1-6.

Caron, M. 1989, Potential use of mycorrhizae in control of soil borne diseases, *Can. J. Plant Pathol.* 11 : 177-179.

Caron, M., Fortin, J. A. and Richard, C. 1986, Effect of inoculation sequence on the interaction between *Glomus intraradices* and *Fusarium oxysporum* f. sp. *radicis - lycopersici* in tomatoes, *Can. J. Plant Pathol.* 8 : 12-16.

149

Catska, V. 1994, Interrelationships between vesicular arbuscular mycorrhizae and rhizosphere microflora in apple replant disease, *Biol. Plant.* 36 : 99-104.

Chabbra, M. L. Bhatnagar, M. K. and Sharma, M. P. 1992, Influence of a vesicular-arbuscular (VA) mycorrhizal fungus on important diseases of maize, *Indian Phytopath.* 45 : 235-236.

Chakravarthy, P. and Misra, P. R. 1986a, The influence of VA mycorrhizae on wilting of *Albizia procera* and *Dalbergia sissoo, Eur. J. For. Pathol.* 16 : 91-97.

Chakravarthy, P. and Misra, P. R. 1986b, Influence of endotrophic mycorrhizae on the fusarial wilt of *Cassia tora* L., *J. Phytopath.* 115 : 130-133

Chou, L. G. and Schmitthenner, A. F. 1974, Effect of *Rhizobium japonicum* and *Endogone mosseae* on soybean root-rot caused by *Pythium ultimum* and *Phytophthora megasperma* var. *sojae, Pl. Dis. Rep.* 58 : 221-225.

Dare, 1996, Annual Report 1995-96, Department of Agricultural Research and Education, Ministry of Agriculture, Government of India, pp. 69-71.

Davis, R. M. 1980, Influence of *Glomus fasciculatum* on *Thielaviopsis basicola* root rot of citrus, *Plant Dis.* 64 : 839-840.

Davis, R. M. and Menge, J. A. 1980, Influence of *Glomus fasciculatum* and soil phosphorus on *Phytophthora* root rot of citrus, *Phytopath.* 70 : 447-452.

Davis, R. M., Menge, J. A. and Erwin, D. 1979, Influence of *Glomus fasciculatum* and soil phosphorus on *Verticillium* wilt of cotton, *Phytopath.* 69 : 453-456.

Davis, R. M., Menge, J. A. and Zentmyer, G. A. 1978, Influence of vesicular arbuscular mycorrhizae on *Phytophthora* root rot of three crop plants, *Phytopath.* 68 : 1614-1617.

Dehne, H. W. 1977, Untersuchungen uber den Einfluss der endotrophen mycorrhiza auf die *Fusarium*-Welke an Tomato und Gurke, Dessertation, Univ. Bonn., W. Germany, 150 p.

Dehne, H. W. 1982, Interaction between vesicular-arbuscular mycorrhizal fungi and plant pathogens, *Phytopath.* 72 : 1115-1119.

Dehne, H. W. and Schoenbeck, F. 1975, The influence of endotrophic mycorrhiza on the *Fusarium* wilt of tomato, *Z. Pflanzenkr. Pflanzenschutz.* 82 : 630.

Dehne, H. W. and Schoenbeck, F. 1979a, The influence of endotrophic mycorrhiza on plant diseases, 1, Colonization of tomato plants by *Fusarium oxysporum* f. sp. *lycopersici. Phytopath., Z.* 95 : 105-110.

Dehne, H. W. and Schonbeck, F. 1979b, The influence of endotrophic mycorrhiza on plant disease, 2, Phenol metabolism and lignification, *Phytopath. Z.* 95 : 210-216.

Dehne, H. W., Schoenbeck, F. and Baltruschat, H. 1978, The influence of endotrophic mycorrhiza on plant diseases, 3, Chitinase activity and the ornithine cycle, *Z. Pflanzenkrankh. Pflanzenschutz.* 85 : 666-678.

Dumas-Gaudot, E., Guillaume, P., Tahiri-Alaoui, A., Gianinazzi-Pearson, V. and Gianinazzi, S. 1994, Changes in poly peptide patterns in tobacco roots colonized by two *Glomus* species, *Mycorrhiza* 4 : 215-221.

Gianinazzi, S. and Gianinazzi - Pearson, V. 1992, Cytology, histochemistry and immunochemistry as tools for studying structure and function in endomycorrhiza, In : *Techniques for the Study of Mycorrhiza (Methods in Microbiology* Vol. 24), Academic Press, London, pp. 109-139.

Gianinazzi, S. and Schuepp, H. (eds.) 1994, *Impact of Arbuscular Mycorrhizae on Sustainable Agriculture and Natural Ecosystems*, Birkhauser, Basel.

Gianinazzi-Pearson, V. and Gianinazzi, S. 1995, Proteins and protein activities in endomycorrhizal symbiosis, In : *Mycorrhiza : Structure, Function, Molecular Biology and Biotechnology*, eds. B. Hock, and A. Varma, Springer, Heidelbert, pp. 251-266.

Gianinazzi-Pearson, V., Franken, P., Gollotte, A., Lherminier, J., Tisserant, B., Lemoine, M. C., Van Tuinen, D. and Gianinazzi, S. 1995, Cellular and molecular approaches in the characterization of symbiotic events in functional arbuscular mycorrhizal associations, *Can. J. Bot.* 73 : S 526-S 532

Graham, J. H. and Egel, D. S. 1988, *Phytophthora* root rot development on mycorrhizal and phosphorus fertilized non-mycorrhizal sweet orange seedlings, *Plant Dis.* 72 : 611-614

Graham, J. H. and Menge, J. A. 1982, Infleunce of vesicular arbuscular mycorrhizas and soil phosphorus on take all disease of wheat, *Phytopath.* 72: 95-98.

Harley, J. L. and Smith, S. E. 1983, *Mycorrhizal Symbiosis*, Academic Press, London

Harrison, M. J. and Dixon, R. A. 1993, Isoflavonoid accumilation and expression of defense gene transcripts during the establishment of vesicular - arbuscular mucorrhizal associations in roots of *Medicago truncatula, Mol. Plant Microbe Interact.* 6 : 643-654.

Hooker, J. E., Jaizme-vega, M. and Atkinson, D. 1994, Biocontrol of plant pathogens using arbuscular mycorrhizal fingi, In : *Impact of Arbuscular Mycorrhizas on Sustainable Agriculture and Natural Ecosystems*, eds. S. Gianinazzi, and H. Schuepp, Birkhauser, Basel, pp. 191-200.

Hwang, S. F., Chakravarty, P. and Prevost, D. 1993, Effect of rhizobia, metalaxyl and VAmycorrhizal fungi on growth, nitrogen fixation and development of *Pythium* root rot of sainfoin, *Plant Dis.* 77 : 1093-1098.

Hwang, S. F. Chang, K. F. and Chakravarty, P. 1992, Effect of vesicular-arbuscular mycorrhizal fingi on the development of *Verticillium* and *Fusarium* wilts of alfalfa, *Plant Dis.* 76 : 239-243.

Iqbal, S. H. and Nasim, G. 1988, VAmycorrhiza as a deterrent to damping off caused by *Rhizoctonia solani* at different temperature regions, *Biol. Paki.* 34 : 215-221.

Iyer, R. and Sundararaju, P. 1993, Interaction of VAmycorrhiza with *Meloidogyne incognita* and *Pythium aphanidermatum* affecting ginger *(Zingiber officinale* R.), *J. Pl. Crop.* 21 : 30-34

Jalali, B. L. 1983, Biological suppression of root disease fungi by vesicular-arbuscular mycorrhizal system, In : Proc. *4th Intern. Congr. Plant Pathol.* Melbourne (Australia), pp. 46.

Jalali, B. L. and Jalali, I. 1991, Mycorrhiza in plant disease control, In : *Handbook of Applied Mycology,* Vol. 1, *Soil and Plants,* eds. D.K. Arora, B. Rai, K. G. Mukerji, and G. R. Knudsen, Marcel Dekker Inc., New York, USA, pp. 131-154.

Jalali, B. L. and Thareja, M. L. 1981, Suppression of fusarial wilt of chickpea in VA mycorrhiza inoclated soils, *Intern. Chickpea Newsl. ICRISAT,* 4 : 21-23.

Jalali, B. L., Chhabra, M. L. and Singh, R. P. 1990, Interaction between vesicular-arbuscular mycorrhizal endophyte and *Macrophomina phaseolina* in mungbean, *Indian Phytopath.* 43 : 527-530

Joseph, P. J. 1998, Management of rhizome rot of ginger *(Zingiber officinale* R.) using VA mycorrhiza and antagonists, Ph.D. thesis, Kerala Agricultural University, Thrissur, Kerala.

Joseph, P. J. and Sivaprasad, P. 1997, Biocontrol potential of arbuscular mycorrhizal isolates against rhizome rot of ginger *(Zingiber officinale* R.), In : *Abstract of the International Conference on Integrated Plant Disease Management for Sustainable Agriculture,* New Delhi, India, p. 251.

Joseph, P. J., Sivaprasad, P., Lekha, K. and Vijayan, M. 1995, Vesicular arbuscular mycorrhizal colonization in ginger and its influence on the natural incidence of rhizome rot, In : *Mycorrhizae : Biofertilizers for the Future, Proc. Third Natl. Conf. on Mycorrhiza,* eds. A. Adholya and S. Singh, Tata Energy Research Institute, New Delhi, pp. 77-80.

Kaye, J. W., Pfleger, F. L. and Stewart, E. L. 1984, Interaction of *Glomus fasciculatum* and *Pythium ultimum* in green house grown poinsettia, *Can. J. Bot.* 62 : 1575-1579.

Katznelson, H., Rouatt, J. W. and Peterson, E. A. 1962, The rhizosphere effect of mycorrhizal and non-mycorrhizal roots of yellow birch seedlings, *Can. J. Bot.* 40 : 379-382.

Khan, S. and Khan, A. G. 1974, Role of VAM fungus in the resistance of wheat roots to infection by *Urocystis tritici, Proc. Pak. Sci. Cong.* 25 : 7.

Kobayashi, N. 1990, Biological control of soil borne diseases with VAM fungi and charcoal compost, In : *Biological Control of Plant Diseases, Proc. Int. Sem. Biol. Control of Plant Diseases and Virus Vectors,* eds. H. Komadan, K. Kiritani and J. Bay-Peterson, Taipei, Taiwan, pp. 151-160.

Kohl, J. and Schlosser, E. 1989, Effect of two *Trichoderma* spp. on the infection of maize roots by vesicular arbuscular mycorrhizae, *Zeitschrift fur pflanzenkrankheiten und pflanzenschutz* 96 : 439-443.

Krishna, K. R. and Bagyaraj, D. J. 1983, Interaction between *Glomus fasciculatum* and *Sclerotium rolfsii* in pea nuts, *Can J. Bot.* 61 : 2349-2351.

Krishna, K. R. and Bagyaraj, D. J. 1984, Phenols in mycorrhizal roots of *Arachis hypogaea, Experientia* 40 : 85-86.

Kumar, C. P. C., Garibova, L. V., Vellikanov, L. L. and Durinina, E. P. 1993, Biocontrol of wheat root rots using mixed cultures of *Trichoderma viride* and *Glomus epigaeus, Indian J. Pl. Prot.* 21 : 145-148.

Lambias, M. R. and Mehdy, M. C. 1993, Suppression of endochitinase, β-1,3 endo glucanase and chalcone isomerase expression in bean vesicular arbuscular mycorrhizal roots under different soil phosphate conditions, *Mol. Plant Microbe Interact.* 6 : 75-83

Lambias, M. R. and Mehdy, M. C. 1995, Differential expression of defense related genes in arbuscular mycorrhiza, *Can. J. Bot.* 73 : 533-540.

Linderman, R. G. 1994, Role of VAM fungi in biocontrol, In : *Mycorrhizae and Plant Health,* eds. F.L. Pfleger and R.G. Linderman, APS Press Minnisota, USA, pp. 1-26.

Marschner, H. and Dell, B. 1994, Nutrient uptake in mycorrhizal symbiosis, *Plant Soil* 159 : 89-102.

Mc Graw, A. C. and Schenck, N. C. 1981, Effects of two species of vesicular-arbuscular mycorrhizal fungi on the development of *Fusarium* wilt of Tomato, *Phytopath.* 71 : 894.

Melo, I. S., Costa, C. P. and Silveira, A. P. D. 1985, Effect of vesicular arbuscular mycorrhizas on aubergine wilt caused by *Verticillium albo-atrum* Reinke and Berth, *Summa Phytopath.* 11 : 173-179.

Meyer, J. R. and Linderman, R. G. 1986, Selective influence on populations of rhizosphere or rhizoplane bacteria and actinomycetes by mycorrhizas formed by *Glomus fasciculatum, Soil Biol. Biochem.* 18 : 191-196.

Morandi, D., Bailey, J. A. and Gianinazzi-Pearson, V. 1984, Isoflavonoid accumilation in soybean roots infected with vesicular - arbuscular mycorrhizal fungi, *Physiol. Pl. Pathol.* 24 : 357-364.

Morton, J. B. and Benny, G. L. 1990. Revised classification of arbuscular mycorrhizal fungi (Zygomycetes) : a new order, Glomales, two new suborders, Glominae and Gigasporinae and two new

151

families, Acaulosporaceae and Gigasporaceae with an emendation of Glomaceae, *Mycotaxon* 37 : 471-491.

Mukerji, K.G. 1999, Mycorrhizae in control of plant pathogens : Molecular Approaches In : *Biotechnological approaches in Biocontrol of Plant Pathogens,* eds. K.G. Mukerji, B.P. Chamola and R.K. Upadhyay, Kluwer Academic / Phenumu Publishers, New York, USA, pp. 135-156.

Mukerji, K.G., Chamola, B.P. and Sharma, M. 1997, Mycorrhiza in Control of Plant Pathogens, In : *Management of Threataning Plant Diseases of National Importance,* eds. V.P. Agnihotri, A.. Sarbhoy and D.V. Singh, Malhotra Publishing House, New Delhi, pp. 297-314.

Nemec, S. 1974, Population of *Endogone* in strawberry fields in relation to root rot infection, *Trans. Br. Mycol. Soc.* 62 : 45.

Paget, D. K. 1975, The effect of *Cylindrocarpon* on plant growth response to vesicular arbuscular mycorrhizae, In : *Endomycorrhizas,* eds. F.E. Sanders, B. Mosse, and P.B. Tinker, Academic Press, London, pp. 593-606.

Parvathi, K., Venkateswarlu, K. and Rao, A. S. 1985, Influence of root infecting fungi on development of *Glomus mosseae* in groundnut, *Curr. Sci.* 54 : 1006-1007.

Paulitz, T. C. and Linderman, R. G. 1991, Lack of antagonism between the biocontrol agent, *Gliocladium virens* and vesicular arbuscular mycorrhizal fungi, *New Phytol.* 117 : 303-308.

Perrin, R. 1991, Mycorrhizes et protection phytosanitaire, In : *Les Mycorrhizes des Arbres et Plantes Cultives,* ed. D.G. Strulla, Lovosier, Paris, France, pp. 93-130.

Ramirez, B. N. 1974, Influence of endomycorrhizae on the relationship of inoculum density of *Phytophthora palmivora* in soil to infection of papaya roots, M.Sc. Thesis, Univ. Florida, Gainsville, USA.

Read, D. J., Francis, R. and Finlay, R. D. 1985, Mycorrhizal mycelia and nutrient cycling in plant communities, In : *Ecological Interactions in Soils, Plants, Microbes and Animals,* ed. A.H. Fitter, British Ecological Society No. 4, Blackwell Scientific Publications, London.

Robert, C. P. 1998, Management of foot-rot of black pepper (*Piper nigrum* L.) with VAmycorrhizae and antagonists, Ph.D. thesis, Kerala Agricultural University, Thrissur, Kerala, India.

Rosendahl, S. 1985, Interactions between vesicular-arbuscular mycorrhizal fungus, *Glomus fasciculatum.* and *Aphanomyces euteiches* root rot of peas, *Phytopath. Z.* 114 : 31-40.

Rosendahl, S., Dodd, J. C. and Walker, C. 1994, Taxonomy and phylogeny of the Glomales, In : *Impact of Arbuscular Mycorrhizas on Sustainable Agriculture and Natural Ecosystems,* eds. S. Gianinazzi and H. Schuepp, Burkhauser, Basel, pp. 1-12.

Rosendahl, S., Rosendahl, C. N. and Thingstrup, I. 1992, The use of vesicular-arbuscular mycorrhizal (VAM) fungi as a biocontrol agent, *Bulletin* OILB/ SROP 15 : 48-50.

Ross, J. P. 1972, Influence of *Endogone* mycorrhiza on *Phytophthora* rot of soybean, *Phytopath.* 62 : 896-897.

Safir, G. 1968, The influence of vesicular-arbuscular mycorrhizae on the resistance of onion to *Pyrenochaeta terrestris,* M.Sc. thesis, University of Illinois, Urbana, USA.

Sarma, Y. R., Anandaraj, M. and Venugopal, M. V. 1996, Biological control of disease of spices, In : *Biological Control on Spices,* eds. M. Anandaraj, and K.V. Peter, Indian Institute of Spices Research (ICAR), Calicut, Kerala, India, pp. 1-19.

Schenck, N. C. 1981, Can mycorrhizal control root disease? *Plant Dis.* 65 : 230-234.

Schenck, N. C. 1987, Vesicular-arbuscular mycorrhizal fungi and the control of fungal root disease, In : *Innovative Approach to Plant Disease Control,* ed. I. Chet, Wiley, New York, USA, pp. 179-191.

Schenck, N. C. and Kellam, M. K. 1978, The influence of vesicular-arbuscular mycorrhizae on disease development, *Fla. Agric. Exp. Stn. Bull.* p. 799.

Schenck, N. C., Ridings, W. H. and Cornell, J. A. 1977, Interaction of two vesicular-arbuscular mycorrhizal fungi and *Phytophthora parasitica* on two citrus root-stocks (Abstr.), In : *Proc. Third North Am. Cong. Mycorrhizae,* Corvallis, Oregon, USA.

Schoenbeck, F. 1979, Endomycorrrhiza in relation to plant diseases, In : *Soil Borne Plant Pathogens,* eds. B. Schippers and Gams, Academic Press, London, pp. 271.

Schoenbeck, F. and Dehne, H. W. 1977, Damage to mycorrhizal and non-mycorrhizal cotton seedlings by *Thielaviopsis basicola, Pl. Dis. Rep.* 61 : 266-268.

Schoenbeck, F. and Dehne, H. W. 1979, Untersuchungen zum Einfluss der endotrophen mycorrhiza auf planzenkrankhiten, 4, Pilzliche sprossparasiten, *Olpidium brassicae,* T.M.V. Z. *Pflanzenkrankh. Pflanzenschutz.* 86 : 103-112.

Secillia, J. and Bagyaraj, D. J. 1987, Bacteria and actinomycetes associated with pot culture of vesicular-arbuscular mycorrhizae, *Can. J. Microbiol.* 83 : 1069-1073.

Selvaraj, T. and Subramanian, G. 1990, Phenols and lipids in mycorrhizal roots of *Sesamum indicum, Curr. Sci.* 59 : 471-473.

Sharma, M. and Mukherji, K. G. 1992, Mycorrhiza - Tool for biocontrol, In : *Recent Developments in*

Biocontrol of Plant Diseases, eds. K. G. Mukerji, J. P. Tewari, D.K. Arora and Geeta Saxena, Aditya Books Private Ltd., New Delhi, India, pp. 52-80.

Sharma, M. and Mukerji, K.G. 1999, VAmycorrhizae in control of fungal pathogens, In : *From Ethnomycology to fungal Biotechnology-Exploiting Fungi from Natural Resources for Novel Products,* eds. J. Singh and K.R. Aneja, Kluwer Academic / Plenum Publishers, New York, USa, pp. 185-196.

Sharma, M., Mittal, N., Kumar, R.N. and Mukerji, K.G. 1998, Fungi : Tool for plant disease management, In : *Microbes for Health, Wealth and Sustainable Environment,* eds. A. Verma, Malhotra Publishing House, New Delhi, pp. 101-154.

Singh, R., Adholeya, A. and Mukerji, K.G. 2000, Mycorrhiza in control of soil-bome pathogens, In : *Mycorrhizal Biology,* eds. K.G. Mukerji, B.P. Chamola and J. Singh, Kluwer Academic/Plenum Publishers, New York, USA, pp. 173-197.

Singh, S. 1998, Interaction of mycorrhizae with soil microflora and microfauna - I, Interaction with soil microflora (except soil microfauna and free living nitrogen fixers), *Mycorrhiza News* 16 : 1-13.

Sivaprasad, P. 1993, Management of root diseases of important spice crops of Kerala with VAmycorrhiza, Annual Report of DBT Project, 1993, Kerala Agricultural University, Thrissur, Kerala, India.

Sivaprasad, P. 1995, Management of root diseases of important spice crops of Kerala with VA-mycorrhiza, DBT Project Report, 1995, Kerala Agricultural University, Thrissur, Kerala, India.

Sivaprasad, P., Jospeh, P. J. and Balakrishnan, S. 1997, Management of foot- rot of black pepper with arbuscular mycorrhizal fungi (AMF), In : *Abstract of the International Conference on Integrated Plant Disease Management for Sustainable Agriculture,* New Delhi, India p. 205.

Sivaprasad, P., Ramesh, B., Mohankumaran, N., Rajmohan, K. and Joseph, P. J. 1995a, Vesicular-arbuscular mycorrhizae for the establishment of tissue culture plantlets, In : *Mycorhizae Biotertilizers for the Future, Proc. Third Nat. Conf. on Mycorrhiza,* eds. A. Adholeya and S. Singh, Tata Energy Research Institute, New Delhi, India, pp. 281-283.

Sivaprasad, P., Robert, C. P., Vijayan, M. and Joseph, P. J. 1995b, Vesicular arbuscular mycorrhizal colonization in relation to 'foot rot' disease intensity in black pepper, In : *Mycorrhizae : Biofertilizers for the Future, Proc. Thrid Mat. Cong. on Mycorhiza,* eds. A. Adholeya, and S. Shingh, Tata Energy Research Institute, New Delhi, pp. 137-146.

Smith, G. S. 1988, The role of phosphorus nutrition in interactions of vesicular mycorrhizal fungi with soil borne nematodes and fungi, *Phytopath.* 78 : 371-374.

St-Arnaud, M., Hamel, C., Caron, M. and Fortin, J. A. 1994, Inhibition of *Pythium ultimum* in roots and growth substrates of mycorrhizal *Tagets patula* colonized with *Glomus intraradices, Can. J. Pl. Pathol.* 16 : 187-194.

Stewart, E. L. and Pfleger, F. L. 1977, Development of poinsettia as influenced by endomycorrhizae, fertilizer and root rot pathogens, *Pythium ultimum* and *Rhizoctonia solani, Flor. Rev.* 159 : 37.

Sylvia, D. M. and Schenck, N. C. 1983, Germination of chlamydospores of three *Glomus* species as affected by soil matric potential and fungal contamination, *Mycologia* 75 : 30-35.

Thomas, L., Mallesha, B. C. and Bagyaraj, D. J. 1994, Biological control of damping off of cardamom by VA mycorrhizal fungus *Glomus fasciculatum, Microbiol. Res.* 149 : 413-417.

Volpin, H., Elkind, Y., Okon, Y. and Kapulnik, Y. 1994, A vesicular arbuscular mycorrhizal fungus *(Glomus intraradices)* induces a defense response in alfalfa roots, *Pl. Physiol.* 104 : 683-689.

Wick, R. L. and Moore, L. D. 1984, Histology of mycorrhizal and non-mycorrhizal *Ilex crenata* "Helleri" challenged by *Thielaviopsis basicola, Can. J. Pl. Pathol.* 6 : 146-150.

Woodhead, S. H. 1978, Mycorrhizal infection of soybean roots as it influences *Phytophthora* root rot, Ph.D. thesis, University of Illinois, Urbana, USA.

Wyss, P., Boller, T. and Wiemken, A. 1991, Phytoalexin response is elicited by a pathogen (*Rhizoctonia solani*) but not by a mycorrhizal fungus (*Glomus mosseae*) in soybean roots, *Experientia* 47 : 395-399.

Wyss, P., Boller, T. and Wiemken, A. 1992, Testing the effect of biological control agents on the formation of vesicular arbuscular mycorrhiza, *Plant Soil* 147 : 159-162.

Zambolin, L. and Schenck, N. C. 1981, Interactions between a vesicular arbuscular mycorrhiza and root infecting fungi on soybean, *Phytopath.* 71 : 267.

Zambolin, L. and Schenck, N. C. 1983, Reduction of the effect of pathogenic root infecting fungi on soybean by the mycorrhizal fungus, *Glomus mosseae, Phytopath.* 73 : 1402-1405.

FUNGAL ANTAGONISTS OF PHYTONEMATODES

R.K. Walia and R. Vats

Department of Nematology
CCS Haryana Agricultural University
Hisar - 125 004, Haryana, INDIA

1. INTRODUCTION

The devastating ecological considerations inherent in the use of chemical nematicides has prompted research efforts to look for alternative pest management strategies, and much emphasis has been diverted to the exploitation of biological control agents. The soil ecosystem, of which phytonematodes constitute one of the components, is a biologically dynamic system involving a galaxy of microorganisms, interacting among themselves in all possible ways. Identifying potential antagonists of phytonematodes and harnessing their biological attributes in the management of phytonematode populations is a challenging job. Several antagonists of phytonematodes have been identified and these belong to diverse groups i.e., fungi, bacteria, viruses (parasites/ pathogens), mites, predatory nematodes, enchytreids, colleraboles, etc. (predators). Of these, fungal antagonists have received greater attention.

Incidentally, the earliest records of antagonists of phytonematodes pertained to fungi (Kühn, 1877; Lohde, 1874; Zopf, 1888). Fungal antagonism to nematodes has been classified in several ways, but based upon their mode of action , these are generally grouped into - (i) Nematode trapping fungi, (ii) Endozoic fungi, (iii) Parasitic or Opportunistic fungi and (iv) Fungi which are neither parasitic nor predacious, but influence the nematodes indirectly by their metabolites in the rhizosphere. Paradoxically, the research trend involving fungal antagonists of nematodes have followed more or less the same order. The omnipresent nematode trapping fungi were the subject of investigations in the earlier phase of nematode biocontrol. Much information gathered on the nematode capturing devices (adhesive hyphae, adhesive knobs, adhesive network, constricting rings, non constricting rings, etc.), their ecology and identification, culminated in a monograph prepared by Barron in 1977. After a lag period, interest in nematode fungal antagonists was once again surged with the discovery of *Paecilomyces lilacinus* by Jatala in 1978. Numerous reviews published since then (Jatala, 1986; Kerry, 1980, 1987; Mittal *et. al.*, 1999; Morgan-Jones and Rodriguez-Kabana, 1984; Sayre, 1980; Stirling, 1992) reflect the intensity of research activity involving fungal antagonists in general, and parasitic fungi in particular.

155

Tremendous research efforts in the recent years have taken the fungal antagonists very close to their commercial exploitation. The much desired basic information is already available in the valuable reviews cited above. The aim of this article is to short-list the promising parasitic/ predacious fungi of phytonematodes and discuss their prospects for development as biocontrol agents.

2. PARASITIC OR OPPORTUNISTIC FUNGI

2.1. *Paecilomyces lilacinus*

2.1.1. Geographical distribution and nematode hosts

P. lilacinus (Thom) Samson is a common soil hyphomycete with a cosmopolitan distribution, especially in the warmer regions of the world (Samson, 1974), but as a nematode destroying fungus it was reported from Peru parasiting eggs of *Meloidogyne incognita* and *Globodera pallida* (Jatala *et al.*, 1979). It also parasitises eggs of *Tylenchulus, Nacobbus* (Jatala, 1985), adults of *M. incognita* and *G. pallida* (Jatala *et al.*, 1979); reduces egg viability in *Rotylenchulus reniformis* (Reddy and Khan, 1988; Walters and Barker, 1994), and populations of *Radopholus similis* (Generalo and Davide, 1995) and *Pratylenchus* sp. (Gapasin, 1995).

2.1.2. Demonstration of efficacy

Subsequent to the first report on the efficacy of *P. lilacinus* in reducing *M. incognita* populations on potato (Jatala *et al.*, 1979), several workers have recorded reduction in nematode populations, root infection and eggmass production with consequent enhanced plant growth following *P. lilacinus* application in green house/microplot tests (Cabanillas and Barker, 1989; Davide and Zorilla, 1983, 1995a,b; Dube and Smart, 1987; Ekanayake and Jayasundara, 1994; Franco *et al.*, 1981; Jatala *et al.*, 1980; Khan and Saxena, 1996; Rao *et al.*, 1998; Siddiqui and Mahmood, 1992; Strattner, 1979; Trivedi, 1990; Villanueva and Davide, 1984a; Wang *et al.*, 1997;) and field trials (Guevara *et al.*, 1985; Herrera *et al.*, 1984, 1985; Jatala *et al.*, 1980, 1981; Salas and Jatala, 1980) especially against *T. semipenetrans*. Walters and Barker (1994) reported that numbers of *R. reniformis* were reduced by 46-48% in a green house test but were suppressed by 59 and 36% at mid season and harvest, respectively in microplots infested with *P. lilacinus*, preventing impairment of shoot and fruit growth of tomato. In field trials against *M. incognita* the yields of okra and tomato were higher in plots treated with *P. lilacinus* than in untreated plots, and the differences between treatments widened at each harvest data. The fungus provided the same level of nematode suppression as the nematicide (Noe and Sasser, 1995).

2.1.3. Mode of action

The suppression of sedentary endoparasitic nematode populations by *P. lilacinus* is attributed to (i) parasitisation of adult females (Jatala *et al.*, 1979), (ii) disintegration of embryos (Fitters and Belder, 1993; Jatala *et al*, 1985) and (iii) inhibition of hatching (Zaki, 1994). The infection process begins with the growth of fungal hyphae in the gelatinous matrix; eventually the eggs of *Meloidogyne, Tylenchulus* or *Nacobbus* species are engulfed by the mycelial network, which become prostate and spiral over the smooth egg surface. In cysts of *Globodera* species the fungus penetrates the vulva or the broken and exposed neck region. After entering the cysts, the fungus grows saprophytically on the mucilagenous body content surrounding the eggs during or before its parasitism of the eggs (Jatala, 1986). However, colonisation appears to occur by simple penetration of the egg cuticle by individual hyphae aided by mechanical and/or enzymatic activities (such as chitinase). Dunn *et al.* (1982) reported that colonisation of the eggs occasionally seemed to be associated with a specialised structure, possibly an appresorium.

2.1.4. Fungal density v/s nematode control

Adiko (1984) obtained 42 and 70% suppression of *M. incognita* on tomato with 20 or 80 g of fungus-infested rice kernels per 15 cm d pot, respectively. Cabanillas and Barker (1989) reported that the best protection against *M. incognita* was attained with 10 and 20 g of fungus-infested wheat kernels per microplot which resulted in a three fold and four fold increase in tomato yield respectively, compared with control. Khan *et al.*(1992) used spores of *P. lilacinus* in polyvenyl alcohol and gelatin with vegetable charcoal mixtures separately for dressing seeds of *Luffa aegyptica* in pot experiments. The medium and high doses of *P. lilacinus* (50 and 75 mg spores, respectively) significantly reduced root galling by *M. incognita acrita*. Walia and Bansal (1992) found that a minimum of 1×10^4 spores were required to initiate significant suppression of nematode populations. Zaki (1994) cultured *P. lilacinus* on gram seed and found that 4 g of fungus coated seed per kg soil was optimum for the effective reduction in the galling (69%) and J-2 (86%) of *M. javanica* in tomato with an optimum eggmass infection (58%) and egg destruction (66%). Eguiguren-Carrion (1995) determined that under green house conditions, the LD 50 for *M. incognita* and *N. aberrans* was 45.3 and 55.3 mg colonised rice per 500 cc soil, respectively, for a populatiojn density of 25 J-2 per cc soil.

2.1.5. Method and time of fungus introduction

Jatala (1986) advocated the use of wheat or rice grains for mass multiplication and application of *P. lilacinus*. Greatest protection against *M. incognita* on tomato was attained when *P. lilacinus* was delivered into soil 10 days before planting and again at planting. *P. lilacinus* infected eggmasses were maximum in plots treated at mid season or at mid season plus an early application, compared with plots treated with the fungus 10 days before planting and /or at planting time (Cabanillas and Barker, 1989). Since *P. lilacinus* is an oviparasite, the stage of embryonic development is probably important when timing the delivery of the fungus into the soil. In a field infested with *M. incognita* and *R. reniformis* when *P. lilacinus* was added either one week before planting or at planting of *Capsicum annum* , significantly more and heavier fruits were obtained in fungus (1 week before planting) treated plots than from the check (Vincente and Acosta, 1992). Zaki and Maqbool (1991) also found that *P. lilacinus* was most effective when applied one week before *M. javanica* inoculation. Rao and Reddy (1994) indicated that neem cake extracts (5 and 10%) are very useful for carrying *P. lilacinus* spores. This method of application of spores facilitates rational integration of biocontrol agent and botanical leading to the exploitation of beneficial effects of both components in the management of *M. incognita*. Results also indicated that the neem cake extracts support the growth of *P. lilacinus*. Similarly soil application of *P. lilacinus* at 3-5% (w/w) on a neem cake base was useful in controlling *M. javanica* on groundnut (Patel *et al.*, 1995). Bare root dip treatment of eggplant seedlings in neem based formulations of *P. lilacinus* for 20 mins gave significant increase in plant growth and reduction in root galling, number of eggs per eggmass and final nematode population (soil + root) of *M. incognita* (Rao *et al.*, 1998). Davide and Zorilla (1995 a) evaluated the efficacy of *P. lilacinus* when applied as tuber dip, soil mix and combination of both on potato against *G. rostochiensis*. All the methods of application resulted in increase in yield but their efficacy varied. Tuber dip method of *P. lilacinus* application for 10 min in spore and mycelial suspension before planting, generally gave better control of *G. rostochiensis* than the soil mix method where the fungus in rice hull substrates was mixed with organic fertilizers, chicken manure and soil at planting (Davide and Zorilla, 1995b). Use of substrates of *ipil-ipil*, rice hulls as well as banana leaves and bracks improved effectiveness of *P. lilacinus* against *Radopholus similis* on banana as compared to the direct use of spores and mycelial suspension applied as soil drench (Generalo and Zorilla, 1995). These substrates supported abundant growth of fungi.

2.1.6. Mass multiplication

For experimental field applications, Jatala (1983) has suggested that the fungus should be cultured on cereal grains. Potato dextrose agar (broth) has been used for culturing *P. lilacinus* for *in vitro* experiments (Bansal *et al.*, 1988; Cabanillas *et al.*, 1989a; Dunn *et al.*, 1982). Wheat, rice grains and gram seeds have been used for mass culturing of the fungus for green house and field experimental applications (Amoncho and Sasser, 1995; Cabanillas and Barker, 1989; Jatala, 1986; Jonathan *et al.*, 1995; Rana and Dalal, 1995; Zaki and Bhatti, 1990). Culturing the fungus on cereal grains for field application requires large amount of grains which is not an economic method. Some workers have tried to culture *P. lilacinus* on agro-industrial waste materials (Bansal *et al.*, 1988), neem leaves (Zaki and Bhatti, 1988), saw dust and cotton seed shell (Wang *et al.*, 1997) with promising results. By and large delivery of *P.lilacinus* into soil has been limited to traditional methods such as infested grains or in aqueous spore suspensions. Commercial production of a material of this fungus named BIOCON has been initiated in Philippines (Jatala, 1986).

2.1.7. Population dynamics and establishment of fungal spores in soil

Villanueva and Davide (1984 b) found that different isolates of *P. lilacinus* had varying growth and sporulation responses to different pH and temperature levels as well as to different light conditions . Bansal *et al.* (1989a) concluded that *P. lilacinus* survives and resumes normal growth after continuous exposure to 40°C for 10 days, 45°C for 6 days and 50°C for 1 day. Further, *P. lilacinus* could grow and sporulate best at initial pH 5 (in buffered medium), and its growth was significantly suppressed as the initial pH was increased (Bansal *et al.*, 1989b). Cabanillas *et al.* (1989b) investigated the potential of 13 *P. lilacinus* isolates from various geographic regions against *M. incognita*. Maximum fungal growth as determined by dry weight of the mycelium occurred from 24 - 30°C; least growth was at 12 and 36°C. As soil temperature increased from 16 to 28° C , both root-knot damage caused by *M. incognita* and percentage of eggmasses infected by *P. lilacinus* increased. The greatest residual *P. lilacinus* activity on *M. incognita* was obtained with a mixture of fungal isolates. In a study involving influence of carrier and storage of *P.lilacinus* on its survival and related protection of tomato against *M. incognita* , Cabanillas *et al.* (1989 a) reported that fungal viability was high in wheat and diatomaceous earth granules, intermediate in alginate pellets and low in soil and chitin amended soil stored at 25 ± 2° C. In field microplots *P. lilacinus* resulted in about a 25%increase in tomato yield and 25% gall suppression compared with nematodes alone in first year. Greatest suppression of egg development occurred in plots treated with *P. lilacinus* in pellets, wheat grain and granules. In the second year, carry over protection of tomato against *M. incognita* resulted in about a 3 times increase in tomato fruit yield and 25% suppression of gall development, compared with control. Higher numbers of fungus -infected eggmasses occurred in plots treated with pellets (32%) than in those treated with chitin amended soil (24%), wheat (16%), granules (12%) or soil (7%). Number of CFU per g soil in plots treated with pellets were 10 times greater than initial levels estimated at planting time in the second year.

158

2.1.8. Compatibility with other management practices

In a field microplot study, *P. lilacinus* and *Pasteuria penetrans* applied singly or in combination reduced population densities of *M. javanica* and increased tomato yield by 13-24%, 2-5% and 8% respectively (Dube, 1989). Siddiqui and Mahmood (1993) found that the combined inoculation of *P. lilacinus* and *Bacillus subtilis* improved dry shoot weight of chickpea significantly when plants were simultaneously inoculated either with *M. incognita* or *Macrophomina phaseolina* or with both. *P.lilacinus* was found highly effective when used in combination with green manuring of maize and *Sesbania aculeata* for the management of *R. reniformis* on pigeon pea (Mahmood and Siddiqui, 1993). The use of *P. lilacinus* with *Bradyrhizobium japonicum* significantly reduced *M. javanica* root-knot indices on chickpea and increased plant growth in microplot experiments (Ehteshamul-Haque *et al.*, 1984). Al-Raddad (1995) noted that inoculation of tomato plants with *P. lilacinus* and *Glomus mosseae* together or separately resulted in similar shoot and plant heights. The highest root development was achieved when mycorrhizal plants were inoculated with *P. lilacinus* to control *M. javanica*. Mittal *et al.* (1995) observed that combination of *P. lilacinus* with chitin enhanced suppression of *M. incognita* more than using them alone in *Solanum melongena*, *Lycopersicon esculentum* and *Cicer arietinum*. *P.lilacinus* at 1 or 2 g per kg soil together with seed treatments with carbosulfon at 0.05 % (w/w) were applied to *Vigna radiata* for control of *Heterodera cajani* in pot trials. All treatments receiving combined applications of nematicide and fungus had significantly lower *H. cajani* populations and significantly higher growth and yield compared to control (Rana and Dalal, 1995). However, Geraldo *et al.* (1996) reported that *P.lilacinus* in combination with phenamiphos did not give additional control of *Meloidogyne* sp. in *Luffa* plants as compared to either fungus or chemical alone. Khan *et al.* (1997) indicated that combined application of *P. lilacinus* and *Trichoderma harzianum* resulted in significantly more reduction in damage caused by *M. incognita* + *Fusarium solani* disease complex and increased plant growth of papaya compared to individual application of either antagonist.

2.2. *Verticillium chlamydosporium*

2.2.1. Geographical distribution and nematode hosts

V.chlamydosporium Goddard is a soil fungus of world wide distribution (Barron and Onions, 1966; Domsch and Gams, 1972; Gams, 1971). Its hosts include cyst nematodes (*Heterodera* spp. and *Globodera* spp.) and root-knot nematodes (*Meloidogyne* spp.) (Crump, 1991; de Leij *et al.*, 1993; Freire and Bridge, 1985; Kerry, 1975; Kerry *et al.*, 1995; Morgan-Jones *et al.*, 1981; Saifullah and Thomas, 1997; Willcox and Tribe, 1974). It was found to be the principal egg pathogen of *Heterodera schachtii* (Bursnall and Tribe, 1974) and *H. avenae* (Kerry, 1975). It is present in soybean growing areas of USA and *H. avenae* infested regions in Australia (Stirling and Kerry, 1983) and India (Bhardwaj and Trivedi, 1996). An isolate of the fungus which was able to colonise the rhizoplane of tomato, especially sites where galls formed, was an effective biological control agent for *Meloidogyne* spp. (de Leij and Kerry, 1990, 1991).

2.2.2. Demonstration of efficacy

Suppression in nematode population and multiplication has been reported following applications of *V. chlamydosporium* in several green house and field tests (de Leij and Kerry, 1991; Freire and Bridge, 1985; Kerry *et al.*, 1995; Morgan-Jones *et al.*, 1981; Owino *et al.*, 1993; Siddiqui and Mahmood, 1996b; Willcox and Tribe, 1974). In a green house test *V. chlamydosporium* gave a 75% control of the first generation of *Heterodera schachtii*. Control

159

was also shown in field trials (Crump, 1991). De Leij *et al.* (1993) reported that in a microplot experiment on sandy loam soil *V. chlamydosporium* controlled populations of *M. hapla* on tomato by more than 90%. Bhardwaj and Trivedi (1996) reported significant reduction in *H. avenae* populations and increase in wheat plant growth in *V. chlamydosporium* treated pots.

2.2.3. Mode of action

V.chlamydosporium is a facultative parasite. All stages of the fungus (hyphae, conidia, chlamydospores) occur in soil and actively growing mycelium infects eggs and females (Davies *et al.*, 1991). The fungus forms close contact with the egg shell and produces specialised perforations causing disintegration of egg shell, vitelline layer and partial dissolution of chitin and lipid layer (Lopez-Llorca and Duncan, 1988; Saifullah and Thomas, 1997; Stirling, 1991). Egg hatching in the presence of fungus was inhibited probably due to toxins secreted by the fungus (Meyer *et al.*, 1990; Morgan-Jones *et al.*, 1983). Generally young white females of cyst nematode are not attacked but those in which eggs have been formed become susceptible to invasion (Bursnall and Tribe, 1974).

2.2.4. Fungal density v/s nematode control

Using different application rates of *V. chlamydosporium* and *M. incognita,* de Leij *et al.* (1992 b) concluded that colonisation of the rhizoplane is dependent on initial fungal inoculum and on galling caused by *M. incognita*; higher fungal and nematode inoculum levels resulted in greater colonisation of the roots. Although galled roots were most extensively colonised by *V. chlamydosporium*, the fungus was least effective in controlling *M. incognita* at high nematode densities, presumably because many eggmasses stayed embedded in the gall tissue and were, therefore, protected from fungal attack. Bhardwaj and Trivedi (1996) used wheat bran multiplied fungus in three dosages (4, 8 and 12 g /kg soil/pot). All the three doses reduced the *H. avenae* population and increased plant growth as compared to control. The rate of 8 g/kg soil significantly reduced cyst number as well as improved the growth of wheat.

2.2.5. Method and time of fungus introduction

The best control of *M. javanica* was obtained by inoculating the soil with *V. chlamydosporium* colonised on rice medium at the rate of 30 g /kg soil. The reduction in number of galls and mature females to the tune of 50 and 84.3%, respectively was observed as compared to control. Introducing the fungus two weeks before the nematode inoculation significantly increased the control of *M. javanica* (Mousa *et al.*, 1995).

2.2.6. Population dynamics and establishment of fungal propagules in soil

The establishment of *V. chlamydosporium* in soil is dependent on the initial fungal inoculum rate used. Increase in numbers of CFU in soil were greater from a small initial inoculum than from a large one, thus suggesting that *V. chlamydosporium* depends partly on external nutrients in soil for its establishment (de Leij *et al.*, 1992 b). *V. chlamydosporium* multiplied in peaty sand from inoculum rates of 500, 1000, 5000 or 10000 chlamydospores/g soil to a maximum of 5.5 x 10^4 CFU/g soil and survived for at least 8 weeks in pots planted with tomato plants inoculated with 1000 J-2 of *M. incognita*. The fungus survived but did not multiply in loamy sand or sand. Establishment of *V. chlamydosporium* on the rhizoplane of tomato was greater in peaty sand than in loamy sand or sand. Nematode control was, in general, greater in peaty sand than in other two soil types. In a microplot experiment on sandy loam, *V. chlamydosporium* controlled populations of *M. hapla* on tomato by more than 90%. The fungus multiplied and

160

survived in soil for at least 123 days, mostly in rhizospheric soil. Kerry *et al.* (1993) estimated that the numbers of CFU in soils in which *Heterodera* spp. multiplication was suppressed, was greater than in those in which the nematode multiplied. Some isolates of *V. chlamydosporium* proliferated in soil and survived in considerable numbers for at least 3 months. Temperatures above 25°C reduced the proportion of eggs of root-knot nematodes parasitized by *V. chlamydosporium*. At this temperature, many of the eggs in the eggmasses would have completed their embryonic development and hatched before they could be colonised by the fungus (de Leij *et al.*, 1992a).

2.2.7. Compatibility with other management practices

Combining *V. chlamydosporium* with an aldicarb treatment equivalent to 2.8 kg a.i./ha did not affect the activity of the fungus, and gave better control (98%) of *M. hapla* than a treatment with *V. chlamydosporium* or aldicarb alone (90%) (de Leij *et al.*, 1993). Use of organic amendments viz., castor, marigold, and neem stimulated *M. javanica* egg parasitisation by *V. chlamydosporium*, while benomyl and mustard significantly inhibited it (Owino *et al.*, 1993). Parasitisation of *Heterodera cajani* by *V. chlamydosporium* on pigeonpea was reduced in the presence of *Trichoderma harzianum*. However, the highest reduction in nematode multiplication was observed when *V. chlamydosporium, T. harzianum* and *Glomus mosseae* were used together (Siddiqui and Mahmood, 1996 b). Kerry and Bourne (1996) suggested that *V. chlamydosporium* in conjunction with other methods, such as the rotation of the poor host, may provide adequate control but the reliability of such approaches needs extensive testing.

2.3. *Hirsutella rhossiliensis*

H. rhossiliensis Minter and Brady parasitises and kills nematodes *in vitro* (Cayrol and Frankowski, 1986; Jaffee and Zehr, 1983; Sturhan and Schneider, 1980; Timper and Kaya, 1989), suppresses nematode numbers in green house tests (Eayre *et al.*, 1987; Muller, 1985), and parasitises high numbers of nematodes in growers fields (Jaffee *et al.*, 1985; Lackey *et al.*, 1993; Muller, 1985). It produces spores that adhere to and penetrate the nematode cuticle and assimilate the body contents prior to the emergence and sporulation (Jaffee and Zehr, 1985). Penetration of cabbage roots by *Heterodera schachtii* was suppressed 50-77% in loamy sand naturally infested with *H. rhossiliensis*, when *H. schachtii* was incubated in the suppressive soil without plants for 2 days, 40-63% of the juveniles had *H. rhossiliensis* spores adhering to their cuticles. Of those with spores, 82-92% were infected. Infected nematodes were killed and filled with hyphae within 2-3 days (Jaffee and Muldoon, 1989). The palletised hyphae of *H. rhossiliensis* suppressed nematodes/cm of root averaged 42 or 98% in soil infested with eggmasses or juveniles of *M. javanica* infecting tomato, and 83 or 98% in soil infested with cysts or juveniles of *H. schachtii* infecting cabbage. *H. rhossiliensis* did not suppress *M.javanica* in eggmass-infested loamy sand when nematode density as determined by bioassay exceeded 1200/100 cc of soil but suppressed *H. schachtii* in cyst-infested loamy sand and loam regardless of nematode density. The fungus was always more effective in soil infested with juveniles than in soil infested with eggmasses or cysts (Lackey *et al.*, 1994). In a green house test, soil inoculation with mycelial colonies of *H. rhossiliensis* with a potential spore production of 25 x10^5/g soil resulted in a suppression of *Globodera pallida* penetration in potato roots by 37 and 51% after 5 and 6 weeks, respectively (Velvis and Kamp, 1996). Field observations, theory, and experiments demonstrated that, (i) parasitisation of nematodes by *H. rhossiliensis* is dependent on nematode density, (ii) local populations of the fungus will go extinct unless supplied with some minimum number of nematodes (the host threshold density) and (iii) natural epidemics of this fungus in populations of nematodes develop slowly and only after long periods of host density (Jaffee, 1992).

2.4. *Dactylella oviparasitica*

D. oviparasitica Stirling and Mankau was first isolated from *Meloidogyne* eggmasses collected from peach orchards in California (Stirling and Mankau, 1978a). Laboratory studies indicated that the fungus was parasitic on *Meloidogyne* eggs and suggested that it contributed to the biological control of root-knot nematodes in the field (Stirling and Mankau, 1978b). Stirling and Mankau (1978a) described method for its isolation and culturing. *D. oviparasitica* exists as different isolates which vary in their pathogenicity and other aspects of biology (Stirling and Mankau, 1979). The fungus grew rapidly between 24-27°C but more eggs were parasitised at lower temperatures; eggs were destroyed in about 9 days at 27°C and 30 days at 15°C (Stirling, 1979). Hyphae of *D. oviparasitica* proliferated rapidly through *Meloidogyne* eggmasses, and appresoria formed when they contacted eggs. The fungus probably penetrated egg shells mechanically, although chitinase production detected in culture suggested that enzymatic penetration was also possible. In soil, *D. oviparasitica* invaded eggmasses soon after they were deposited on the root surface and eventually parasitised most of the first eggs laid. Occasionally the fungus parasitised *Meloidogyne* females, halting egg production prematurely. The fungus parasitised eggs in the gelatinous matrix or eggs freed from the matrix and placed on agar or in soil. *D. oviparasitica* parasitised eggs of four *Meloidogyne* spp., *Acrobeloides* spp., *Heterodera schachtii* and *Tylenchulus semipenetrans*. In tests in a growth chamber, parasitism by *D. oviparasitica* suppressed galling on *M. incognita* infected tomato plants (Stirling and Mankau, 1979). *D. oviparasita* occurred in 20-60% of eggs throughout the year in orchard and as infected eggs were rapidly destroyed, parasitism was probably underestimated. *Meloidogyne* produced fewer eggs on peach and more were parasitised than on tomatoes or vines. It was concluded that on peach, parasitism by *D. oviparasitica* effectively limited *Meloidogyne* multiplication but was less effective with better host plants (Stirling *et al.*, 1979). In the field, where conditions are not always ideal for the nematodes, and *Meloidogyne* populations need only be kept below economic thresholds, parasitism by *D. oviparasitica* may reduce *Meloidogyne* populations sufficiently to be of some economic benefit.

2.5. *Cylindrocarpon destructans*

First reported as a parasite of *Heterodera avenae* by Goffart (1932) from Germany, *C. destructans* is found at rather low frequency. It has also been found in cysts of *H. schachtii* (Bursnall and Tribe, 1974; Crump, 1987; Kerry and Crump, 1977; Tribe, 1979), eggs (Morgan- Jones and Rodriguez-Kabana, 1986) and females (Crump and Flynn, 1992) of *G. rostochiensis* and *G. pallida*. Infected eggs become filled with mycelium and are not distinctive, though often slightly brownish in colour (Tribe, 1977). Jain (1979) found that the fungus was very effective in destroying all eggs, larvae and cysts of *H. avenae* in a short time (Goffart, 1932). Under field conditions against *H. avenae* a single application of *C. destructans* causes 81% reduction in the number of nematode infected plants and two applications caused 93% reduction (Kondakova and Tikhonova, 1981).

3. ENDOZOIC FUNGI

3.1. *Nematophthora gynophila*

Initially identified as an *Entomophthora*-like fungus, *N. gynophila* Kerry and Crump is an obligate zoospore-forming oomycete, and was found in 97% fields in southern England where it attacks females of several species of cyst nematodes, viz., *H. goettingiana, H. trifolii, H. carotae, H. schachtii, H. cruciferae* (Kerry and Crump, 1977), and prevents cyst formation(Kerry and Crump, 1980). The fungus is widespread in soils infested with *H. avenae* and can prevent nematode multiplication on susceptible cereal crops (Kerry and Crump, 1977; Kerry *et al.*, 1980). It parasitises *H. glycines* in USA (Crump *et al.*, 1983) and Germany (Knuth and Dietermann, 1983). In India, a *Nematophthora* sp. was isolated from females of *H. avenae, H. cajani, H. graminis, H. mothi, H. sorghi* and *H. zeae*. It destroyed the nematode females but the eggs and larvae already developed prior to infection were not parasitised by the fungus (Sharma and Swarup, 1988). *N. gynophila* parasitises both the eggs and adults of *H. avenae* (Sayre, 1986). The cuticle of parasitised females is destroyed within about 7 days at 13° C leaving a mass of resting spores which are readily dispersed in the ·soil (Crump and Kerry, 1977). The fungal ability to preferentially attack and destroy adult females exerted great pressure on nematode populations by virtually destroying their reproductive capacity. The emerging adult females on cereal roots were parasitised by the motile zoospores of *N. gynophila* and were effectively controlled. The zoospores were able to move readily in wet than dry soils. Use of formalin killed spores of *N. gynophila* and hence nematode populations increased considerably (Kerry *et al.*, 1980). *N. gynophila* infected more females of *H. schachtii* at 10° C and 15° C than at 20° C (Crump and Kerry, 1982).

N.gynophila produces an extensive branching mycelium which becomes septate and often disarticulates into short segments. These segments of vegetative hyphae may give rise to long discharge tubes and form sporangia which penetrate the disrupted nematode cuticle and protrude into the soil. The cytoplasm within sporangia divides into biflagellate zoospores which at maturity are released rapidly after the tip of the discharge tube ruptures. After a short active period (3-60 min) they encyst, eventually producing either a germ tube or a second motile stage. Each nematode can give rise to 200 zoosporangia and each sporangium may produce upto 120 zoospores. The remaining hyphae give rise to resting spores which are produced laterally, ·one spore per hyphal segment. The infected females which are eventually filled with about 3000 thick-walled resting spores (Kerry, 1980), are fragile and difficult to extract, thus the kill is underestimated (Kerry *et al.*, 1982).

Being an obligate parasite, *N. gynophila* cannot be increased saprophytically in soil and warrants inundative treatments if significant nematode control is to be achieved. However, resting spores of this fungus can be extracted quantitatively from soil (Crump and Kerry, 1981). Though *N. gynophila* has been cultured axenically (Graf and Madelin, 1989), its growth is extremely slow, hyphae are fragile and colonies are short-lived. Mass culturing of the fungus is difficult (Stirling, 1991). This may restrict its use to transplanted crops in which only small volumes of soil would require treatment. The resting spores of *N. gynophila* can survive in soil for at least 5 years in the absence of nematode females (Kerry, 1984) and could be used to control cyst nematodes on crops which demand a rotation.

3.2. *Catenaria* spp.

Catenaria auxiliaris (Kuhn) Tribe was first described in 1877 by Kuhn in the females of *H. schachtii*. He named it as *Tarichium auxiliare*. A century later the fungus was redescribed as *C. auxiliaris* after it was found to produce posteriorly uniflagellate zoospores (Tribe, 1977 b).The taxonomy and biology of *C. auxiliaris* have been studied in detail (Kerry, 1979, 1980). *C. auxiliaris* has commonly been found associated with populations of *Heterodera* species in UK, Australia, USA and India (Crump *et al.*, 1983; Kerry and Mullen, 1981; Sharma and Swarup, 1988; Stirling, 1980). Zoospores of *C. auxiliaris* infect females of *H. schachtii* in 3-4 days. Females become infected after they rupture the root cortex and are exposed to soil. The resting sporangia are more frequently recorded and each develops internally a single yellow-brown resting spore (Sharma and Swarup, 1988; Tribe, 1977b). In Britain about 15% of soils infested with *H. avenae* contain *C. auxiliaris*, but the fungus is more widespread in fields infested with *H. schachtii* (Tribe, 1977b). The fungus caused more than 70% reduction in the

163

eggs and larvae production in the infected females of *H. cajani* and *H. zeae* in India (Sharma and Swarup, 1988). The fungus has not been cultured *in vitro* and, therefore, infected nematodes must be washed or picked directly from roots (Kerry, 1974). Although the fungus grows on mucilage formed from degenerating nematode tissue but has not been cultured axenically (Sharma and Swarup, 1988; Stirling, 1991).

Catenaria vermicola is an endoparasite on different life cycle stages of cyst nematodes. Sharma and Swarup (1988) reported *C. vermicola* parasitising cysts of *H. avenae, H. cajani, H. zeae, H. graminis, H. mothi* and *H. sorghi* in India. The white females /cysts infected with *C. vermicola* did not turn brown and their entire content was replaced by zoosporangia with discharge tubes breaking through the cyst cuticle. Dhawan (1985) showed that *C. vermicola* infected *H. avenae* juveniles *in vitro* and completed its life cycle in 2-3 days. The fungus also infected juveniles of *H. sorghi* and *H. cajani*. The resting sporangia observed in juveniles and zoosporangia in eggs, and the polyphagous nature of the fungus indicated that it is well adapted for survival within cysts. The fungus can be cultured on artificial media (Sharma and Swarup, 1988).

Catenaria anguillulae Sorokin, a widespread nematophagous endoparasite (Barron, 1977), is a weak parasite. Boosalis and Mankau (1965) stated that it mainly colonises dead and injured nematodes. However, pathogenicity studies added 13 genera and 9 species of nematodes to the host range of the fungus (Esser and Ridings, 1973). The influence of temperature, pH, salt concentration, and nature of substrate on the incidence of *C. anguillulae* has been studied well (Sayre and Keeley, 1969). The antagonistic potential of different isolates of *C. anguillulae* varied against *H. schachtii* (Voss *et al.*, 1992). The fungus can be cultured easily on agar medium (Sayre and Keeley, 1969).

4.   TRAPPING FUNGI

4.1. *Arthrobotrys* spp.

Species of *Arthrobotrys* vary in their growth and ability to form traps at different pH but *A. oligospora* was little affected by pH 4.9-8.1 (Mankau, 1964). *A. irregularis* was less effective against *Meloidogyne* sp. on tomatoes in soils below pH 6.5. It was also suggested that *A. irregularis* should be added 4 weeks before planting to ensure its establishment in soil (Cayrol and Frankowski, 1979). It can survive temperature extremes of 18° C for 8 days and 40° C in dry conditions for only 1 day (Cayrol and Frankowski, 1980). *A. irregularis* reduced damage to tomatoes when applied on rye grains at 140g/sq. m. (1.4 t/ha) to soil infested with *Meloidogyne* (Cayrol and Frankowski, 1979). This fungus is commercially available as Royal 350 in France. *Arthrobotrys conoides* reduced damage to the mushroom (*Agaricus bisporus*) caused by nematodes, particularly *Ditylenchus myceliophagus*, when introduced into the compost on rye grain at a rate of 1 % (w/w) with the mushroom spawn (Cayrol *et al.*, 1978). It grew rapidly in the compost and unlike some other trapping species did not inhibit the growth of mushroom. It was developed as Royal 300 for commercial use against nematode problem of mushroom in France. In one trial , yields were increased by 20% and nematode populations reduced by 40% but many nematodes remained in the compost and a second trial gave a smaller response (Al Hazmi *et al.*, 1982a). Soil temperature influenced the efficacy of the fungus and although applications on vermiculite equivalent to 300 m3 /ha reduced number of eggs, there were many nematodes left at the end of experiment (Al Hazmi *et al.*, 1982b).

Reproduction of *M. incognita* and the incidence of root galling on corn were reduced by addition of *A. conoides* and/or green alfalfa. Numbers of juveniles were reduced by as much as 84%, and eggs were fewest in early to mid season soil samples from microplots (Al Hazmi *et al.*, 1982a).

In Philippines *Arthrobotrys cladodes* gave 49-79% control of *M. incognita*. This fungus can be mass produced in bottles with local substrates such as grit, rice hull, chopped water lily, meshed potato and *ipil-ipil* leaves (Villanueva and Davide, 1984).

The application of *A. oligospora* preparation lowered the development of root galling due to *Meloidogyne* spp. by 46.5-81.9% and was more effective than nematicides (Slepetiene *et al.*, 1993). In pot trials with tomatoes, inoculations of the substrate with the endoparasitic fungi, *A. oligospora* and *A. dactiloides* reduced the population of *M. incognita* (Arndt and Leuprecht, 1994). Green house trials to compare the effects of *A. conoides, A. musiformis, A. robusta, A. irregularis* and *A. thaumasia* on the control of *M incognita* revealed that when ground maize was used, there was a significant reduction in the number of tomato root galls in the presence of *A. irregularis* in the first crop, and *A. thaumasia* in the following crop. Reduction in the number of galls was not observed in the first crop, when the suspension of conidia was used, however, *A. robusta* and *A. thaumasia* significantly reduced gall numbers in the second crop. No significant differences in the number of eggmasses and juveniles was evident (Dias and Ferraz, 1994).

4.2. *Monacrosporium* spp.

Among the nematophagous hyphomycetes *Monacrosporium* is a well known predator. In India, several species of this fungus have been reported as predators of nematodes, which snare and kill nematodes by adhesive hyphal meshes (Sachidananda and Ramakrishnan, 1971; Siddiqui and Mahmood, 1996; Srivastava, 1985; Srivastava and Dayal, 1985). *M. javanica* is more susceptible to *Monacrosporium cionopagum* than *H. schachtii* (Jaffee and Muldoon, 1995). Out of six *Monacrosporium* species tested under green house conditions, two species were found to be most efficient in reducing the number of galls, number of eggmasses and number of juveniles of *M. incognita* in the soil (Pria and Ferraz, 1996). Calcium alginate pellets containing hyphae of *M. cionopagum, M. ellipsosporum* repelled *M. incognita* J-2, repulsion of *R. reniformis* was less pronounced (Robinson and Jaffee, 1996).

4.3. *Dactylaria* spp.

Dactylaria candida produces adhesive knobs which are very effective nematode trapping agents and nematodes are often attacked at more than one infection site (Barron, 1977). *D. candida* possesses both unicellular adhesive knobs borne on short 2-3 celled stalks and non-constricting rings. Adhesive knobs produce an infection peg and infection bulb. Infection pegs in *D. candida* are aseptate while in *D. brochopaga* and *D. scaphoides* are septate. The infection hyphae proliferate within the nematode body and are efficient in digesting and assimilating the contents of nematode body. Infection hyphae of *D. candida* are stouter than vegetative hyphae and are capable of parasitizing eggs of *Aphelenchus avenae* (Dowsett and Reid, 1977 a). *D. scaphoides* produces three dimensional adhesive network. The nematodes entangled in the network become quiescent due to penetration by appresorium. Subsequently, formation of an infection bulb and infection hyphae takes place which digest the contents of nematode body (Dowsett and Reid, 1979). The infection processes of *D. candida, D. brochopaga* and *D. scaphoides* are analogous to one another (Dowsett and Reid, 1977a,b, 1979). Another species *D. dasguptae* has been found to reduce the nematode population by 35% during mushroom (*Agaricus bisporus*) cultivation (Koning *et al.*, 1996).

5. CONCLUSIONS

It is clear from the above account that fungal antagonists are abound and contribute significantly in the natural control of phytonematodes in different ways. The nematode trapping fungi, though omnipresent, are basically saprophytes and do not exhibit specificity towards phytonematodes. Moreover, their nematode trapping activity is governed by certain specific edaphic conditions. Though many of these can be mass multiplied with ease on synthetic media, yet their practical use for nematode suppression is still limited for want of information on the factors which induce significant predacious behaviour.

The zoospore forming fungi (*Catenaria*, etc.) have their own limitations viz., difficulty of mass multiplication and requirement of optimum soil moisture levels at critical periods to enhance their activity. The choice of most promising fungal antagonists for field scale use thus falls on egg and female parasites such as *P. lilacinus* and *V. chlamydosporium*. They have wide host ranges among phytonematodes, can dwell in diverse agroclimatic conditions and it is possible to mass multiply them. However, virulence is variable among isolates, and there is need to study their long-term sustenance in edaphic systems to maintain nematode suppressive activity. Our experience with *P. lilacinus* shows that competition with resident microflora and fauna may be a limiting factor and may, therefore, warrant inundative treatments. Augmentation of soil organic matter may enhance fungal soil population at the cost of nematode suppression. Human hazardous factors also need to be ruled out before commercial prospects are realised. Short-term approach may end up in a fiasco. Ecological considerations should be extensively worked out for a long-term commercial viability of the product.

REFERENCES

Adiko, A. 1984, Biological control of *Meloidogyne incognita* with *Paecilomyces lilacinus*, M.S. thesis, North Carolina State University, Raleigh, USA

Al-Hazmi, A.S., Schmitt, D.P. and Sasser, J.N. 1982a, Population dynamics of *Meloidogyne incognita* on corn grown in soil infested with *Arthrobotrys conoides*, *J. Nematol.* 14 : 44-50.

Al-Hazmi, A.S., Schmitt, D.P. and Sasser, J.N. 1982b, The effect of *Arthrobotrys conoides* on *Meloidogyne incognita* population densities in corn as influenced by temperature, fungus inoculum density, and time of fungus introduction into the soil, *J. Nematol.* 14 : 168-174.

Al-Raddad, A.M. 1995, Interaction of *Glomus mosseae* and *Paecilomyces lilacinus* on *Meloidogyne javanica* of tomato, *Mycorrhiza* 5 : 233-236.

Amoncho, A. and Sasser, J.N. 1995, Biological control of *Meloidogyne incognita* with *Paecilomyces lilacinus*, *Biocont.* 1 (4) : 51-61.

Arndt, M. and Leuprecht, B. 1994, Trials on alternatives for control of nematodes in vegetable crops, *Gartenbau Magazine* 3 (5) : 24-25.

Bansal, R.K., Walia, R.K. and Bhatti, D.S. 1988, Evaluation of some agro-industrial waste for mass propagation of the nematode parasitic fungus *Paecilomyces lilacinus*, *Nematol. Medite.* 16 : 135-136.

Bansal, R.K., Walia, R.K. and Bhatti, D.S. 1989a, Effect of exposure to high temperature on the survival of nematode destroying fungus, *Paecilomyces lilacinus*, In : *Proc. 76th Indian Scie. Cong.*, (Abstr.), Part III, Section X : 139-140.

Bansal, R.K., Walia, R.K. and Bhatti, D.S. 1989b, Effect of pH on the growth of nematode destroying fungus, *Paecilomyces lilacinus*, In : *Proc. 76th Indian Science Congress*, (Abstr.), Part III, Section X : 140-141.

Barron, G.L. 1977, *The Nematode Destroying Fungi*, Canadian Publications Ltd., Guelph, Canada, 140 pp.

Barron, G.L. and Onions, A.H.S. 1966, *Verticillium chlamydosporium* and its relationship to *Diheterospora, Stemphyliopsis* and *Paecilomyces*, *Can. J. Bot.* 44 : 861-869.

Bhardwaj, P. and Trivedi, P.C. 1996, Biological control of *Heterodera avenae* on wheat using different inoculum levels of *Verticillium chlamydosporium*, *Ann. Pl. Prot. Sci.* 4 : 111-114.

Boosalis, M.G. and Mankau, R. 1965, Parasitism and predation of soil microorganisms, In: *Ecology of Soil-borne Plant Pathogens*, eds. K.F. Baker and W.C. Snyder, Univ. of California Press, Berkeley, USA, pp. 374-391.

Bursnall, L.A. and Tribe, H.T. 1974, Fungal parasitism in cysts of *Heterodera*, II, Egg parasites of *H. schachtii*, *Trans. Bri. Mycol. Soc.* 62 : 595-601.

Cabanillas, E. and Barker, K.R. 1989, Impact of *Paecilomyces lilacinus* inoculum level and application time on control of *Meloidogyne incognita* on tomato, *J. Nematol.* 21 : 115-120.

Cabanillas, E., Barker, K.R. and Nelson, L.A. 1989a, Survival of *Paecilomyces lilacinus* in selected carriers and related effects on *Meloidogyne incognita* on tomato, *J. Nematol.* 21: 121-130.

Cabanillas, E., Barker, K.R. and Nelson, L.A. 1989b, Growth of isolates of *Paecilomyces lilacinus* and their efficacy in biocontrol of *Meloidogyne incognita* on tomato, *J. Nematol.* 21 : 164-172.

Cayrol, J.C. and Frankowski, J.P. 1979, Une methode de lutte biologique centre les nematodes a galles des racines appartenant au genre *Meloidogyne, Pepinieristes, Horticulteurs, Maratchers - Revue Horticole* 193 : 15-23.

Cayrol, J.C. and Frankowski, J.P. 1980, Connaissances nouvelles sur le champignon nematophage *Arthrobotrys irregularis* (Royal 350), *Pepinieristes, Horticulteurs, Maratchers-Revue Horticole* 203 :33-38.

Cayrol, J.C. and Frankowski, J.P. 1986, Influence of the number of parasitising conidia of *Hirsutella rhossiliensis* on the mortality of *Ditylenchus dipsaci, Revue de Nematologie* 9 : 411-412.

Cayrol, J.C., Frankowski, J.P., Laniece, A., D'Hardemore, G. and Talon, J.P. 1978, Centre les nematodes en champignouniere. Mise au point d'une methode de lutte biologique a' l'aide d'um Hyphomycete predateur : *Arthrobotrys robusta* souche "antipolis" (Royal 300), *Pepinieristes, Horticulteurs, Maratchers - Revue Horticole* 184 : 23-30.

Crump, D.H. 1987, A method for assessing the natural control of cyst nematode populations, *Nematol.* 33 : 232-243.

Crump, D.H. and Flynn, C.A. 1992, Biological control of the potato cyst nematode using parasitic fungi, *Asp. App. Biol.* 33 : 161-165.

Crump, D.H. and Kerry, B.R. 1977, Maturation of females of the cereal cyst nematode on oat roots and infection by an *Entomophthora*-like fungus in observation chambers, *Nematol.* 23 : 398-402.

Crump, D.H. and Kerry, B.R. 1981, A quantitative method for extracting resting spores of two nematode parasitic fungi, *Nematophthora gynophila* and *Verticillium chlamydosporium* from soil, *Nematol.* 27 : 330-338.

Crump, D.H. and Kerry, B.R. 1982, Pathogens of cyst nematodes. In : *Rothamsted Report for 1981*, Part I, p.165.

Crump, D.H., Sayre, R.M. and Young, L.D. 1983, Occurrence of nematophagous fungi in cyst nematode populations, *Plant Dis.* 67 : 63-64.

Davide, R.G. and Zorilla, R.A. 1983, Evaluation of a fungus, *Paecilomyces lilacinus* (Thom)Samson, for the biological control of the potato cyst nematode *Globodera rostochiensis* Woll. as compared with nematicides, *Philipp. Agricul.* 66 : 397-404.

Davide, R.G. and Zorilla, R.A. 1995a, Evaluation of three nematophagous fungi and some nematicides for the control of potato cyst nematode *Globodera rostochiensis, Biocont.* 1 (3) : 45-55.

Davide, R.G. and Zorilla, R.A. 1995b, Evaluation of *Paecilomyces lilacinus* and some nematicides for the control of potato cyst nematode *Globodera rostochiensis, Biocont.* 1(3) : 69-76.

Davies, K.G., De Leij, F.A.A.M. and Kerry, B.R. 1991, Microbial agents for the biological control of plant parasitic nematodes in tropical agriculture, *Trop. Pest Manag.* 37 : 303-320.

De Leij, F.A.A.M. and Kerry, B.R. 1990, Influence of temperature and nematode species on the efficacy of the fungus, *Verticillium chlamydosporium*, as a biological control agent of root-knot nematodes, *Nematol.* 36 : 367.

De Leij, F.A.A.M. and Kerry, B.R. 1991, The nematophagous fungus *Verticillium chlamydosporium* as a potential biological control agent for *Meloidogyne arenaria, Rev. de Nematol.* 14 : 157-164.

De Liej, F.A.A.M., Denneby, J.A. and Kerry, B.R. 1992a, The effect of temperature and nematode species on interactions beetween the nematophagous fungus, *Verticillium chlamydosporium* and *Meloidogyne* spp., *Nematol.* 38 : 65-79.

De Leij, F.A.A.M., Kerry, B.R. and Denneby, J.A. 1992b, The effect of fungal application rate and nematode density on the effectiveness of *Verticillium chlamydosporium* as a biological control agent for *Meloidogyne incognita, Nematol.* 38 : 112-122.

De Leij, F.A.A.M., Kerry, B.R. and Denneby, J.A. 1993, *Verticillium chlamydosporium* as a biological control agent for *Meloidogyne incognita* and *M. hapla* in pot and microplot tests, *Nematol.* 39 : 115-126.

Dias, W.P. and Ferraz, S. 1994, Evaluation of species of *Arthrobotrys* for the control of *Meloidogyne incognita, Fitopatologia Brasileira* 19: 189-192.

Dhawan, S.C. 1985, Mode of infection and life cycle of *Catenaria vermicola* infecting juveniles of *Heterodera avenae, Indian Phytopath.* 38 : 380-381.

Domsch, K.H. and Gams, W. 1972, *Fungi in Agricultural Soils*, Longman, London, 290 pp.

Dowsett, J.A. and Reid, J. 1977a, Light microscope observations on the trapping of nematodes by *Dactylaria candida, Can. J. Bot.* 55: 2956-2962.

Dowsett, J.A. and Reid, J. 1977b, Transmission and scanning electron microscope observations on the trapping of nematodes by *Dactylaria candida, Can. J. Bot.* 55: 2963-2970.

Dowsett, J.A. and Reid, J. 1979, Observations on the trapping of nematodes by *Dactylaria scaphoides* using optical transmission and scanning electron microscope technique, *Mycologia* 71: 379-391.

167

Dube, B.N. 1989, Biological control of *Meloidogyne javanica* by *Paecilomyces lilacinus* and *Pasteuria penetrans*, In : *Proc. Integrated Pest Management in Tropical and Subtropical Cropping Systems '89*, Vol.2, pp. 639-645.

Dube, B.N. and Smart, G.C. Jr. 1987, Biological control of *Meloidogyne incognita* by *Paecilomyces lilacinus* and *Pasteuria penetrans*, *J. Nematol.* 19 : 222-227.

Dunn, M.T., Sayre, R.M., Carrell, A. and Wergin, W.P. 1982, Colonization of nematode eggs by *Paecilomyces lilacinus* (Thom) Samson as observed with SEM, *Scan. Elec. Microscopy* 3 : 1351-1357.

Eayre, C.G., Jaffee, B.A. and Zehr, E.I. 1987, Suppression of *Criconemella xenoplax* by the nematophagous fungus *Hirsutella rhossiliensis*, *Plant Dis.* 71 : 832-834.

Eguiguren-Carrion, R. 1995, Control of *Meloidogyne incognita* and *Nacobbus* sp. with *Paecilomyces lilacinus* in the green house and effect of nematicides on the fungus, *Biocont.* 1 (4) : 41-49.

Ehteshamul-Haque, S., Zaki, M.I., Abid, M. and Ghaffar, A. 1994, Use of biocontrol agents with *Bradyrhizobium japonicum* in the control of root-knot nematode in chickpea, *Pakistan J. Nematol.* 12 : 149-154.

Ekanayake, H.M.R.K. and Jayasundara, N.J. 1984, Effect of *Paecilomyces lilacinus* and *Beauveria bassiana* in controlling *Meloidogyne incognita* on tomato in Sri Lanka, *Nematol. Mediterranea* 22 : 87-88.

Esser, R.P. and Ridings, W.H. 1973, Pathogenicity of selected nematodes by *Catenaria anguillulae*, In : *Proc Soil and Crop Science Society of Florida*, USA 33: 60-64.

Fitters, P.F.L. and Belder, E. Den. 1993, A time-lapse technique to study the effect of fungal products on embryogenesis of nematode eggs, *Mededlingen van de Faculteit Landbouw-wetenschappen, Universiteit Gent.* 58 (2 B) : 751-756.

Franco, J., Jatala, P. and Bocangel, M. 1981, Efficiency of *Paecilomyces lilacinus* as a biocontrol agent of *Globodera pallida*, *J. Nematol.* 13 : 438-439.

Freire, F.C.O. and Bridge, J. 1985, Parasitism of eggs, females and juveniles of *Meloidogyne incognita* by *Paecilomyces lilacinus* and *Verticillium chlamydosporium*, *Fitopatologia Brasileira* 10 : 577-596.

Gams, W. 1971, *Cephalosporium- artige Schimmelpilze* (Hyphomycetes), Fischer, Stuttgart, 262 pp.

Gapasin, R.M. 1995, Evaluation of *Paecilomyces lilacinus* (Thom) Samson for the control of *Pratylenchus* sp. in corn, *Biocont.* 1(4) : 35-39.

Generalo, L.C. and Davide, R.G. 1995, Evaluation of biological control efficiency of three fungi grown in different substrates against *Radopholus similis* on banana, *Biocont.* 1 (3) : 35-43.

Giraldo, M., Leguizamon, J.E. and Bernardo Chavez, C. 1996, Biological control of *Meloidogyne* spp. Goeldi with the fungus *Paecilomyces lilacinus* (Thom) Samson in luffa plants *Luffa cylindrica, L. aegyptiaca, Fitopatologia Colombiana* 20 (1/2): 20-25.

Goffart, H. 1932, Untersuchungen am Hafernematoden *Heterodera schachtii* Schm. Unter besonderer Berucksichtigung der Schleswig-Golstarnischen verhaltnisse, *Arab. Bill. Reich Aust. Lend-u Forstw,* 20: 1-28.

Graf, N.J. and Madelin, M.F. 1989, Axenic culture of the cyst nematode parasitizing fungus, *Nematophthora gynophila, J. Inverete. Pathol.* 53: 301-306.

Guevara, E., Jatala, P., Salas, R. and Herrera, E. 1985, Eficiencia de algunos nematicides y *Paecilomyces lilacinus* en el control de *Meloidogyne* spp. en olivos, Resumenes IX Congr. Peru, Fitopatol., Huanuco, Peru, p. 63.

Herrera, E., Jatala, P. and Canicoba, R. 1984, Estudio de la dinamica poblacional de *Tylenchulus semipenetrans* Cobb 1913, mediante aplicaciones de nematicidas y *Paecilomyces lilacinus* (Thom) Samson.

Herrera, E., Jatala, P. and Canicoba, R. 1985, Respuesta al control quimico y biologico del nematodo de los citricos. Inst. Nac. Invest. Promoc. Agropecu. (INIPA), Informe Especial, Year 4, 2: 16 pp.

Jaffee, B.A. 1992, Population biology and biological control of nematodes, *Can. J. Microbiol.* 38: 359-364.

Jaffee, B.A., Gaspard, J.T. and Ferris, H. 1989, Density dependent parasitism of the soil-borne nematode *Criconemella xenoplax* by the nematophagous fungus *Hirsutella rhossiliensis, Microb. Ecol.* 17 : 193-200.

Jaffee, B.A. and Muldoon, A.E. 1989, Suppression of cyst nematode by natural infestation of a nematophagous fungus, *J. Nematol.* 21: 505-510.

Jaffee, B.A. and Muldoon, A.E. 1995, Susceptibility of root-knot and cyst nematode to the nematode trapping fungi *Monacrosporium ellipsosporum* and *M. cionopagum, Soil Biol. Biochem.* 27: 1083-1090.

Jaffee, B.A. and Zehr, E.I. 1983, Effects of certain solutes, osmotic potential, and soil solutions on parasitism of *Criconemella xenoplax* by *Hirsutella rhossiliensis, Phytopath.* 73: 544-546.

Jaffee, B.A. and Zehr, E.I. 1985, Parasitic and saprophytic abilities of the nematode-attacking fungus *Hirsutella rhossiliensis, J. Nematol.* 17: 341-345.

Jain, D.K. 1979, Occurrence of nematophagous fungi and their mode of action on cereal cyst nematode (*Heterodera avenae*), *Thes. Abst.* 5 : 88-89.

Jatala, P. 1983, Biological control with the fungus *Paecilomyces lilacinus*, In : *Proc. 3rd Research and Planning Conference on Root-knot nematodes Meloidogyne* spp., Coimbra, Portugal, Region VII, pp. 183-187.

Jatala, P. 1985, Biological control of nematodes, In : *An Advanced Treatise on Meloidogyne-Biology and Control*, eds. J.N.Sasser and C.C.Carter, Vol. 1, North Carolina State University, Graphics, pp. 303-308.

Jatala, P. 1986, Biological control of plant parasitic nematodes, *Ann. Rev. Phytopath.* 24 : 453-489.

Jatala, P., Franco, J., Gonzalez, A. and O'Hara, C.M. 1985, Hatching stimulation and inhibition of *Globodera pallida* eggs by enzymatic and exopathic toxic compounds of some biocontrol fungi, *J. Nematol.* 17: 501.

Jatala, P., Kaltenbach, R. and Bocangel, M. 1979, Biological control of *Meloidogyne incognita acrita* and *Globodera pallida* on potatoes, *J. Nematol.* 11: 303.

Jatala, P., Kaltenbach, R., Bocangel, M., Devaux, A.I. and Campos, R. 1980, Field application of *Paecilomyces lilacinus* for controlling *Meloidogyne incognita* on potatoes, *J. Nematol.* 12: 226-227.

Jatala, P., Salas, R., Kaltenbach, R. and Bocangel, M. 1981, Multiple application and long-term effect of *Paecilomyces lilacinus* in controlling *Meloidogyne incognita* under field applications, *J. Nematol.* 13: 445.

Jonathan, E.I., Padmanabhan, D. and Ayyamperumal, A. 1995, Biological control of root-knot nematode on betelvine, *Piper betle*, by *Paecilomyces lilacinus*, *Nematologia Mediterranea* 23: 191-193.

Kerry, B.R. 1974, A fungus associated with young females of the cereal cyst nematode, *Heterodera avenae*, *Nematol.* 20: 259-260.

Kerry, B.R. 1975, Fungi and the decrease of cereal cyst nematode populations in cereal monoculture, 3rd meeting on Pathological Factors of the Monoculture of Cereals, Gembloux, Belgium, 14 pp.

Kerry, B.R. 1979, Fungal parasites of females of cyst nematodes, *J. Nematol.* 12 : 304-305.

Kerry, B.R. 1980, Biocontrol : fungal parasites of female cyst nematodes, *J. Nematol.* 13 : 253-259.

Kerry, B.R. 1984, Nematophagous fungi and the regulation of nematode populations in soil, *Helminthological Abstracts*, Series B 53 : 1-14.

Kerry, B.R. 1987, Biological control, In : *Principles and Practices of Nematode Control in Crops*, eds. R.H. Brown and B.R. Kerry, Academic Press, New York, USA, pp. 233-263.

Kerry, B.R. and Bourne, J.M. 1996, The importance of rhizosphere interactions in the biological control of plant parasitic nematodes - a case study using *Verticillium chlamydosporium*, *Pesti. Sci.* 47: 69-75.

Kerry, B.R. and Crump, D.H. 1977, Observations on fungal parasites of females and eggs of the cereal cyst nematode, *Heterodera avenae*, and other cyst nematodes, *Nematol.* 23: 193-201.

Kerry, B.R. and Crump, D.H. 1980, Two fungi parasitic on females of cyst nematodes (*Heterodera* spp.), *Trans. Brit. Mycol. Soc.* 74: 119-125.

Kerry, B.R., Crump, D.H. and Irving, F. 1995, Some aspects of biological control of cyst nematodes, *Biocont.* 1(4): 5-14.

Kerry, B.R., Crump, D.H. and Mullen, L.A. 1980, Parasitic fungi, soil moisture and multiplication of the cereal cyst nematode, *Heterodera avenae*, *Nematol.* 26: 57-68.

Kerry, B.R., Crump, D.H. and Mullen, L.A. 1982, Studies on the cereal cyst nematode, *Heterodera avenae* under continuous cereals, 1975-78, II, Fungal parasitism of nematode females and eggs, *Ann. Appl. Biol.* 100: 489-499.

Kerry, B.R., Kirkwood, I.A., De Leij, F.A.A.M., Barba, J., Leijdens, M.B. and Brookes, P.C. 1993, Growth and survival of *Verticillium chlamydosporium* Goddard, a parasite of nematodes in soil, *Biocont. Sci. Technol.* 3: 355-365.

Kerry, B.R. and Mullen, L.A. 1981, Fungal parasites of some plant parasitic nematodes, *Nematropica* 2: 187-189.

Khan, H.A., Khan, S.A., Qamar, F. and Seema, N. 1992, Preliminary studies on seed dressing of *Luffa aegyptica* with *Paecilomyces lilacinus* against *Meloidogyne incognita* var. *acrita* during the germination of seed, *Sarhad J. Agric.* 8: 227-230.

Khan, T.A., Khan, S.T., Fazal, M. and Siddiqui, Z.A. 1997, Biological control of *Meloidogyne incognita* and *Fusarium solani* disease complex in papaya using *Paecilomyces lilacinus* and *Trichoderma harzianum*, *Internat. J. Nematol.* 7: 127-132.

Khan, T.A. and Saxena, S.K. 1996, Comparative efficacy of *Paecilomyces lilacinus* in the control of *Meloidogyne* spp. and *Rotylenchulus reniformis* on tomato, *Pakistan J. Nematol.* 14: 111-116.

Knuth, P. and Deitermann, H. 1983, *Nematophthora gynophila* reported from GFR, Nachr BP. dt. *PflSchutz-dienst. Braun.*, 35: 68-69.

Kondakova, E.I. and Tikhonova, L.V. 1981, The effect of parasitic fungi on the infestation of wheat with *Heterodera avenae* Woll. Byull. *Vses. Inst. Gelmint. im. K.I. Skr.*, 31: 33-35.

Koning, G., Hamman, B. and Eicker, A. 1996, The efficacy of nematophagous fungi on predaceous nematodes in soil compared with saprophagous nematodes in mushroom compost, *South African J. Bot.* 62 : 49-53.

Kuhn, J. 1877, Vorlaufiger Bericht uber die bisherigen Ergebnisse der Seit dem Jahre 1875 im Aftrage des vereins for Rubehzucher-Industrie ausgegiihrten versuche Zur Ermittelung der ursacho der Rubenmudiqueit des Bodens and Zur Er forschung der Natur der Nematoden, Z. Ver. Ruben. Ind. Dent. Reich. (Ohne Bond), 452-457.

Lackey, B.A., Jaffee, B.A. and Muldoon, A.E. 1994, Effect of nematode inoculum on suppression of root-knot and cyst nematodes by the nematophagous fungus *Hirsutella rhossiliensis*, *Phytopath.* 84: 415-420.

Lackey, B.A., Muldoon, A.E. and Jaffee, B.A. 1993, Alginate pellet formulation of *Hirsutella rhossiliensis* for biological control of plant parasitic nematodes, *Biolo. Cont.* 3: 155-160.

Lohde, G. 1874, Einige neuen parasitischen plize, Tageblatt der 47, Versammlung deutscher naturaforcher.

Lopez-Llorca, L.V. and Duncan, J.M. 1986, New media for the estimation of fungal infection in eggs of the cereal cyst nematode, *Heterodera avenae* Woll, *Nematol.* 32: 486-490.

Mahmood, I. and Siddiqui, Z.A. 1993, Integrated management of *Rotylenchulus reniformis* by green manuring and *Paecilomyces lilacinus*, *Nematologia Mediterranea* 21: 285-287.

Mankau. R. 1964, Ecological relationships of predacious fungi associated with the citrus nematode, *Phytopath.* 54: 1435.

Mankau, R. 1980, Biocontrol : Fungi as nematode control agents, *J. Nematol.* 12: 244-252.

Meyer, S.L.F. 1994, Effects of a wild type strain and a mutant strain of the fungus *Verticillium lecanii* on *Meloidogyne incognita* populations in green house studies, *Fundam. App. Nematol.* 17: 563-567.

Meyer, S.L.F., Huettel, R.N. and Sayre, R.M. 1990, Isolation of fungi from *Heterodera glycines* and *in vitro* bioassays for their antagonism to eggs, *J. Nematol.* 22: 532-537.

Mittal, N., Saxena, G. and Mukerji, K.G. 1995, Integrated control of root-knot disease in three crop plants using chitin and *Paecilomyces lilacinus*, *Crop Protec.* 14 : 647-651.

Mittal, N., Saxena, G. and Mukerji, K.G. 1999, Biological control of root knot nematode by nematode - destroying fungi, In : *From Ethnomycology to Fungal Biotechnology Exploiting from Natural Resources for Novel Products*, eds. J. Singh and K.R. Aneja, Kluwer Academic / Plenum Publishers, New York, USA, pp. 143-172.

Morgan-Jones, G., Godoy, G. and Rodriguez-Kabana, R. 1981, *Verticillium chlamydosporium*, fungal parasite of *Meloidogyne arenaria* females, *Nematropica* 11: 115-119.

Morgan-Jones, G. and Rodriguez-Kabana, R. 1984, Species of *Verticillium* and *Paecilomyces* as parasites of cysts and root-knot nematodes, *Phytopath.* 74: 831.

Morgan-Jones,G. and Rodriguez-Kabana, R. 1986, Fungi associated with cysts of potato cyst nematodes in Peru, *Nematropica* 16: 21-31.

Morgan-Jones, G., White, J.F. and Rodriguez-Kabana, R. 1983 Phytonematode pathology: ultrastuctural studies, 1, Parasitism of *Meloidogyne arenaria* eggs by *Verticillium chlamydosporium*, *Nematropica* 13: 245-260.

Mousa, E.M., Basiony, A.M. and Mahdy, M.E. 1995, Control of *Meloidogyne javanica* by *Verticillium chlamydosporium* on tomatoes, *Afro-Asian J. Nematol.* 25: 113-115.

Muller, J. 1985, The influence of two pesticides on fungal parasites of *Heterodera schachtii*, *Les Colloques de l'INRA* 31: 225-231.

Noe, J.P. and Sasser, J.N. 1995, Evaluation of *Paecilomyces lilacinus* as an agent for reducing yield losses due to *Meloidogyne incognita*, *Biocont.* 1(3): 57-67.

Owino, P.O., Waudo, S.W. and Sikora, R.A. 1993, Biological control of *Meloidogyne javanica* in Kenya : effect of plant residues, benomyl and decomposition products of mustard (*Brassica campestris*), *Nematol.* 29: 127-134.

Patel, D.J., Vyas, R.V., Patel, B.A. and Patel, R.S. 1995, Bioefficacy of *Paecilomyces lilacinus* in controlling *Meloidogyne javanica* (Pathotype 2)on groundnut, *Internat. Arachis Newsl.* No. 15: 46-47.

Pria, M.D. and Ferraz, S. 1996, Biological control of *Meloidogyne incognita*, race 3, by six species of the fungus *Monacrosporium* spp. alone or in combination with *Verticillium chlamydosporium*, *Fitopatologia Brasileira* 21: 30-34.

Rana, B.P. and Dalal, M.R. 1995, Management of *Heterodera cajani* in mungbean with *Paecilomyces lilacinus* and carbosulfan, *Ann. Pl. Protec. Scie.* 3: 145-148.

Rao, M.S. and Reddy, P.P. 1994, A method for conveying *Paecilomyces lilacinus* to soil for the management of root-knot nematodes on eggplant, *Nematologia Mediterranea* 22 : 265-267.

Rao, M.S., Reddy, P.P., and Nagesh, M. 1998, Use of neem based formulation of *Paecilomyces lilacinus* for the effective management of *Meloidogyne incognita* infecting eggplant, In : *Proc. 3rd International Symposium of Afro-Asian Society of Nematologists*, (Abst.) Coimbatore, p. 73.

Reddy, P.P. and Khan, R.M. 1988, Evaluation of *Paecilomyces lilacinus* for the biological control of *Rotylenchulus reniformis* infecting tomato compared with carbofuran, *Nematologia Mediterranea* 16 : 113-115.

Robinson, A.F. and Jaffee, B.A. 1996, Behavioural responses of root-knot and reniform nematodes to alginate pellets containing nematophagous fungi, In : *Proc. Beltwide Cotton Conferences*, Vol. 1 Nashville, TN, USA, 282 pp.

Sachidanand, J. and Ramakrishnan, K. 1971, Nematophagous fungi of agricultural soil I, *Mycopathologia et Mycologia Applicata* 43: 235-241.

Saifullah and Thomas, B.J. 1997, Parasitism of *Globodera rostochiensis* by *Verticillium chlamydosporium* Low temperature scanning electron microscopy and freeze fracture study, *Internat. J. Nematol.* 7: 30-34.

Salas, R. and Jatala, P. 1980, Capacidad nematofaga del hongo *Paecilomyces lilacinus* (Thom) Samson para el control del nematodo del nudo *Meloidogyne incognita*, Resumenes *13th Conv. Nac. Entomol.*, Huacho, Peru. pp. 20-21.

Samson, R.A. 1974, *Paecilomyces* and some allied hyphomycetes, *Stud. Mycol.* 6: 1-119.

Sayre, R.M. 1986, Pathogens for biological control of nematodes, *Crop Protec.* 5: 268-276.

Sayre, R. M. 1980, Promising organisms for biocontrol of nematodes, *Pl. Dis.* 64: 526-532.

Sayre, R.M. and Keeley, L.S. 1969, Factors influencing *Catenaria anguillulae* infections in a free-living and a plant-parasitic nematode, *Nematol.* 15: 492-502.

Sharma, R. and Swarup, G. 1988, *Pathology of Cyst Nematodes*, Malhotra Publishing House, New Delhi, India, 88 pp.

Siddiqui, Z.A. and Mahmood, I. 1992, Biological control of root-rot disease complex of chickpea caused by *Meloidogyne incognita* race 3 and *Macrophomina phaseolina*, *Nematologia Mediterranea* 20 : 199-202.

Siddiqui, Z.A. and Mahmood, I. 1993, Biological control of *Meloidogyne incognita* race 3 and *Macrophomina phaseolina* by *Paecilomyces lilacinus* and *Bacillus subtilis* alone and in combination on chickpea, *Fundamen. App. Nematol.* 16: 215-218.

Siddiqui, Z.A. and Mahmood, I. 1996a, Biological control of plant parasitic nematodes by fungi : a review, *Bioresou. Technol.* 58 : 229-239.

Siddiqui, Z.A. and Mahmood, I. 1996b, Biological control of *Heterodera cajani* and *Fusarium udum* on pigeon pea by *Glomus mosseae*, *Trichoderma harzianum* and *Verticillium chlamydosporium*, *Israel J. Pl. Sci.* 44 : 49-56.

Slepetiene, J., Mackevic, N. and Tepliakova, T. 1993, A comparative charateristic of the effect of predatory fungi and nematicides on soil nematodes, *Acta Parasitol. Lituanica* 24: 47-57.

Srivastava, S.S. 1985, Ecology and distribution of predatory nematophagous fungi in IARI campus, New Delhi, India. *Indian J. Nematol.* 16: 128-130.

Srivastava, S.S. and Dayal, R. 1985, Fungal predator of nematodes - *Monacrosporium*, *Indian Phytopath.* 35: 650-653.

Stirling, G.R. 1979, Effect of temperature on parasitism of *Meloidogyne incognita* eggs by *Dactylella oviparasitica*, *Nematol.* 25: 104-110.

Stirling, G.R. 1980, Parasites and predators of cereal cyst nematode (*Heterodera avenae*) in South Australia, In : *Proc. 4th National Conference*, Australian Pathology Society, Perth, Austalia, p. 33.

Stirling, G.R. 1991, *Biological Control of Plant Parasitic Nematodes : Progress, Problems and Prospects*, CABI, Wallingford, pp. 106.

Stirling, G.R. 1992, Biological control of plant parasitic nematodes: Progress, problems and prospects, *Nematol.* 38: 392-394.

Stirling, G.R. and Kerry, B.R. 1983, Antagonists of the cereal cyst nematode, *Heterodera avenae* Woll. in Australian soils, *Austral. J. Experim. Agric. Anim. Husban.* 23: 318-324.

Stirling, G.R. and Mankau, R. 1978a, *Dactylella oviparasitica*, a new fungal parasite of *Meloidogyne* eggs, *Mycologia* 70: 774-783.

Stirling, G.R. and Mankau, R. 1978b, Parasitism of *Meloidogyne* eggs by a new fungal parasite, J. *Nematol.* 10: 236-240.

Stirling, G.R. and Mankau, R. 1979, Mode of parasitism of *Meloidogyne* and other nematode eggs by *Dactylella oviparasitica*, J. *Nematol.* 11: 282-288.

Stirling, G.R., McKenry, M.V. and Mankau, R. 1979, Biological control of root-knot nematodes (*Meloidogyne* spp.) on peach, *Phytopath.* 69: 806-809.

Strattner, A. 1979, Biological control of nematodes, *C.I.P. Circular* 7 (3): 3.

Sturhan, D. and Schneider, R. 1980, *Hirsutella heteoiderae*, ein neuer nematodenparasitarer Pilz, *Phytopathologische Zeitschrift* 99: 105-115.

Timper, P. and Kaya, H.K. 1989, Role of the second-stage cuticle of entomogenous nematodes in preventing infection by nematophagous fungi, *J. Invert. Pathol.* 54: 314-321.

Tribe, H.T. 1977a, Pathology of cyst nematodes, *Biol. Rev.* 52: 477-507.

Tribe, H.T. 1977b, A parasite of white cysts of *Heterodera : Catenaria auxiliaris*, *Trans. Bri. Mycol. Soc.* 69: 367-376.

171

Tribe, H.T. 1979, Extent of disease in populations of *Heterodera* with special reference to *H. schachtii*, *Ann. App. Bio.* 92: 61-72.

Trivedi, P.C. 1990, Evaluation of a fungus, *Paecilomyces lilacinus* for the biological control of root-knot nematode, *Meloidogyne incognita* on *Solanum melongena*, In : *Proc. 3rd International Conference on Plant Protection in the Tropics*, eds. F.A.C. Ooi, G.S. Lim and P.S. Tang, Vol. 6, Genting Highlands, Malaysia, pp. 29-33.

Velvis, H. and Kamp, P. 1996, Suppression of potato cyst nematode root penetration by the endoparasitic nematophagous fungus *Hirsutella rhossiliensis*, *Europ. J. Pl. Pathol.* 102: 115-122.

Vicente, N.E. and Acosta, N. 1992, Biological and chemical control of nematodes in *Capsicum annum* L. *J. Agricul. Univ. Puerto Rico* 76 (3/4): 171-176.

Villanueva, L.M. and Davide, R.G. 1984a, Evaluation of several isolates of soil fungi for biological control of root-knot nematodes, *Philipp. Agricul.* 67: 361-371.

Villanueva, L.M. and Davide, R.G. 1984b, Influence of pH, temperature, light and agar media on the growth and sporulation of a nematophagous fungus, *Paecilomyces lilacinus* (Thom) Samson, *Philipp. Agricul.* 67: 223-231.

Vonyoukalon, E. 1993, Effect of *Arthrobotrys irregularis* on *Meloidogyne arenaria* on tomato plants, *Fundament. App. Nematol.* 16: 321-324.

Voss, B., Utermohl, P. and Wyss, U. 1992, Variation between strains of the nematophagous endoparasitic fungus *Catenaria anguillulae* Sorokin, II, Attempts to achieve parasitism of *Heterodera schachtii* Schmidt in pot trials, *Zeitschrift fur Pflanzenkrankheiten und Pflanzenschutz* 99 (3): 311-318.

Walia, R.K. and Bansal, R.K. 1992, Factors governing efficacy of nematode-oviparasitic fungus, *Paecilomyces lilacinus* : effect of fungus inoculum level, *Afro-Asian Nematol. Network* 1: 9-11.

Walters, S.A. and Barker, K.R. 1994, Efficacy of *Paecilomyces lilacinus* in suppressing *Rotylenchulus reniformis* on tomato, *J. Nematol.* 26 : 600-605.

Wang, C.J., Song, C.Y., Zhang, X. de, Xie, Y.Q., Liu, Y.Z. and Wang, M.Z. 1997, Sustainable control efficacy of *Paecilomyces lilacinus* against *Heterodera glycines*, *Chinese J. Biol. Cont.* 13 :26-28.

Willcox, J. and Tribe, H.T. 1974, Fungal parasitism in cysts of *Heterodera*, I, Preliminary investigations, *Trans. Brit. Mycol. Soc.* 62: 585-594.

Zaki, F.A. 1994, Dose optimization of *Paecilomyces lilacinus* for the control of *Meloidogyne javanica* on tomato, *Nematologia Mediterranea* 22 : 45-47.

Zaki, F.A. and Bhatti, D.S. 1988, Economical method for mass culturing of *Paecilomyces lilacinus* (Thom) Samson, *Curr. Scie.* 57 (3): 153.

Zaki, F.A. and Bhatti, D.S. 1990, Effect of castor (*Ricinus communis*) and the biocontrol fungus *Paecilomyces lilacinus* on *Meloidogyen javanica*, *Nematol.* 36: 114-122.

Zaki, M.J. and Maqbool, M.A. 1991, *Paecilomyces lilacinus* controls *Meloidogyne javanica* in chickpea, *Internat. Chickpea Newsl. No.* 25 : 22-23.

Zopf, W. 1888, Zur kenntnis der Infections-krankheiten niederer Thiere und Pflanzen, *Nova Acta Academiae Caesariae Leopoldino-Carolinae* 52: 314-376.

BACTERIAL ANTAGONISTS OF PHYTONEMATODES

R.K. Walia[1], S.B. Sharma[2] and R. Vats[1]

[1]Department of Nematology
CCS Haryana Agricultural University
Hisar-125 004, Haryana, INDIA

[2]International Crops Research Institute for the
Semi-Arid Tropics (ICRISAT), Patancheru - 502 324
Andhra Pradesh, INDIA

1. INTRODUCTION

Soil is a dynamic natural resource. It forms a thin cover of unconsolidated minerals and organic matter on earth's surface and functions to maintain the ecosystems on which all life depends (Doran *et al.,* 1996). A myriad of living organisms such as insects, nematodes, bacteria, algae, fungi, earthworms coexist in the soil. The microbial populations of the soil has enormous diversity and as many as 10,000 different species may be found in a single gram of soil (Torsvik *et al.,* 1990). They are responsible. for smooth operation of the biogeochemical cycles in the soil ecosystem. Nematodes are the most abundant metazoans in soil and they are next only to arthropods in species diversity. All these organisms coexist with numerous diverse functions, interrelations, and interactions. Plant parasitic nematodes or phytonematodes are important constraints to agricultural production all over the world. These soil pests are popularly known as the 'hidden enemies' of farmers because of their microscopic size and subterranean habitat. Of all the microorganisms which are known to occur in soil along with nematodes only a few have been identified as potential biocontrol agents of phytonematodes. The published literature indicates that some species of fungi and bacteria are the most common parasites of nematodes (Davies, 1998). Some bacteria are potent antagonists of phytonematodes. These nematode antagonistic bacteria are broadly of two types: (i) bacteria that are pathogenic to nematodes or the nematode disease producing bacteria, and (ii) bacteria whose secretions or metabolic products are harmful to nematodes or the nematode toxin producing bacteria.

The *Pasteuria* group of gram-positive, endospore-forming bacteria are hyperparasites of nematodes and they are an example of disease producing bacteria. All the economically important genera of phytonematodes have an association with these bacteria. The second type includes strains of *Agrobacterium radiobacter, Azotobacter chroococcum, Bacillus thuringiensis, B. cereus, Clostridium* spp., *Pseudomonas* spp. and *Streptomyces* spp. These bacteria produce toxins that are harmful to plant parasitic nematodes. In this article we have presented the status of information on *Pasteuria* spp. as well as on the nematode toxin producing bacteria.

2. NEMATODE PATHOGENIC BACTERIA

The *Pasteuria* group of bacteria are hyperparasites of nematodes and water fleas (*Daphnia* spp.). A population of *Pasteuria* sp. was first observed by Cobb in 1906 and later Thorne (1940) described it as a protozoan (*Duboscqia penetrans*). Afterwards, based on electron microscopic studies, the prokaryotic nature of this organism was established and it was designated as a bacterium, *Bacillus penetrans* (Mankau, 1975a,b). Presently, populations of these bacteria are grouped in the genus *Pasteuria* (Sayre and Starr, 1985). Some isolates have the potential for use in the management of cyst, and root-knot nematodes (Sharma and Swarup, 1988; Stiriding, 1991).

2.1. Taxonomy and Classification

The taxonomy of *Pasteuria* still remains unclear but it is probably made up of a number of species and isolates which differ in their host ranges and virulence. The life cycles, host ranges, and spore morphologies are considered as important characters in classifying these bacteria (Sayre *et al.*, 1991). Using these criteria, three species have been described on phytonematodes. These are: *P. penetrans* (Thorne) (Sayre and Starr, 1985) parasitic on root-knot nematodes (*Meloidogyne* spp.); *P. thornei* (Starr and Sayre, 1988) from root-lesion nematodes (*Pratylenchus* spp.); and *P. nishizawae* (Sayre, *et al.*, 1991) which is a parasite of the cyst nematodes (*Heterodera* sp. and *Globodera* sp.). Some other different types of populations of *Pasteuria* have been found on *Heterodera avenae* (Davies *et al.*, 1990) and *H. goettingiana* (Sturhan *et al.*, 1994). There is a wide yet unexplored natural diversity in the genepool of *Pasteuria.*

Sharma and Davies (1996a) think that the current parameters used for characterization of *Pasteuria* into different species are unsatisfactory as they are prone to variability. Ciancio *et al.* (1994) have also concluded that no definite criteria are available to separate *Pasteuria* populations into species or pathotypes. There are no methods available to isolate the *Pasteuria* populations directly from the soil, other than with a bait nematode. This procedure selects a sample of the original population and is not a true representative of the natural diversity present in the soil. Host specificity of the *Pasteuria* population could be an artefact of the method of isolation (Sharma and Davies, 1996) and culturing the bacterium on a particular host nematode further reduces the variability present in the original population (Davies *et al.*, 1994). The *Pasteuria* populations can be divided into two groups: those that complete their life cycles in the juveniles (*P. thornei* group; PT group) and those that do not produce spores until the nematode has matured (*P. penetrans* group; PP group). *Pasteuria nishizawae* belongs to PP group while populations reported on *H. avenae* (Davies *et al.*, 1990) and *H. goettingiana* (Sturhan *et al.*, 1994) belong to PT group. Further classification of *Pasteuria* requires more information particularly analysis of their DNA.

2.2. Distribution and Host Range

Pasteuria spp. are present in diverse ecosystems in more than 51 countries. *Pasteuria* spores have been observed on several groups of soil nematodes including tylenchids, dorylaims, mononchids, rhabditids, aphelenchids (Sturhan, 1985). There are more than 205 nematode species within 96 genera and 10 orders that are hosts of *Pasteuria* (Sturhan, 1988).

Some of the recent reports on root-knot nematodes are from Pakistan (Maqbool and Zaki, 1990), Portugal and Italy (Abrantes and Vovlas, 1988); Iraq (Fattah *et al.*, 1989); on *Xiphinema bakeri* and *X. brasiliensis* in Peru (Ciancio and Mankau, 1989); on *Trophonema okamotoi* in USA (Inserra *et al.*, 1992) and on *Helicocotylenchus pseudorobustus* and *Rotylenchus capensis* in Greece (Vovlas *et al.*, 1993). Ciancio *et al.* (1994) found *P. penetrans* associated with 52 species of plant parasitic and saprophytic nematodes; 31 species were new host records of *P. penetrans*. They reported *Pasteuria* from 11 new geographical regions. Random surveys of agricultural fields in Haryana state in northern India have also revealed widespread occurrence of *Pasteuria* spp. on *Meloidogyne* sp.; *Heterodera cajani; H. mothi; Verutus mesoangustus; Mesodorylaimus japonicus; Aporcellaimellus* sp.; *Ecumenicus monhystera; Discolaimus tenax; Paralongidorus sali* and *Xiphinema brevicolle* (Walia *et al.*, 1994).

2.3. Host Specificity or Diversity

There are conflicting opinions on the host specificity of *Pasteuria* spp. because of large polymorphism in the host range of different populations. Spores of *Pasteuria* sp. originating from a particular nematode host generally attach to taxonomically close species of the original nematode host (Ciancio *et al.*, 1992; Inserra *et al.*, 1992; Oostendorp *et al.*, 1990; Silva *et al.*, 1994). The populations of *P. penetrans* can be genetically heterogenous with respect to host specificity (Channer and Gowen, 1992). Isolates of *Pasteuria* collected from *H. cajani* and *H. mothi* differ in their host ranges in India. Spores collected from *H. cajani* attach to *V. mesoangustus* and Ambala population of *H. avenae*; while those from *H. mothi* attach to *H. graminis* and Mahendergarh population of *H. avenae* in Haryana (Bajaj *et al.*, 1997). Spores of an isolate collected from *H. cajani* in Haryana do not adhere to *M. incognita* and *M. javanica* (Walia *et al.*, 1990). Another isolate of *P. penetrans* originally collected from *H. mothi* in New Delhi (northern India) also parasitized populations of *H. cajani, H. graminis, H. sorghi, H. zeae, H. avenae,* and *M. incognita* (Sharma, 1985). Another field population of *P. penetrans* parasitic on *H. cajani, M. javanica,* and *Rotylenchulus reniformis* was found in southern India (Sharma and Sharma, 1989). In the United Kingdom, an isolate of *Pasteuria* collected from juveniles of *H. avenae,* also adhered to the cuticles of *H. schachtii, H. glycines, Globodera pallida, G. rostochiensis* and *M. javanica* (Davies *et al.*, 1990). The attachment of endospore to the nematode cuticle may also vary with geographic distribution of the bacterial population. Stirling (1985) found that Australian populations of *P. penetrans* were more host specific than populations from USA. It is possible that the host specificity of the Australian bacteria developed because of lengthy association with nematode hosts of limited genetic diversity. Sharma and Davies (1996b) found two sympatric *Pasteuria* populations that were totally different in their host preferences. These sympatric populations are either the consequence of artificial selection pressure or they represent two distinct genetically isolated populations. The latter situation can arise in parasites where their hosts have been geographically isolated previously. Western blot analysis using a

175

polyclonal antibody showed large differences in the antigenic profiles between the spore extracts of these two sympatric populations. Although culturing of each of these two populations reduced the original diversity, these populations were still able to attack nematodes of different genera representing a wide geographical region.

2.4. Pathogenesis

2.4.1. Attachment of endospore to the nematode cuticle

The endospores of *Pasteuria* spp. adhere to the cuticle of nematodes in soil. This is the first important step in the pathogenicity of the bacterial population on nematodes. Greater number of spores attach along the lateral fields on nematode body because these are slightly raised areas and have more chances of coming in contact with the spores as the nematodes move along in the soil. The extent of spore encumbrance is influenced by the populations of bacterial endospores and nematodes (Davies *et al.*, 1991), temperature (Hatz and Dickson, 1992; Sekhar and Gill, 1990; Stirling *et al.* 1990;), and soil texture (Singh and Dhawan, 1992). Spores heated to 60, 80, 100 and 120°C may attach to nematodes but do not invade or develop inside them (Ginnakou *et al.*, 1997).

The spores generally adhere randomly to the nematode cuticle while on some nematodes the spores attach to anterior or posterior part of the nematodes in greater number (Sharma and Davies, 1996a). Sometimes there are large differences in number of spores attached to the cuticle of different individuals in a nematode population. This could either be due to heterogeneity in the nematode cuticle or in the epitopes on spore surface. In the laboratory, sonication of spores for 5-30 minutes disrupts the sporangial wall and significantly enhances the rate at which spores attach to the second stage juveniles (living or dead) of *M. javanica* (Stirling *et al.*, 1986). The spore surface is covered in fine fibers (adhesins) that are involved in the attachment of spore to the nematode cuticle (Persidis *et al.*, 1991). The cuticle surface of nematodes has an important role in the ability of this hyperparasitic infection (Davies and Danks, 1992). The surface of the spore is negatively charged and a balance between electrostatic and hydrophobic interactions is important in deciding whether or not the spores attach to the nematode cuticle (Afolabi *et al.*, 1995). The spores have highly glycosylated proteins with N-acetyl glucosamine (Persidis *et al.*, 1991) and glucosyl residues occurring on the spore surface attach to the lectins (carbohydrate binding proteins) on the nematode surface (Davies and Danks, 1993). Differences in the amount and nature of proteins on the surface of spores of different populations of *P. penetrans* (Davies *et al.*, 1992) and amount of surface associated proteins in different species of nematodes (Davies and Danks, 1992) may have an important role in the observed host specificity. Speigel *et al.* (1996) have found that several nematode surface coat components, such as carbohydrate residues, carbohydrate recognition domains, and a 250 kDa antigen are involved in spore attachment to the surface of *M. javanica*.

Polyclonal antibodies raised to the surface of nematode cuticle reduce the ability of spores to bind to the nematode cuticle. It suggests that the antibody blocks the receptor on cuticle surface for spore attachment (Davies *et al.*, 1992). Sharma and Davies (1997) found that a monoclonal antibody (HC/145) raised to the cuticle of second stage juvenile of *H. cajani* increased the spore attachment by 124%. It indicates that the components of the cuticle involved in inhibiting the spore attachment may be masking the *Pasteuria* receptor present on the cuticle.

2.4.2. Endospore germination

The endospore attached to the nematode cuticle germinates only after the root-knot nematode establishes feeding sites in the host root, and it takes about 8 days after root invasion (Sayre and Starr, 1988). The germ tube of the endospore penetrates the cuticle and it appears to be enzymatic action since no displacement of cuticular or hypodermal components is involved (Stirling *et al.*, 1986). After entering the hypodermal tissue, the germ tube develops into a spherical vegetative colony. Some isolates of *Pasteuria* complete their life cycles in the second stage juveniles of cyst nematodes (Davies *et al.*, 1990; Sturhan *et al.*, 1994). These reports assume significance in view of their deviation from already known life-cycle development of *Pasteuria* sp. among sedentary endoparasitic nematodes. These isolates belong to PT group of *Pasteuria* (Sharma and Davies, 1996a). It may be possible that the spore-encumbered juveniles feed ectoparasitically prior to root penetration and subsequently fail to invade the roots.

2.4.3. Vegetative stages

The development of bacterial life stages has been described in detail by Mankau (1975a,b) and Sayre and Starr (1985). The microcolonies break away from the colonised hypodermis and proliferate within the pseudocoelom of the nematode. Daughter colonies continue to form upon lysis of intercalary cells in the microcolony, as the nematode develops to adult stage. There is apparently no toxin production by the bacterium inside the nematode host. The infected nematode continues to feed on the giant cells (Bird, 1986). The reproductive system in the diseased nematodes does not develop and resources inside nematode body are utilized for the growth of the bacterium. Eventually, quartets of developing sporangia predominate in the nematode pseudocoelom, and they give rise to doublets and finally to the single endospores. The development of the bacterium is synchronised with that of the nematode.

2.4.4. Endospore formation

The endosporogenesis proceeds by formation of a septum in the anterior of the endospore mother cell, condensation of a forespore from the anterior protoplast, formation of multilayered walls about the forespore, lysis of the old sporangial wall, and release of the endospore.

The onset of sporogenesis coincides with the initiation of reproductive phase in nematode. The nematode reproductive hormones might provide a signal to the bacterium to initiate sporogenesis. Consequently, there is no or little egg production by the root-knot female and it ultimately turns into a carcass full of bacterial spores. The endospores are finally released into the soil when plant roots and nematodes decompose. A root-knot nematode female may contain 2×10^6 spores. Temperature has a profound effect on the duration of life cycle of bacterium and its pathogenicity (Ahmed and Gowen, 1991).

2.4.5. Soil phase

Spores of *P. penetrans* are non-motile and they are resistant to heat and desiccation (Oostendorp *et al.*, 1990). Their dispersal in soil is largely dependent upon the rate of water percolation (Oostendorp *et al.*, 1990), the size of soil pore openings, tillage practices, and soil invertebrates (Sayre and Starr, 1988).

177

2.5. Biocontrol Potential

2.5.1. Efficacy

Prasad (1971) was perhaps the first to find that the greenhouse tomatoes inoculated with *M. incognita* had fewer galls on the roots when grown is soil infested with *P. penetrans*. Application of *P. penetrans* in several greenhouse and microplot tests has consistently suppressed the nematode-induced root galling and egg production with consequent enhanced plant growth (Daudi *et al.*, 1990; Vergas *et al.*, 1992; Walia, 1994b; Zaki and Maqbool, 1992a). Sayre (1984) obtained significantly higher yield of cucumber in plots where *P. penetrans* was also present than in the plots infested only with nematodes. In microplot tests, application of *P. penetrans* substantially increased the yields of soybean (212%) and winter vetch (219%); this yield increase was comparable to that often achieved with nematicides (Brown *et al.*, 1985). Walia and Dalal (1994) recorded 18-20 % increase in tomato yield by treatment of *M. javanica* infested nursery soil with *P. penetrans*.

2.5.2. Mode of action

The infection of *P. penetrans* reduces the invasion of roots by the juveniles and even those who invade the roots do not lay eggs. The reduction in root penetration occurs when the nematodes have 15-21 spores attached (Davies *et al.*, 1988), but as many as 50 spores per nematode may be needed for marked reductions in the root invasion by the juveniles (Sell and Hansen, 1987). The movement of juveniles is significantly reduced when they are encumbered with an average of seven spores per juvenile (Davies *et al.*, 1991). Adhesion of bacterial spores on the nematode cuticle across the striae may obstruct bending of nematode body, rendering it less mobile, thus leading to reduced root invasion. The reproductive system fails to develop in the infected females and they do not lay eggs. This leads to marked reduction in the secondary infection by the second or subsequent generation juveniles. The effective control of nematodes by *P. penetrans* is dependent on the density of spores per unit soil (Chen *et al.*, 1996b; Ciancio and Bourijate, 1995; Singh and Dhawan, 1994).

Stirling *et al.* (1990) have attempted to predict the concentration of *P. penetrans* spores needed to control root-knot nematodes in a field. The nematodes are rarely able to initiate galls after moving more than 2 cm through soil containing 1×10^5 spores per g soil. Their data suggest that spore concentrations of 1×10^5 and 2.2×10^5 spores per g soil, respectively would be required at 27 °C to ensure that an average of 20 and 50 spores are attached per nematode. At such spore concentrations, many juveniles would not be able to invade the roots and infected females would not produce any eggs. The above predictions are also supported by the results of the field experiments of Stirling (1984) with dried root preparation of *P. penetrans*. The spore concentration of the inoculum used in this work was estimated at 1×10^8 to 1×10^9 spores per g (Stirling and Wachtel, 1980). Since plants growing in soil containing 424 to 600 mg dried root preparation per kg soil were almost free of galls, control appears to have been obtained with spore concentration between 0.5 and 5×10^5 spores per g soil. Although these predictions of the concentrations of *P. penetrans* spores required to control the root-knot nematode are based on limited data and need further verification, they are indicative of the inoculum quantity needed for use of *Pasteuria* as a biological control agent. If the number of spores are in the vicinity of 1×10^{10} spore per ml of females, about 20 L of inoculum per ha would be required to achieve the required spore concentration in the top 15 cm of soil profile.

2.5.3. Method and time of bacterium inoculation

Soil infested with endospores of *P. penetrans* is a source of inoculum and can be applied in the field where nematode control is needed. Root powder consisting of *Pasteuria* infected nematodes also serves as a good source of endospore inoculum.. The efficacy of endospore infested soil or root powder is a function of endospore concentration and effective nematode control is achieved when at least 80% of·the bioassayed juveniles were encumbered with 10 or more spores. However, there are no standardized methods to achieve uniform results all the times (Adiko and Gowen, 1994). Application of spores 2.5 cm deep in soil is effective for the control of *M. incognita* on tomato (Ahmad *et al.*, 1994). Presently there are no techniques to mass produce the endospores for large scale field applications. However, it is possible to treat small areas in the nursery beds, spot application, and seed coating. Vegetable nurseries invariably provide a continuous source of nematodes which can move long distances along with the seedlings. If *Pasteuria* spores are added to such nematode-infested nursery soils, the bacterium will also spread to the vegetable fields along with nematode-infected seedlings. This may aid in gradual control of the root-knot nematodes in the field. In our preliminary trials, we found that application of aqueous spore suspension (1500 spores per g nursery soil) to nematode-infested soil improved the growth of egg plant by 30% and reduced the root damage by *M. javanica* by 36% to when the seedlings were to planted in a nematode infested soils (Walia *et al.*, 1992). Application of endospores in root powder preparation was also very effective in reducing the nematode-caused damage to tomato. Addition of spore concentration of 1×10^4 per g nematode infested soil resulted in 30% parasitization of *M. javanica* females and it increased to 67% at 1×10^5 spore density (Walia, 1994b). These seedlings, when transplanted to the field, produced 18-20% higher yield of tomato, and had low levels of nematode infection. At-harvest of tomato crop, about 18% and 50% of the juveniles in soil had spores adhering to their body in the treatments of 1×10^4 and 1×10^5 spores, respectively (Walia and Dalal, 1994).

2.5.4. Mass production

Pasteuria are obligate parasites. Attempts to culture the *Pasteuria* on artificial medium have been unsuccessful. In the 1970s, spore-infested soils were used as a source of parasite (Mankau, 1972). In the 1980s, Stirling and Wachtel (1980) have devised another method in which tomato roots heavily infected with *Meloidogyne* females parasitized by *P. penetrans* are obtained by inoculating juveniles with spores adhered to their body. These roots are air-dried and ground to make a powder which is easy to handle and store (Gowen and Channer, 1988; Sharma and Stirling, 1991). Stirling and Wachtel (1980) developed a bioassay to enable comparison of the potency of different preparations of the parasite. Verdejo and Jaffee (1988) have described a three component system for gnotobiotic production of spores. The root-knot nematode juveniles, with or without *P. penetrans* spores, are added to plates containing *Agrobacterium rhizogenes*-transformed tomato or potato roots on solid Gamborg's B5 medium. After about six weeks, the females are examined for spore production. Chen *et al.* (1996a) compared six methods for quantification of the endospore concentrations of *P. penetrans* from tomato roots. Mortar disruption and machine disruption methods gave the highest estimations (endospores per g root) of 83 and 79 million, respectively. These methods were significantly superior to incubation bioassay (48 million), enzymatic disruption (32 million) and enzymatic disruption + floatation (26 million) methods. A centrifugation bioassay method gave the lowest estimation of 13 million. If the nematode-bacterium infected roots are dried, powdered and then passed through 100 mesh (150 µm pore size) sieve, the fine root powder which passes through the sieve has more concentration of spores. This powder can be used for coating of seeds with aid of a sticker. A very small quantity (50 mg in 10 ml water) of the root powder can be used to count spores directly with a haemacytometer.

In vitro cultures of *P. penetrans* were attempted by inoculating either spore or mycelial bodies on a diverse range of simple and complex media and incubating them in aerobic, reduced oxygen, anaerobic and increased CO_2 environments. Signs of spore germination or growth of vegetative stages were never observed (Williams *et al.*, 1989). Bishop and Ellar (1991) also screened several previously published media without any success. Two defined media were then formulated-medium one maintained inoculated ball-mycelia of *P. penetrans* in an apparently viable state for one month giving low yields of mature spores; and medium two gave a small increase in the number of inoculated ball-mycelia but lysis resulted. Limited *in vitro* cultivation of *P. nishizawae* has recently been achieved (Reise *et al.*, 1991). A complex undefined medium containing 111 ingredients has been developed to propagate this organism. All stages of life cycle of this organism (doublets, triplets, cauliflower, sporangia, endospores and immature and mature spores) appeared to be produced and maintained for up to six transfers over an 8 month period.

2.5.5. Population dynamics

Cropping frequency and host nematode density in soil influence the populations of *P. penetrans* (Ostendorp *et al.*, 1991; Verdejo-Lucas, 1992). The population densities of *P. penetrans* increase gradually over time from the relatively low levels of spores initially. It can establish and build up gradually to levels which may keep the nematode population below damaging levels in the long run. Thus, even if large scale production of inoculum of *P. penetrans* proves to be impossible or too costly, the use of this antagonist in integrated nematode management schemes may be feasible.

2.5.6. Compatibility with other control practices

Applications of nematicides and *Pasteuria* endospores in soil together generally have a synergistic effect on nematode control. Commonly used pesticides (nematicides) do not have any noticeable adverse effect on the bacterial population (Maheswari *et al.*, 1987; Sekhar and Gill, 1991). The low dosages of nematicides may enhance the infection of *Pasteuria* (Brown and Nordmeyer, 1985) and it is possible to develop treatment protocols which encourage the natural build up of spore concentrations in soil over repeated crop cycles (Daudi and Gowen, 1992). Combined application of *P. penetrans* and oxamyl have additive effects in suppressing root galling and egg production on nematode susceptible tomato variety. Oxamyl treatment enhances the efficacy of *P. penetrans* and the nematode suppressive effect lasts for more than one crop season. Soil solarization also enhances the efficacy of *P. penetrans* (Tzortzakakis and Gowen, 1994). Application of *Paecilomyces lilacinus* and *P. penetrans* together is more effective in controlling the root-knot nematodes and enhancing the crop yield than application of either of the organisms alone on vetch (Dube and Smart, 1987); mung bean (Shahzad *et al.*, 1990); okra (Zaki and Maqbool, 1991) and egg plant (Zaki and Maqbool, 1992b). *Verticillium chlamydosporium* and *P. penetrans* together complement each other and give 92% control of *M. incognita* population on tomato in pot experiments (Leij *et al.*, 1992).

180

3. NEMATODE TOXIN PRODUCING BACTERIA

Various types of bacteria colonize the plant rhizosphere. The metabolic products of these bacteria influence not only the plant growth but also the neighboring microorganisms including the plant parasitic nematodes. Some bacteria decompose the organic matter and the decomposition products such as volatile fatty acids, hydrogen sulphide and ammonia inhibit the nematode populations. Some plant health promoting bacteria such as *Rhizobium, Bradyrhizobium,* and heterotrophic bacteria like *Azotobacter* and *Azospirillum* have nematode antagonistic traits (Chahal and Chahal, 1988; Huang, 1987; Ramakrishnan *et al.*, 1996; Vats and Dalal, 1998). The plant health promoting rhizobacteria may adversely influence the intimate relationship between the plant parasitic nematodes and their hosts by regulating the nematode behavior during the early root penetration phase of parasitism. Two mechanisms of action may be responsible for reduction in nematode infection : (i) production of metabolites which reduce hatching and host attraction, and (ii) degradation of specific root exudates which control nematode behavior (Sikora and Hoffman-Hergarten, 1993).

Of various rhizobacteria that have been isolated so far, about 7-10 % have antagonistic potential against cyst and root-knot nematodes (Sikora, 1992). In greenhouse tests, *Agrobacterium radiobacter* and *Bacillus sphaericus* cause 41% reduction in root invasion by *Globodera pallida* in potato. In field trials the blend of two bacteria caused a 31 % and *B. sphaericus*-alone 29 % reduction in root invasion by the nematode. Tuber yield increased by 18 % and 22 % in the field trials when tubers were treated with *A. radiobacter* or the blend, respectively (Racke and Sikora, 1992). Application of *Bacillus subtilis, Bradyrhizobium japonicum,* singly or in combination, with mycorrhiza results in increased shoot dry weight, number of nodules, phosphorus content and reduced nematode (*Heterodera cajani*) multiplication and *Fusarium*-wilt incidence on pigeonpea (Siddiqui and Mahmood, 1995).

Strains of *Bacillus thuringiensis* (*Bt*) are toxic to plant parasitic nematodes. Jacq *et al.* (1977) found that a thermostable toxin of *Bt* was toxic to populations of *Meloidogyne, Panagrellus* and *Aphelenchus*, and prevented *M. incognita* larvae from forming galls on tomato roots. DIPEL and SAN strains of *Bt* sub sp. *kurstaki* suppressed the populations of both *M. javanica* and *Tylenchulus semipenetrans* (Osman *et al.*, 1988). Zuckerman *et al.* (1993) reported that a strain of *Bt* (CR-371) caused significant reduction in galls on tomato in a greenhouse trial. In field trials in Puerto Rico, *Bt* treated tomatoes and pepper had significantly fewer root galls due to *M. incognita*. CR-371 treated strawberry plant also had smaller populations of *Pratylenchus penetrans* in a greenhouse trial in USA.

4. CONCLUSIONS

The nematode parasitic bacteria (*Pasteuria* group) have undoubtedly generated a great interest in the biological control of plant parasitic nematodes (Walia, 1994a). Presently, nematologists in several countries are working on this hyperparasite. The taxonomic reorganization at species level among the members of *Pasteuria* group remains confusing, as reports of new nematode hosts and isolates with a typical life cycle patterns are becoming available. The work on *Pasteuria* group has also generated new knowledge on nematode cuticle surface.

The major attributes which favour *P. penetrans* as a potential biological control agent are - demonstrable nematode control potential, long viability of spores, resistance to heat and desiccation, persistence in field soils, compatibility with chemical nematicides, non toxicity to plants, other soil biota and human kind, and easy storage. Presently, use of

181

Pasteuria for nematode control is very limited. The major hurdle is lack of techniques for its mass production *in vitro*. It is important to identify synthetic medium for culturing of *Pasteuria* to expand the use of this organism as a commercially viable biocontrol agent of phytonematodes. The compatibility of *P. penetrans* with chemical nematicides and similar prospects with other management practices make it a suitable candidate for development as a component of integrated nematode management systems. However, it is already about 25 years of across nations research on *Pasteuria* spp. to develop a commercially viable biological control agent for plant parasitic nematodes. There is not yet any classical example of large scale effective control of a nematode pest using *Pasteuria*. There is a need for much greater funding support to make it happen.

The rhizobacteria hold a good promise in reducing the damage caused by phytonematodes. These bacteria are easy to culture and produce in large quantity. However, research on use of rhizobacteria for nematode control has a long way to go to decide whether or not these bacteria provide a practical option for nematode management. We need to identify and select the most effective strains, study their mode of action, and demonstrate their efficacy in farmers fields.

REFERENCES

Abrantes, I.M.De O. and Vovlas, N. 1988, A note on the parasitism of the phytonematodes *Meloidogyne* sp. and *Heterodera fici* by *Pasteuria penetrans, Can. J. Zool.* 66 : 2852-2854.

Adiko, A. and Gowen, S.R. 1994, Comparison of two inoculation methods of root-knot nematodes for the assessment of biocontrol potential of *Pasteuria penetrans, Afro Asian J. Nematol.* 4 : 32-34.

Afolabi, P., Davies, K,G. and O'Shea, P.S. 1995, The electrostatic nature of the spore of *Pasteuria penetrans,* the bacterial parasite of root-knot nematodes, *J. App. Bacteriol.* 79 : 244-249.

Ahmad, R., Abbas, M.K., Khan, M.A., Inam-ul-Haq, M., Javed, N. and Sahi, S.T. 1994, Evaluation of different methods of application of *Pasteuria penetrans* for the biocontrol of root-knot of tomato (*Meloidogyne incognita*), *Pakistan J. Nematol.* 12 : 155-160.

Ahmad, R. and Gowen, S.R. 1991, Studies on the infection of *Meloidogyne* spp. with isolates of *Pasteuria penetrans, Nematologia Mediterranea* 19 : 229-233.

Bajaj, H.K., Dabur, K.R., Walia, R.K. and Mehta, S.K. 1997, Host specificity among isolates of bacterial parasite, *Pasteuria* spp. from *Heterodera cajani* and *H. mothi, Internat. J. Nematol.* 7 : 227-228.

Bird, A.F. 1986, The influence of the actinomycete, *Pasteuria penetrans*, on the host-parasite relationship of the plant - parasitic nematode, *Meloidogyne javanica, Parasitol.* 93 : 571-580.

Bird, A.F. and Brisbane, P.G. 1988, The influence of *Pasteuria penetrans* in field soils on the reproduction of root-knot nematodes, *Revue de Nematologie* 11 : 75-81.

Bishop, A.H. and Ellar, D.J. 1991, Attempts to culture *Pasteuria penetrans in vitro, Biocon. Scie. Technol.* 1 : 101-114.

Brown, S.M., Kepner, J.L. and Smart, G.C. Jr. 1985, Increased crop yields following applications of *Bacillus penetrans* to field plots infested with *Meloidogyne incognita, Soil Biol. Biochem.* 17 : 483-486.

Brown, S.M. and Nordmeyer, D. 1985, Synergistic reproduction in root galling by *Meloidogyne javanica* with *Pasteuria penetrans* and nematicides, *Revue de Nematologie* 8 : 285-286.

Chahal, P.P.K. and Chahal, V.P.S. 1988, Biological control of root-knot nematode of brinjal (*Solanum melongena* L.) with *Azotobacter chroococcum*, In : *Advances in Plant Nematology*, eds. M.A. Maqbool, A.M. Golden, A. Ghaffar and L.R. Krusberg, Proc. U.S.-Pakistan International Workshop on Plant Nematology, Karachi, Pakistan, pp. 257-264.

Channer, A.G. and Gowen, S.R. 1992, Selection for increased host resistance and increased pathogen specificity in *Meloidogyne-Pasteuria penetrans* interaction, *Fundam. App. Nematol.* 15 : 331-339.

Chen, Z.X., Dickson, D.W. and Hewlett, T.E. 1996a, Quantification of endospore concentrations of *Pasteuria penetrans* in tomato root material, *J. Nematol.* 28 : 50-55.

Chen, Z.X., Dickson, D.W. Mc Sorley, R., Mitchell, D.J. and Hewlett, T.E. 1996b, Suppression of *Meloidogyne arenaria* race-1 by soil application of endospores of *Pasteuria penetrans, J. Nematol.* 28 : 159-168.

Ciancio, A., Bonsignore, R., Vovlas, N. and Lamberti, F. 1994,Records and spore morphometrics of *Pasteuria penetrans* group parasites of nematodes, *J. Inverteb. Pathol.* 63 : 260-267.

Ciancio, A. and Bourijate, M. 1995, Relationship between *Pasteuria penetrans* infection levels and density of *Meloidogyne javanica*, *Nematologia Mediterranea* 23 : 43-49.

Ciancio, A. and Mankau, R. 1989, Note on *Pasteuria* sp. parasitic in longidorid nematodes, *Nematropica* 19 : 105-109.

Ciancio, A. Mankau, R. and Mundo-Ocampo, M. 1992, Parasitism of *Helicotylenchus lobus* by *Pasteuria penetrans* in naturally infested soil, *J. Nematol.*, 24 : 29-35.

Cobb, N.A. 1906, Fungus maladies of the sugarcane, with notes on associated insects and nematodes. (Second edition), Hawaiian Sugar Planters Assoc. Exp. Sta. Div, *Path. Physiol. Bull.* 5 : 163-195.

Daudi, A.T., Channer, A.G., Ahmed, R. and Gowen, S.R. 1990, *Pasteuria penetrans* as a biocontrol agent of *Meloidogyne javanica* in the field in Malawi and in microplots in Pakistan, In : *Proc. British Crop Protection Conference, Pests and Diseases* 1 : 253-257.

Daudi, A.T. and Gowen, S.R. 1992, The potential for managing root-knot nematode by use of *Pasteuria penetrans* and oxamyl, *Nematologia Mediterranea* 20 : 241-244.

Davies, K.G. 1998, Natural parasite and biological control, In: *Cyst Nematodes*, ed. S. B. Sharma, Chapman and Hall, London, UK.

Davies, K.G. and Danks, C. 1992, Interspecific differences in the nematode surface coat between *Meloidogyne incognita* and *M. arenaria* related to the adhesion of the bacterium *Pasteuria penetrans*, *Parasitol.* 105 : 475-480.

Davies, K.G. and Danks, C. 1993, Carbohydrate/protein interactions between the cuticle of infective juveniles of *Meloidogyne incognita* and spores of the obligate hyperparasite *Pasteuria penetrans*, *Nematologica* 39 : 53-64.

Davies, K.G., Flynn, C.A., Laird, V. and Kerry, B.R. 1990, The life cycle, population dynamics and host specificity of a parasite of *Heterodera avenae*, similar to *Pasteuria penetrans*, *Revue de Nematologie* 13 : 303-309.

Davies, K.G., Kerry, B.R. and Flynn, C.A. 1988, Observations on the pathogenicity of *Pasteuria penetrans*, a parasite of root-knot nematodes, *Ann. App. Biol.* 112 : 491-501.

Davies, K.G., Laird, V. and Kerry, B.R. 1991, The motility, development and infection of *Meloidogyne incognita* encumbered with spores of obligate hyperparasite, *Pasteuria penetrans*, *Revue de Nematologie* 14 : 611-618.

Davies, K.G., Leij, F.A.A.M. de, and Kerry, B.R. 1991, Microbial agents for the biological control of plant parasitic nematodes in tropical agriculture, *Trop. Pest Manag.* 37 : 303-320.

Davies, K.G., Redden, M. and Pearson, T.K. 1994, Endospore heterogeneity in *Pasteuria penetrans* related to adhesion to plant parasitic nematodes, *Lett. App. Microbiol.* 19 : 370-373.

Davies, K.G., Robinson, M.P. and Laird, V. 1992, Proteins involved in the attachment of a hyperparasite, *Pasteuria penetrans*, to its plant parasitic nematode host, *Meloidogyne incognita*, *J. Inverteb. Pathol.* 59 : 18-23.

Doran, J.W., Serratonio, M. and Liebig, M..A. 1996, Soil health and sustainability, *Adv. Agron.* 56: 1-45.

Dube, B. and Smart, G.C. Jr. 1987, Biological control of *Meloidogyne incognita* by *Paecilomyces lilacinus* and *Pasteuria penetrans*, *J. Nematol.* 19 : 222-227.

Fattah, F.A., Saleh, H.M. and Aboud, H.M. 1989, Parasitism of the citrus nematode, *Tylenchulus semipenetrans* by *Pasteuria penetrans* in Iraq, *J. Nematol.*, 21 : 431-433.

Ginnakou, I.O., Pembroke, B., Gowen, S.R. and Davies, K.G. 1997, Effects of long term storage and above normal temperatures on spore adhesion of *Pasteuria penetrans* and infection of the root-knot nematode, *Meloidogyne javanica*, *Nematol.* 43 : 185-192.

Gowen, S.R. and Channer, A.G. 1988, The production of *Pasteuria penetrans* for the control of root-knot nematodes, In : *Proc. Brighton Crop Protection Conference, Pests and Diseases*, Brighton, U.K. pp.1215-1220.

Hatz, B. and Dickson, D.W. 1992, Effect of temperature on attachment, development and interactions of *Pasteuria penetrans* on *Meloidogyne arenaria*, *J. Nematol.* 24 : 512-521.

Huang, J.S., 1987, Interactions of nematodes with rhizobia, In: *Vistas on Nematology*, eds. J.A. Veech and D.W. Dickson, SON, Hyattsville, MD, USA, pp. 310-306.

Inserra, R.N., Oosterndorp, M. and Dickson, D.W. 1992, *Pasteuria* sp. parasitizing *Trophonema okamotoi* in Florida, *J. Nematol.* 24 : 36-39.

Jacq, V.A., Ignoffo, C.M. and Dropkin, V.H. 1977, Deleterious effects of the thermostable toxin of *Bacillus thuringiensis* on species of soil-inhabiting, myceliophagus, and plant - parasitic nematodes, *J. Kans. Entomol. Soc.* 50 : 394.

Leij, F.A.A.M. de, Davies, K.G. and Kerry, B.R. 1992, The use of *Verticillium chlamydosporium* Goddard and *Pasteuria penetrans* (Thorne) Sayre and Starr alone and in combination to control *Meloidogyne incognita* on tomato plants, *Fundam. Appl. Nematol.* 15 : 235-242.

183

Maheswari, T.U., Mani, A. and Rao, P.K. 1987, Combined efficacy of the bacterial spore parasite *Pasteuria penetrans* (Thorne) and nematicides in the control of *Meloidogyne javanica* on tomato, *J. Biol. Cont.* 1 : 53-57.

Mankau, R. 1972, Utilization of parasites and predators in nematode pest management ecology, In : *Proc. Tall Timbers Conference on Ecological Animal Control by Habitat Management* 4 : 129-143.

Mankau, R. 1975a, *Bacillus penetrans* n. comb. causing a virulent disease of plant parasitic nematodes, *J. Inverteb. Pathol.* 26 : 333-339.

Mankau, R. 1975b, Prokaryote affinities of *Duboscquia penetrans* Thorne, *J. Protozool.* 21 : 31-34.

Maqbool, M.A. and Zaki, M.J. 1990, Occurrence of *Pasteuria penetrans* on mature females of *Meloidogyne javanica* and *M. incognita* root-knot nematodes in Pakistan, *Pakistan J. Nematol.* 8 : 13-15.

Oostendorp, M., Dickson, D.W. and Mitchell, D.J. 1990, Host range and ecology of isolates of *Pasteuria* spp. from the southern United States, *J. Nematol.* 22 : 525-531.

Oostendorp, M., Dickson, D.W. and Mitchell, D.J. 1991, Population development of *Pasteuria penetrans* on *Meloidogyne arenaria*, *J. Nematol.* 23 : 58-64.

Osman, G.Y., Salem, F.M. and Ghattas, A. 1988, Bioefficacy of two bacterial insecticide strains of *Bacillus thuringiensis* as a biological control agent in comparison with a nematicide, Nemacur, on certain parasitic nematoda, *Anzeiger fur Schadlingskunde Pflanzenschutz Umweltschutz* 61 : 35-37.

Persidis, A., Lay, J.G., Manousis, T., Bishop, A.H. and Ellar D.J. 1991, Characterisation of potential adhesions of the bacterium, *Pasteuria penetrans* and of putative receptors on the cuticle of *Meloidogyne incognita*, a nematode host, *J. Cell Scie.* 100 : 613-622.

Prasad, N. 1971, Studies on the biology, ultrastructure, and effectiveness of a sporozoan endoparasite of nematodes, Ph.D. dissertation, University of California, Riverside, USA.

Racke, J. and Sikora, R.A. 1992, Influence of the plant health promoting rhizobacteria *Agrobacterium radiobacter* and *Bacillus sphaericus* on *Globodera pallida* root infection of potato and subsequent plant growth, *J. Phytopath.* 134 : 198-208.

Ramakrishnan, S., Gunasekaran, C.R. and Vadivelu, S. 1996, Effect of biofertilizers *Azolla* and *Azospirillum* on root-knot nematode *Meloidogyne incognita* and plant growth of okra, *Indian J. Nematol.* 26 : 127-130.

Reise, R.W., Hackett, K. and Huettel, R.N. 1991, Limited *in vitro* cultivation of *Pasteuria nishizawae*, *J. Nematol.* 23 : 547-548.

Sayre, R.M. 1984, *Bacillus penetrans* : A biocontrol agent of *Meloidogyne incognita* on cucumber, In : *Proc. 1st Inernational Congress of Nematolgy*, Guelph, Canada, p. 81.

Sayre, R.M. and Starr, M.P. 1985, *Pasteuria penetrans* (ex Thorne, 1940) nom. rev., comb. n., sp. n., a mycelial and endospore-forming bacterium parasitic in plant parasitic nematodes, In : *Proc. Helminthological Society of Washington*, USA 52 : 149-165.

Sayre, R.M. and Starr, M.P. 1988, Bacterial diseases and antagonisms of nematodes, In: *Diseases of Nematodes* Vol. I, eds. G.O. Poinar Jr. and H.B. Jansson, CRC Press, Boca Raton, Florida, USA, pp. 69-101.

Sayre, R.M., Starr, M.P., Golden, A.M., Wergin, W.P. and Endo, B.Y. 1988, Comparison of *Pasteuria penetrans* from *Meloidogyne incognita* with a related mycelial and endospore-forming bacterial parasite from *Pratylenchus brachyurus*, In : *Proceedings of Helminthological Society of Washington*, USA 55 : 28-49.

Sayre, R.M., Wergin, W.P., Nishizawa, T. and Starr, M.P. 1991, Light and electron microscopical study of a bacterial parasite from the cyst nematode *Heterodera glycines*, In : *Proc. Helminthological Society of Washington*, USA, 58 : 69-81.

Sayre, R.M., Wergin, W.P., Schmidt, J.M. and Starr, M.P. 1991, *Pasteuria nishizawae* sp. nov., a mycelial and endospore forming bacterium parasitic on cyst nematodes of genera *Heterodera* and *Globodera*, *Res. Microbiol.* 142 : 551-564.

Sekhar, N.S. and Gill, J.S. 1990, Effect of temperature on attachment of *Pasteuria penetrans* spores to second - stage juveniles of *Meloidogyne incognita*, *Indian J. Nematol.* 20 : 57-59.

Sekhar, N.S. and Gill, J.S. 1991, Efficacy of *Pasteuria penetrans* alone and in combination with carbofuran in controlling *Meloidogyne incognita*, *Indian J. Nematol.* 21 :61-65.

Sell, P. and Hansen, C. 1987, Beziehungen zwischen Wurzelgallennematoden und ihrem naturlichen Gegenspieler *Pasteuria penetrans*, *Mededelingen Faculteit Landbouwwetenschappen Rijksuniversiteit Gent.* 52 : 607-615.

Shahzad, S., Ehteshamul-Haque, S. and Ghaffar, A. 1990, Efficacy of *Pasteuria penetrans* and *Paecilomyces lilacinus* in the biological control of *Meloidogyne javanica* on mungbean, *Internat. Nematol. Network Newsl.* 7 : 34-35.

Sharma, R. 1985, Studies on pathology of some cyst nematodes of the genus *Heterodera*, Ph. D. thesis, Indian Agricultural Research Institute, New Delhi, India.

184

Sharma, R. and Sharma, S.B. 1989, Sticky swarm disease of *Heterodera cajani* caused by *Pasteuria penetrans*, *Internat. Pigeonpea Newsl.* 10: 26-27.

Sharma, R. and Swarup, G. 1988, *Pathology of Cyst Nematodes*, Malhotra Publishing House, New Delhi, India.

Sharma, R.D. and Stirling, G.R. 1991, *In vivo* mass production system for *Pasteuria penetrans*, *Nematol.* 37 : 483-484.

Sharma, S.B. and Davies, K.G. 1996a, Characterization of *Pasteuria* isolated from *Heterodera cajani* using morphology, pathology and serology of endospores, *System. App. Microbiol.* 19: 106-112.

Sharma, S.B. and Davies, K.G. 1996b, Comparison of two sympatric *Pasteuria* populations isolated from a tropical vertisol soil, *World J. Microbiol. Biotechnol.* 12: 361-366.

Sharma, S.B. and Davies, K.G. 1997, Modulation of spore adhesion of the hyperparasitic bacterium *Pasteuria penetrans* to nematode cuticle, *Lett. App. Microbiol.* 25: 426-430.

Siddiqui, Z.A. and Mahmood, I. 1995, Biological control of *Heterodera cajani* and *Fusarium udum* by *Bacillus subtilis*, *Bradyrhizobium japonicum* and *Glomus fasciculatum* on pigeonpea, *Fundam. App. Nematol.* 18 : 559-566.

Sikora, R.A. 1992, Management of the antagonistic potential in agricultural eco-systems for the biological control of plant parasitic nematodes, *Ann. Rev. Phytopathol.* 30 : 245-270.

Sikora, R.A. and Hoffman-Hergarten, S. 1993, Biological control of plant parasitic nematodes with plant health promoting rhizobacteria, pest management : biologically based techniques, In : *Proc. Beltsville Symposium XVIII*, Agricultural Research Services, U.S. Department of Agriculture, Beltsville, Maryland, USA, pp. 166-172.

Silva de, M.P., Gowen, S.R. and De Silva, M.P. 1994, Attrempts to adapt a population of *Pasteuria penetrans* originating from *Meloidogyne javanica* to its less susceptible host *M. arenaria*, *Afro-Asian J. Nematol.* 1 : 40-43.

Singh, B. and Dhawan, S.C. 1992, Effect of soil texture on attachment of bacterial spores of *Pasteuria penetrans* to the second stage juveniles of *Heterodera cajani*, *Indian J. Nematol.* 22 : 72-74.

Singh, B. and Dhawan, S.C. 1994, Effect of *Pasteuria penetrans* on the penetration and multiplication of *Heterodera cajani* in *Vigna unguiculata* roots, *Nematologia Mediterranea* 22 : 159-161.

Speigel, Y., Mor, M. and Sharon, E. 1996, Attachment of *Pasteuria penetrans* endospores to the surface of *Meloidogyne javanica* second stage juveniles, *J. Nematol.* 28 : 328-334.

Starr, M.P. and Sayre, R.M. 1988, *Pasteuria thornei* sp. nov. and *Pasteuria penetrans* sensu stricto emend., mycelial and endospore - forming bacteria parasitic, respectively, on plant-parasitic nematodes of the genera *Pratylenchus* and *Meloidogyne*, *Ann. Inst. Pasteur/Microbiol.* 139 : 11-31.

Stirling, G.R. 1984, Biological control of *Meloidogyne javanica* with *Bacillus penetrans*, *Phytopath.* 74 : 55-60.

Stirling, G.R. 1985, Host specificity of *Pasteuria penetrans* with the genus *Meloidogyne*, *Nematol.* 31 : 203-209.

Stirling, G.R. 1991, *Biological Control of Plantparasitic Nematodes*, CAB International, Wallingford, UK.

Stirling, G.R., Bird, A.F. and Cakurs, A.B. 1986, Attachment of *Pasteuria penetrans* spores to the cuticles of root-knot nematodes, *Revue de Nematologie* 9 : 251-260.

Stirling, G.R., Sharma, R.D. and Perry, J. 1990, Attachment of *Pasteuria penetrans* spores to the root-knot nematode *Meloidogyne javanica* in soil and its effects on infectivity, *Nematol.* 36 : 246-252.

Stirling, G.R. and Wachtel, M.F. 1980, Mass production of *Bacillus penetrans* for the biological control of root-knot nematodes, *Nematol.* 26 : 308-312.

Sturhan, D. 1985, Untersuchungen uber Verbreitung und Wirte des Nematoden-parasiten *Bacillus penetrans*, *Mitt. Biol. Bundesanst., Land Forstwirtsch., Berlin-Dahlem*, 226 : 75.

Sturhan, D. 1988, New host and geographical records of nematode-parasitic bacteria of the *Pasteuria penetrans* group, *Nematol.* 34 : 350-356.

Sturhan, D., Winkelheide, R., Sayre, R.M. and Wergin, W.P. 1994, Light and electron microscopical studies of the life-cycle and developmental stages of a *Pasteuria* isolate parasitising the pea cyst nematode, *Heterodera goettingiana*, *Fundam. App. Nematol.* 17 : 29-42.

Thorne, G. 1940, *Duboscqia penetrans* n. sp. (Sporozoa, Microsporidia, Nosematidae), a parasite of the nematode, *Pratylenchus penetrans* (de Man) Filipjev, In : *Proc. Helminthological Society of Washington*, USA, 7 : 51-53.

Torsvik, V., Goksoy, J. and Daae, F. L. 1990, High diversity in DNA of soil bacteria, *App. Environ. Microbiol.* 56: 782-787.

Tzortzakakis, E.A. and Gowen, S.R. 1994, Evaluation of *Pasteuria penetrans* alone and in combination with oxamyl, plant resistance and solarization for control of *Meloidogyne* spp. on vegetables grown in green houses in Crete, *Crop Prot.* 13 : 455-462.

185

Vats, R. and Dalal, M.R. 1998, Interrelationship between *Rotylenchulus reniformis* and *Rhizobium leguminosarum* on *Pisum sativum*, III, In : *Proc. International Symposium of Afro-Asian Society of Nematologists*, Coimbatore, India, p. 38.

Verdejo, S. and Jaffee, B.A. 1988, Reproduction of *Pasteuria penetrans* in a tissue-culture system containing *Meloidogyne javanica* and *Agrobacterium rhizogenes*-transformed roots, *Phytopath.* 78 : 1284-1286.

Verdejo-Lucas, S. 1992, Seasonal population fluctuation of *Meloidogyne* spp. and the *Pasteuria penetrans* group in Kiwi orchards, *Pl. Dis.* 76 : 1275-1279.

Vergas, R., Acosta, N., Monllor, A. and Betancourt, C. 1992, Control of *Meloidogyne* species with *Pasteuria penetrans* (Thorne) Sayre and Starr, *J. Agric. Univ. Puerto Rico* 76 : 63-70.

Vovlas, N., Ciancio, A. and Vlachopoulas, E.G. 1993, *Pasteuria penetrans* parasitizing *Helicotylenchus pseudorobustus* and *Rotylenchus capensis* in Greece, *Afro-Asian J. Nematol.* 3 : 39-42.

Walia, R.K. 1994a, Biological control of phytonematodes : Principles and prospects, In : *Nematode Pest Management in Crops,* eds. D.S. Bhatti and R.K. Walia, CBS Publishers and Distributors, Delhi, India, pp.228-238.

Walia, R.K. 1994b, Assessment of nursery treatment with *Pasteuria penetrans* for the control of *Meloidogyne javanica* on tomato, in green-house, *J. Biol. Cont.* 8 : 68-70.

Walia, R.K., Bajaj, H.K. and Dalal, M.R. 1994, Records of bacterial parasite, *Pasteuria* spp. on phyto and soil nematodes in Haryana (India), *Curr. Nematol.* 5 : 223-225.

Walia, R.K., Bansal, R.K. and Bhatti, D.S. 1990, A new bacterial parasite (*Pasteuria* sp.) isolated from pigeonpea cyst nematode, *Heterodera cajani, Internat. Nematol. Network Newsl.* 7 : 30-31.

Walia, R.K., Bansal, R.K. and Bhatti, D.S. 1992, Efficacy of bacterium (*Pasteuria penetrans*) and fungus (*Paecilomyces lilacinus*), alone and in combination against root-knot nematode (*Meloidogyne javanica*) infecting brinjal (*Solanum melongena*), In : *Current Trends in Life Sciences*, Vol. XXI. *Recent Developments in Biocontrol of Plant Pathogens,* eds. K. Manibhushan Rao and A. Mahadevan, Today and Tomorrow's Printers and Publishers, New Delhi, India, pp. 119-124.

Walia, R.K. and Dalal, M.R. 1994, Efficacy of bacterial parasite, *Pasteuria penetrans* application as nursery soil treatment in controlling root-knot nematode, *Meloidogyne javanica* on tomato, *Pest Manag. Eco. Zool.* 2 : 19-21.

Williams, A.B., Stirling, G.R., Hayward, A.C. and Perry, J. 1989, Properties and attempted culture of *Pasteuria penetrans* , a bacterial parasite of root-knot nematode (*Meloidogyne javanica*), *J. App. Bacteriol.* 67 : 145-156.

Zaki, M.J. and Maqbool, M.A. 1991, Combined efficacy of *Pasteuria penetrans* and other biocontrol agents on the control of root-knot nematode on okra, *Pakistan J. Nematol.* 9 : 49-52.

Zaki, M.J. and Maqbool, M.A. 1992a, Effect of spore concentrations of *Pasteuria penetrans* on the attachment of *Meloidogyne* larvae and growth of okra plants, *Pakistan J. Nematol.* 10 : 69-73.

Zaki, M.J. and Maqbool, M.A. 1992b, Effects of *Pasteuria penetrans* and *Paecilomyces lilacinus* on the control of root-knot nematodes in brinjal and mung, *Pakistan J. Nematol.* 10: 75-79.

Zuckerman, B.M., Dicklow, M.B. and Acosta, N., 1993, A strain of *Bacillus thuringiensis* for the control of plant-parasitic nematodes, *Biocont. Scie. Technol.* 3 : 41-46.

HORSE PURSLANE (*TRIANTHEMA PORTULACASTRUM* L.) AND ITS BIOCONTROL WITH FUNGAL PATHOGENS : AN OVERVIEW

K.R. Aneja, S.A. Khan and S. Kaushal

Department of Microbiology
Kurukshetra University
Kurukshetra - 136 119, Haryana, INDIA

1. INTRODUCTION

Trianthema portulacastrum L. (Family: Aizoaceae) commonly known as horse purslane, blackpigweed, carpetweed, gudbur, hogweed, itcit and santha is an important terrestrial weed, indigenous to South Africa, but now is also found in tropical and subtropical areas throughout the world (Balyan and Bhan, 1986; Balyan and Malik, 1989). It is one of the most troublesome terrestrial weeds not only of Northwest India, but of many parts of the world and needs urgent attention (Balyan and Malik, 1989; Brar *et al.*, 1995).It has been reported to be widely distributed in India, Srilanka, West Asia, Africa and Tropical America (Balyan and Bhan, 1986; Duthie, 1960).

In India, it is a very common weed of various farm crops, non-croplands, in grasslands and wastelands. It has been observed as a problematic weed in various agricultural crops in the states of Uttar Pradesh, Punjab, Haryana, Rajasthan and Delhi (Singh and Prasad, 1994). Heavy infestations of *Trianthema* has been reported in black gram (Mohammed Ali and Durai, 1987); cotton (Brar *et al.*, 1995; Tiwana and Brar, 1990), mungbean (Gupta *et al.*, 1990; Sandhu *et al.*, 1993); onion (Singh *et al.*, 1992); pearl millet (Balyan *et al.*, 1993; Rathee *et al.*, 1992), pigeonpea (Chauhan *et al.*, 1995) and sugarcane (Chauhan and Singh, 1993; Phogat *et al.*, 1990). It is also reported to be a major weed of gardenland representing 85% of weed population (Sankaran and Rethinam, 1974).

It has been found to be a very aggressive weed in mustard, maize, arhar, soybean, potato and onion crops in the states of Haryana and Punjab. Upto 60-70% infestation of this weed has been reported in arhar and soybean fields and 80-90% in maize and brassica fields. Because of its infestation in various agricultural crops of this region especially during the rainy season, horse purslane should be referred to as number one problematic weeds of agricultural crops in Northern India (Aneja and Kaushal, 1999).

187

Horse purslane, a prostrate herbaceous weed, grows successfully in varied habitats, such as along roadsides, in railway tracks, pastures, margins of cultivated fields, wastelands, gentle and steep slopes and deeply shaded gardens. *Trianthema* plants bloom and set seeds sequentially throughout the growing season. A much branched glabrous herb, it thrives well in neutral or alkaline soils, having low percentage of calcium and magnesium content. Leaves are fleshy with dilated and membranous petiole. *Trianthema portulacastrum* is characterized by having more than 15 stamens and a single style. There are two varieties of *T. portulacastrum*, one with red flowers and other with white flowers (Chandra and Sahai, 1979). Seeds are reniform and dull black with raised concentric lines on their surfaces. They essentially have no dormancy and can germinate soon after they mature (Balyan and Bhan, 1986). Seeds germinate between a temperature range of 20 and 45°C but maximum germination occurs at 30°C. The maximum seedling emergence takes place during June and July, while rapid and vigorous growth occurs during July and August (rainy season) when environmental conditions i.e. both temperature and relative humidity are the best suited for its germination and growth. Plant grows rapidly and reaches peak growth within 40-45 days of its emergence. Node formation starts 10-12 days after sowing and increases gradually with plant age. Seed development and maturation period being only 7 days, thousands of seeds are produced in one season. The production of flowers and seeds starts 20 to 30 days after sowing . During the span of 4 months in the kharif season, 3-4 flushes of this weed emerge, keeping the competition of this weed continuously with the associated crops. Seeds are hard coated hence persist in the soil for many years and infest the crops raised subsequently (Umarani and Selvaraj, 1995).

2. CONTROL

T. portulacastrum, because of its highly competitive and aggressive behavior amongst various crops and adaptability under wide range of environmental conditions has drawn the attention of agriculturists, plant pathologists and weed control scientists all over the world and attempts are being made to control this weed by all possible means i.e. mechanical, chemical and biological.

Hand weeding and hoeing are common practices of controlling this weed in the developing countries of the world; but this method is quite expensive and time consuming thus ineffective as new weed seeds germinate after every hoeing and reinfest the crop, thus using maximum soil nutrients. Moreover, hoeing is not possible during rainy season and labour shortage due to paddy transplanting at that time further accentuates the problem (Brar *et al.*, 1995).

Chemical control of this weed in different crops is been practised either using them singly or in combinations such as Acifluorfen, Acifluorfen + bentazon; Acifluorfen + 2,4 DB, Alachoral, Atrazine, Fluchloralin, Formesafen Isoproturon, Lactofen Metolachlor, Oxyfluorfen, Paraquat, Pendimethalin, Pyrivate + 2,4 DB and Trifluralin. It is true that the chemical herbicides are the most effective and immediate solution to the most weed problems but they are not the only or necessarily the best solution due to the following reasons :

(i)   Such control is only feasible in cultivation or in small areas of pasture and not when the weed covers very large areas because of the expenditure involved.

(ii) Frequent use of chemicals has posed certain problem of pollution hazard and has been the subject of growing concern for both environmentalist and public health authorities through out the world.

(iii) Chemicals affect non target organisms too.

(iv) Chemicals are effective only for a short period.

(v)   Chemicals contaminate soil and ground water, and

(iv) Most importantly chemicals are costly too.

During the last two decades biological control of weeds has received considerable attantion (Green *et al.*,1998) because the excessive use of chemical herbicides results in (i) an increasing number of resistant or tolerant weeds; (ii) environmental contamination increasing detrimental affects on nontarget organisms, contamination of soil, ground water and food; and (iii) strong problem criticizes due to health conceives from such contamination. Therefore, biological control of weeds, a part of integrated approach to weed management, has developed to reduce the dependency on chemical herbicides (Auld and Morin, 1995; Boyette *et al.*, 1991; Green *et al.*, 1998; Heap, 1996; Jobidon, 1991).

3. BIOLOGICAL WEED CONTROL

Biological control of weeds with plant pathogens is an effective, safe, selective and practical means of weed management that has gained considerable importance during recent years (Aneja, 1999). Goal of biological weed control is the deliberate use of mainly host specific arthropods and pathogens, to reduce the population density of a weed below its economic or ecological injury level (Schroeder and Muller-Scharer, 1995). It is considered to be the best long-term solution of weed problem (Aneja, 1997, 1999; Charudattan, 1991, 1997; Evans,1995a; McWhorter and Chandler, 1982) and is considered to be the safe, economic and environmentally sustainable solution (McFadyen, 1998).

Biological weed control is the use of mainly host specific pathogens to reduce the population density of a weed to below its economic or ecological level (Schroeder and Muller-Scharer, 1995; Watson, 1994). Numerous comprehensive reviews describe in detail the fundamentals, methodology and progress of biological weed control (Aneja,1999; Bhan *et al.*, 1998; Bruzzese *et al.*,1997; Charudattan, 1990a,b, 1991; Green *et al.*, 1998; Hasan and Ayers, 1990; Mortensen,1998; Muller-Scharer and Frantzen,1996; Schroeder, 1983; Schroeder and Muller-Scharer,1995; TeBeest *et al.*,1992; TeBeest and Templeton, 1985; Wapshere, 1982; Watson, 1989; Watson *et al.*, 1997). Advantages of using plant pathogens for the control of weeds have been summarized by several workers (Aneja,1997,1999; Auld and Morin, 1995; Charudattan 1990a,b; Evans, 1995a; Freeman,1980; Freeman and Charudattan,1985; Sands and Miller, 1993; Templeton,1982; Zettler and Freeman, 1972). All classes of plant pathogens *viz.* fungi, bacteria, nematodes, viruses or mycoplasmas, are being used in the biocontrol of weeds, but the number of fungi considered far exceeds that of other pathogen groups, because most of the pathogens of higher plants are fungi. Other biological attributes to their pre-eminence in biological weed control are: They are easy to identify than bacteria and viruses; their taxonomic position is well defined and their ability for active invasion of plant tissues (Charudattan, 1990a; Hasan, 1983; TeBeest and Templeton, 1985). Number of deuteromycetous fungi studied substantially exceeds that of other classes as this group is among the more commonly encountered classes of fungi, as pathogens on higher plants, and the members of this group can be easily cultured and induced to sporulate.

3.1. Strategies

Biological weed control can be practised in four different ways: (i) The Classical (Inoculative) method aiming at the reduction and long term stabilization of naturalized weeds by the introduction and establishment of control organisms from the weed's native distribution area. This method is most frequently used and results in permanent control of naturalized weeds outside cultivation; (ii) The Conservation approach aiming to enhance the effect of existing control organisms by environmental manipulation; so far rarely used; (iii) The

Augmentative approach based on collection and redistribution of native control organisms and; (iv) The Bioherbicidal (Inundative) approach using periodical releases of native control organisms to suppress weed population in crop production systems (Charudattan, 1991; Julien, 1992; Schroeder, 1983,1994; Schroeder and Muller-Scharer, 1995); (v) Another approach to the biological weed control, *viz.* "System management approach has been proposed by Muller-Scharer and Frantzen (1996), related to the conservation and augmentative approach. The aim of this system management approach is to shift the balance between host and pathogen in favour of the pathogen mainly by stimulating the build up of a disease epidemic on target weed population. It emphasizes its qualitative aspects related to cautious manipulation of a weed pathosystem; (vi) Beside these approaches there is an another approach of weed control i.e. Integrated approach. In this strategy, plant pathogens are combined with compatible herbicides or with mechanical or cultural practices. The combined effect of plant pathogens and insects for the biological control of weeds has also been suggested (Charudattan, 1986; Singh *et al.*, 1992). The integrated weed management system combines use of multiple-pest resistant, high yielding, well adapted crop cultivars that also resist weed competition and precise placement, use of judicious irrigation practices, planned crop rotations; field sanitation, harvesting method that do not spread weed seeds and use of biological control agents such as pathogens, insects. Of the various approaches, two approaches *viz.* Classical (inoculative) and Bioherbicidal (inundative) exclusively used for the control of weeds have been practised widely all over the world (Mortensen,1998).

3.1.1. Classical (Inoculative) approach

This strategy involves the importation and release of one or more natural enemies that attack the target weed in its native ranges into areas where the weed is introduced and is troublesome and where natural enemies are absent. Objective of classical biological weed control is generally not the eradication of weed species, but self perpetuating regulation of the weed population at an acceptable level (Evans, 1995a; Watson, 1991). The pathogens used in this strategy are generally self disseminating (e.g. rust, smut and certain dry spored foliar fungi). They are spread by spores or other infective propagules that are disseminated by wind, water or insect vectors (Charudattan 1990b; Hasan, 1988). Basic criteria for the development of classical biological weed control project include:
(i) Determination of suitability of the weed for classical biological control.
(ii) Conducting surveys for suitable plant pathogens in target weed's native range.
(iii) Host specific biocontrol agents.
Classical biological control agents have been used against exotic weeds growing in areas that are relatively unmanaged, inaccessible or of low economic return. The best examples of this strategy include: use of *Puccinia chondrillina* Bubak and Syd. from Mediterranean (S. Europe) for control of *Chondrilla juncea* L. in Australia (Cullen *et al.*, 1973; Evans, 1995a,b; Hasan and Wapshere, 1973; Wapshere, 1985); use of *Phragmidium violaceum* Schultz from Europe for the control of *Rubus constrictus* and *R. ulmifolius* L. (Bruzzese and Hasan,1987; Oehrens, 1977); Use of *Entyloma compositarum* Farlow for control of *Ageratina riparium* (Regel.) K. and R. in Hawaii (Trujillo, 1985; Trujillo *et al.*, 1988). Control of *Cardus thoermeri* Weinm (musk thistle) by *Puccinia carduorum* Jacky in Northeastern USA (Politics and Bruckart, 1986; Politics *et al.*, 1984); Control of *Parthenium hysterophorus* L. in Australia with *Puccinia abrupta* Diet and Holw. var. *partheniicola* (Jackson) Parmelee (Evans, 1987; Evans and Tomley, 1994) and control of *Cryptostegia grandiflora* Roxb. ex R.Br. by *Maravalia cryptostegiae* (Cumm.) Ono. in Australia (Evans, 1993) are other successful examples. In addition to these, *Uromycladium tapperianum* (Sacc.) McAlp, a native gall rust on *Acacia*

saligna (Labill.) Wendl. in Australia, has been introduced into South Africa in 1987 to control the shrub *A. saligna* (Post Jackson Willow) (Morris, 1991, 1997; Mortensen, 1998).

Sands and Miller (1993) stated that there will always be a room for a classical approach to the problem ending in either public or private releases of a host specific exotic pathogen. The strength of the classical approach is that it can result in the control of a single species of pest, the development costs are low to moderate and it may have a long term residue and consequently long term control.

3.1.2. Bioherbicidal (Inundative) approach

Bioherbicides are biological control agents applied in similar ways to chemical herbicides to control weeds. The active ingredient in a bioherbicide is a living microorganism and it is applied in inundative doses of propagules. Most commonly the microorganism used is a fungus and its propagules i.e. spores and fragments of mycelium. In this case the bioherbicide is referred as a mycoherbicide and the strategy as a mycoherbicidal strategy (Auld and Morin, 1995). A bioherbicide formulation may contain the viable agent in either a dormant and a metabolically active state, the formulation that which contains the agent in a metabolically active state tend to be less tolerant to environmental stresses, have shorter shelf lives, and require specific packaging to enable gas and moisture exchange. On the other hand, formulation that contains the agent in a state of low metabolic activity (e.g. after drying spores and/or mycelia) tend to have longer shelf lives, are easier to pack, and are more tolerant to environmental stresses such as environmental temerature and relative humidity fluctuations and extremes (Green *et al.,* 1998; Pauu, 1988). According to Powell (1992), the packaging of final formulation must be selected to control gas exchange, movement of moisture, and to prevent contamination of the product. Use of pathogens in a 'product' form and an application technique similar to the chemical tactic are salient features distinguishing the mycoherbicides from classical agents. Applications of pathogens are made to specific areas in volumes and dosages that achieve control of the target weed within specified time and before economic losses are incurred. This strategy emphasizes manipulation of fungal pathogens with respect to establishing and controlling the extent of disease caused in a specific population of the host (TeBeest *et al.,* 1992).

Mycoherbicides have several advantages over conventional chemical herbicides (Aneja,1999; Auld and Morin, 1995; Ayres and Paul, 1990; Charudattan, 1990a). In mycoherbicidal tactic, plant pathogenic fungi are developed and used to control weeds in a way chemical herbicides are used. Mycoherbicides are highly specific disease inducing fungi, isolated from weeds, increased in fermentation tanks and sprayed on fields to biologically control a specific weed without harm to the crop or any non target species in the environment (Templeton *et al.,* 1988). In a few cases mycoherbicide could be used when the target weed dose not have chemical and mechanical control. Presently there are over ten mycoherbicides which are commercially being used in the developed countries of the world (Table 1) (Aneja, 1999). The commercial mycoherbicide first appeared on the USA market in the early 1980s with the release of the product DeVine (a formulation of *Phytophthora palmivora*) in 1981 to control milkweed vine in Florida citrus grooves and in the next year the release of the product Collego (a formulation of *Colletotrichum gloeosporioides* f.sp. *aeschynomene*) to control northern jointvetch, a leguminous weed in rice. Biomal (formulation of *C. gloeosporioides* f. sp. *malvae*) Netherlands, Velgo (*C. coccodes*) for control of *Abutilon theophrasti* in the USA and Canada, Luboa-2 (*C. gloeosporioides* f. sp. *cuscutae*) for *Cuscuta* spp. in China and ABG 5003 (a strain of *Cercospora rodmanii*) for control of *Eichhornia crassipes* (water hyacinth) in the USA (Aneja, 1996, 1997; Charudattan, 1991; Mortenson, 1998).

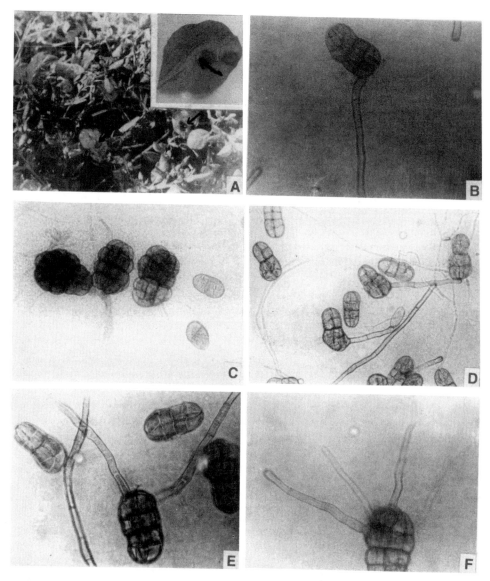

Figure 1. *Gibbago trianthemae* Simmons on *Tranthema portulacastrum* L. (A) a plant population of *T. portulacastrum* showing infection of *G. trianthemae* (arrows) in the field; (B) a conidiophore bearing conidium of *G. trianthemae*; (C) Conidiophore with two conidia; (D) germinating conidia showing the production of secondry conidia; (E-F) germinating conidia of *G. trianthemae*; (E) with two germ tubes; (F) with four germ tubes.

192

Table 1. Mycoherbicides presently being used all around the world (After Aneja, 1999)

Weed(s)	Fungal pathogen	Mycoherbicide	Reference(s)
Morrenia odorata (milkweed vine)	*Phytophthora palmivora*	DEVINE	Ridings(1986); Ridings *et al.* (1976)
Aeschynomene virginica (northern jointvetch)	*Colletotrichum gloeosporioides* f. sp. *aeschynomene*	COLLEGO	TeBeest and Templeton (1985); Templeton (1986); Templeton *et al.*(1984)
Cassia obtusifolia (sickle pod)	*Alternaria cassiae*	CASST	Bannon(1988); Charudattan *et al.* (1986); Walker (1983); Walker and Boyette (1985)
Malva pusilla (round leaf mallow)	*Colletotrichum gloeosporioides* f. sp. *malvae*	BIOMAL	Auld and Morin (1995); Makowski (1987); Mortensen (1988)
Prunus serotina (black cherry)	*Chondrostereum purpureum*	BIOCHON (ECOCLEAR)	de Jong (1997); Mortensen (1998)
Abutilon theophrasti (velvet leaf)	*Colletotrichum coccodes*	VELGO	Wymore *et al.* (1988)
Cuscuta spp. (dodder)	*C. gloeosporioides* f. sp. *cuscutae*	LUBOA	Wang (1990)
Eichhornia crassipes (water hyacinth)	*Cercospora rodmanii*	ABG 5003	Aneja (1996, 1998) Conway (1976); Conway *et al.* (1978); Charudattan (1984,1986); Freeman and Charudattan (1984); Carudattan *et.al.* (1985)

4. STATUS OF RESEARCH ON BIOCONTROL OF HORSE PURSLANE

No significant work has been done on the biocontrol of *T. portulacastrum*. A total of five pathogens, four fungi namely *Cercospora trianthemae* (Chiddarwar, 1962); *Gibbago trianthemae* (Simmons, 1986); *Drechslera (Exserohilum) indica* (=*Bipolaris indica*) (Rao and Rao, 1987; Taber *et al.*, 1988) and *Fusarium oxysporum* (Darshika and Daniel, 1992) and mosaic virus (Sastry, 1980) have been recorded so far on this weed (Table 2). Except *G. trianthemae*, no other fungal pathogen has been evaluated for its biocontrol potential. It was first of all described from the USA, Cuba and Venezuela as a new phaeodictyoconidial genus of Hyphomycetes (Simmons, 1986). The authors Isolated *G. tranthemae* on *Trianthema portulacastrum* in 1994 from Kurukshetra for the first time from India (Aneja and Kaushal, 1999). It has several characteristics similar to genera *Alternaria, Embellisia, Stemphylium* and *Ulocladium* but the unique pattern of secondary sporulation distinguishes the genus from all other phaeodictyoconidial hyphomycetes with some otherwise similar characters (Fig. 1) The conidiophores of *G. trianthemae* are macronematous, 1-4 trans septate, straw coloured upto 150 µm long, 4µm wide, slightly swollen at the apex with a single apical conideogenus locus proliferating sympodially with successive conidiogenus loci quite distinct to each other, producing a single conidium at the apex of each proliferation and retaining a distinct hilum at

193

the locus after the secession of a conidium. Conidia are yellow brown, beakless, muriform, ellipsoid, smooth walled, 3-6 complete or partial transverse septa and 1-6 complete or partial longitudinal septa, 30-68 x 19-38 μm.

Table 2. Pathogens reported on *Trianthema portulacastrum*

Pathogen/s	Symptoms	Country	References
Fungus			
1. *Cercospora trianthemae* Chiddarwar	Leaf spot	India	Chiddarwar(1962)
2. *Drechslera indica* (Rai, Wadhwani and Tewari) Mouchacca (= *Bipolaris indica*) Rai, Wadhwani and Tewari	Leaf spot	India Japan	Rao and Rao (1987), Taber *et al.* (1988)
3. *Fusarium oxysporum* Schlecht		India	Darshika and Daniel (1992)
4. *Gibbago trianthemae* * Simmons	Leaf spot	U.S.A. Cuba, Venezuela, India	Simmons (1986) Aneja and Kaushal (1999)
Virus			
5. *Trianthema* mosaic virus		India	Sastry (1980)

* New Disease for India

Light microscopic observations revealed that conidia are produced by the formation of a minute pore in the wall of the conidiophore showing poroconidial ontogeny. During conidial development, the fertile cell wall of the conidial initial remains intact, suggesting that conidia of this genus are holoblastic, based on Cole and Samson's (1979) classification of development processes in conidial fungi. Mitchell (1988) studied the effectiveness of *G. trianthemae* as a bioherbicide for the control of horse purslane in green house conditions and reported that weed growth was reduced by 50% at the lowest concentration. He emphasized that further studies are still needed on the impact of environment and on application technology of the potential of this pathogen to develop it into a bioherbicide. Work is being carried out in our laboratory to control this weed with biocontrol agents, especially *G. trianthemae* a fungal pathogen, recorded on this weed from this area. *G. trianthemaea* on evaluation for its biocontrol potential on *T. portulacastrum* in experimental pots revealed that the defoliation of the plants started after 20 days of inoculum spraying. Percent infection on leaves ranged between 72% and 84% after 30 days post inoculation with inoculum at the rate of 2.2×10^5 conidia/ml. Other features for the development of *G. trianthemae* as a mycoherbicide for the control of horse purslane recorded are its easy culturability on a nutrient medium, good sporulation capacity and narrow host range.

5. CONCLUSION

T. portulacastrum L. is emerging as a problematic weed in various crops, especially in tropical and subtropical areas of the world. There are two ways to check the nuisance value of a weed (i) converting a problematic weed into a resource through its multifarious uses such as its use as a vegetable, fodder, green manure or medicinal; and (ii) to control it through integrated pest management strategy. Although various pre- and post-emergence chemical herbicides are available to control this weed but keeping in view the pollution hazards created by chemicals, the need of the hour is to intensify research on how to control this weed either through biological control agents or with an integrated approach using chemical plus biological agents. *Gibbago trianthemae*, a fungal pathogen reported on this weed from the USA and India could be developed as a mycoherbicide.

REFERENCES

Aneja, K.R. 1996, Exploitation of fungal pathogens for biocontrol of water hyacinth, In: *Some Facets of Biodiversity*, eds. R.K. Kohali, N. Jerath and D. Batish, SES and SCST Publications, Chandigarh, India, pp. 141-156.

Aneja, K.R. 1997, Discovery and development of mycoherbicides for biological control of weeds, In: *New Approaches in Microbial Ecology*, eds. J.P. Tiwari, G. Saxena, N. Mittal, I. Tewari and B.P. Chamola, Aditya Books Private Limited, New Delhi, India, pp. 517-555.

Aneja, K.R. 1998, Biological suppression of aquatic weeds with fungal pathogens, In: *Biological Suppression of Plant Diseases, Phytoparasitic Nematodes and Weeds*, eds. S.P. Singh and S.S. Hussaini, Project Directorate of Biological Control, Bangalore, India, pp. 174-191.

Aneja, K.R. 1999, Biotechnology for the production and enhancement of mycoherbicide potential, In: *From Ethnomycology to Fungal Biotechnology*, eds. J. Singh and K.R. Aneja, Plenum Publishers, UK, pp. 91-114.

Aneja, K.R. and Kaushal, S. 1999, Occurrence of *Gibbago trianthemae* on horse purslane in India, *Int. J. Biol. Contr.*(in Press).

Auld, B.A. and Morin, L. 1995, Constraints in the development of bioherbicides, *Weed Technol.* 9: 638-652.

Ayres, P. and Paul, N. 1990, Weeding with fungi, *New Scien.* 1 Sept.,1990. pp. 36-39.

Balyan, R.S. and Bhan,V.M. 1986, Emergence, growth and reproduction of horse purslane (*Trianthema portulacastrum*) as influenced by environmental conditions, *Weed Sci.* 34: 516-519.

Balyan, R.S. and Malik, R.K. 1989, Control of horse purslane (*Trianthema portulacastrum*) and barnyardgrass (*Echinochloa crus-galli*) in mung bean (*Vigna radiata*), *Weed Sci.* 37: 695-699.

Balyan, R.S., Kumar, S., Malik, R.K. and Panwar, R.S. 1993, Post-emergence efficacy of atrazine in controlling weeds in pearl millet, *Indian J. Weed Sci.* 25: 7-11.

Bannon, J.S. 1988, CASST herbicide (*Alternaria cassiae*): a case history of a mycoherbicide, *Am. J. Alt. Agric.* 3: 71-76.

Bhan, V.M., Kauraw, L.P. and Chile, A., 1998, Biological suppression of weeds with pathogens- present scenario, In: *Biological Suppression of Plant Diseases, Phytoparasitic Nematodes and Weeds*, eds. S.P. Singh and S.S. Hussaini, Project Directorate of Biological Control, Bangalore, India, pp. 164-173.

Boyette, C.D., Quimby, Jr. P.C., Connick, Jr. W.J., Daigle, D.J. and Fulgham, F.E. 1991, Progress in the production, formulation and application of mycoherbicides, In: *Microbial Control of Weeds*, ed. D.O. TeBeest, Chapman and Hall, New York, USA, pp. 209-222.

Brar, A.S., Thind, R.J.S. and Brar, L.S. 1995, Integrated weed control in upland cotton, (*Gossypium hirsutum* L.), *Indian J. Weed Sci.* 27: 138-143.

Bruzzese, E. and Hasan, S. 1987, Infection of blackberry cultivars by the European blackberry rust fungus, *Phragmidium violaceum, J. Horticult. Sci.* 64: 475-479.

Bruzzese, E., Mclaren, D. and Kwong, R. 1997, Biological control, *Trees Natu. Resou.* 39: 20-24.

Chandra, B. and Sahai, R. 1979, Autecology of *Trianthema portulacastrum* Linn, *Indian J. Ecol.* 6: 17-21.

Charudattan, R. 1984, Microbial control of plant pathogens and weeds, *J. Georgia Entomol. Soc.* 19(2): 40-62.

Charudattan, R. 1986, Integrated control of water hyacinth (*Eichhornia crassipes*) with a pathogen, insects and herbicides, *Weed Sci.* 34: 26-30.

Charudattan, R. 1990a, Prospects for biological control of weeds by plant pathogens, *Fitopatol. Bras.* 15: 13-19.

Charudattan, R. 1990b, Pathogens with potential for weed control, In: *Microbes and Microbial Products as Herbicides,* ed. R.E. Hoagland, American Chemical Society, Washington, DC. USA, pp. 132-154.

Charudattan, R. 1991, The mycoherbicide approach with plant pathogens, In: *Microbial Control of Weeds*, ed. D.O. TeBeest, Chapman and Hall, New York, USA, pp. 24-57.

Charudattan, R. 1997, Status of DeVine and Collego, *IBG News* 6: 8.

Charudattan, R., Linda, S.B., Kluepfel, M. and Osman, Y.A. 1985, Biocontrol efficacy of *Cercospora rodmanii* on water hyacinth, *Phytopath.* 75: 1263-1269.

Charudattan, R., Walker, H.L., Boyette, C.D., Ridings, W.H., TeBeest, D.O., Van Dyke, C.G. and Worsham, A.D. 1986, Evaluation of *Alternaria cassiae* as a mycoherbicide for sicklepod (*Cassia obtusifolia*) in regional field tests, In: *Southern Cooperative Series Bulletin 317,* Alabama Agri. Expt. Station, Auburn University, USA, pp. 1-19.

Chauhan, D.R., Balyan, R.S., Kataria, O.P. and Dhankar, R.S. 1995, Weed management studies in pigeonpea (*Cajanus cajan* L.), *Indian J. Weed Sci.* 27: 80-82.

Chauhan, R.S. and Singh, G.B. 1993, Chemical weed control in spring planted sugarcane, *Indian J. Weed Sci.* 25: 47-50.

Chiddarwar, P.P. 1962, Contribution to our knowledge of the Cercosporae of Bombay State-III, *Mycopath et Mycol. Appl.* 17: 71-78.

Cole, G.T. and Samson, R.A. 1979, *Patterns of Development in Conidial Fungi,* Pitman Press, London, pp. 190.

Conway, K.E. 1976, Evaluation of *Cercospora rodmanii* as a biological control agent of water hyacinth, *Phytopath.* 66: 914-917.

Conway, K.E., Freeman, T.E. and Charudattan, R. 1978, Development of *Cercospora rodmanii* as a biological control agent of water hyacinth, In: *Proc. EWRS Vth Symp. on Aquatic Weeds*, Wageningen, The Netherlands.

Cullen, J.M., Kable, P.F. and Catt, M. 1973, Epidemic spread of a rust imported for biological control, *Nature* 244: 462-464.

Darshika, P. and Daniel, M. 1992, Changes in the chemical content of *Adhatoda* and *Trianthema* due to fungal diseases, *Indian J. Pharmaceu. Scie.* 54:73-75.

de Jong, M.D. 1997, New commercial bioherbicides, *IBG News* 6: 8.

Duthie, J.F.1960(Reprint), *Flora of the Upper Gangetic Plain, P*eriodical Experts, Delhi, India, pp. 500.

Evans, H.C. 1987, Fungal pathogens of some subtropical and tropical weeds and the possibilities for biological control, *Biocon. News Inform.* 8: 7-30.

Evans, H.C. 1993, Studies on the rust *Maravalia cryptostegiae*, a potential biological control agent of rubber-vine weed (*Cryptostegia grandiflora,* Asclepiadaceae :Periplocoideae) in Australia, 1 : Life-cycle, *Mycopathologia* 124: 163-174 .

Evans, H.C. 1995a, Fungi as biocontrol agents of weeds : a tropical perspective, *Can. J. Bot.* 73: S58-S64.

Evans, H.C. 1995b, Pathogen-weed relationships, The practice and problems of host range screening, In: *Proc. VIII Int. Symp. Biol. Contr. Weeds*, Lincoln Univ., New Zealand, eds. E.S. Delfosse and R.R. Scott, DSIR / CSIRO, Melbourne, pp. 539-555.

Evans, H.C. and Tomley, A.J. 1994, Studies on the rust *Maravalia cryptostegiae* a potential biological control agent of rubber-vine weed, *Cryptostegia grandiflora* (Asclepiadaceae: Periplocoideae), in Australia, III: Host range, *Mycopathologia* 126: 93-108.

Freeman, T.E. 1980, Use of conidial fungi in biological control, In: *Biology of Conidial Fungi* II, eds. G.T. Cole and B.R. Kendrick, Academic Press Inc., New York, USA, pp. 143-165.

Freeman, T.E. and Charudattan, R. 1984, *Cercospora rodmanii* Conway, a biocontrol agent for water hyacinth, *Florida Agricultural Experiment Stations Bulletin,* 842, University of Florida, Gainesville, USa, pp. 18.

Freeman, T.E. and Charudattan, R. 1985, Conflicts in the use of plant pathogens as biocontrol of weeds, In: *Proc. VI Int. Symp. Biol. Contr. Weeds,* ed. E.S. Delfosse, Agriculture Canada, Ottawa, Canada, pp. 351-357.

Green, S., Stewart-Wade, S.M., Boland,G.J., Teshler, M.P. and Liu, S.H. 1998, Formulating microorganisms for biological control of weeds, In: *Plant-Microbe Interactions and Biological Control,* eds. G.J. Boland and L.D. Kuykendall, Marcel Dekker Inc., New York, USA, pp. 249-281.

Gupta, Y.K., Katyal, S.K., Panwar, R.S. and Malik, R.K. 1990, Integrated weed management in summer mungbean (*Vigna radiata* (L.) Wilzeck), *Indian J. Weed Sci.* 22: 38-42.

Hasan, S. 1983, Biological control of weeds with plant pathogens-status and prospects, In: *Proc. 10th Int. Cong. Plant Prot.* 2: 759-776.

Hasan, S. 1988, Biocontrol of weeds with microbes, In: *Biocontrol of Plant Diseases,* eds. K.G. Mukerji and K.L. Garg, CRC Press, Boca Raton, Florida, USA, pp. 129-151.

Hasan, S. and Ayres, P.G. 1990, The control of weeds through fungi, principles and prospects, *New Phytol.* 115: 201-222.

Hasan, S. and Wapshere. A.J. 1973, The biology of *Puccinia chondrillina* a potential biological control agent of skeleton weed, *Annu. Appl. Biol.* 74: 325-332.

Heap, I.M. 1996, International survey of herbicide resistant weed, *Weed Sci. Am. Abs.* 8: 9.

Jobidon, R. 1991, Some future directions for biologically based vegetation control in forest research, *For. Chron.*67: 514-519.

Julien, M.H. 1992, *Biological Control of Weeds*: *A World Catalogue of Agents and Their Target Weeds,* Farnham Royal, Commonwealth Agricultural Bureaux, UK.

Makowski, R.M.D. 1987, The evaluation of *Malva pusilla* Sm. as a weed and its pathogen *Colletotrichum gloeosporioides* (Penz.) Sacc. f. sp. *malvae* as a bioherbicide, Ph.D. Dessertation, University of Saskatchewan, Saskatoon, Canada, pp. 225.

McFadyen, R.E.C. 1998, Biological control of weeds, *Annu. Rev. Entom.* 43: 369-393.

McWhorter, C.G. and Chandler, J.M. 1982, Conventional weed control technology, In: *Biological Control of Weeds with Plant Pathogens*, eds. R. Charudattan and H..L. Walker, John Wiley and Sons, New York, USA, pp. 5-27.

Mitchell, J.K. 1988, *Gibbago trianthemae* a recently described Hyphomycete with bioherbicide potential for control of horse purslane (*Trianthema portulacastrum*), *Plant Dis.* 72: 352-355.

Mohammed, Ali, A. and Durai, R. 1987, Control of *Trianthema portulacastrum* in black gram, *Indian J. Weed Sci.* 19: 52-56.

Morris, M.L. 1991, The use of plant pathogens for biological weed control in South Africa, *Agric. Ecosyst. Environ.* 37: 239-255.

Morris, M.L. 1997, Classical biocontrol of weeds with plant pathogens, *IBG News* 6: 11.

Mortensen, K. 1988, The potential of endemic fungus *Colletotrichum gloeosporioides* f. sp. *malvae*, for biocontrol of round-leaved mallow (*Malva pusilla*) and velvateleaf (*Abutilon theopharsti*), *Weed Sci.* 36: 473-478.

Mortensen, K. 1998, Biological control of weeds using microorganisms, In: *Plant-Microbe Interactions and Biological Control,* eds. G.J. Boland and L.D. Kuykendall, Marcel Dekker Inc., New York, USA, pp. 223-248.

Muller-Scharer, H. and Frantzen, J. 1996, An emerging system management approach for biological weed control in crops: *Senecio vulgaris* as a research model, *Weed Res.* 36: 483-491.

Oehrens, E. 1977, Biological control of blackberry through the introduction of rust, *Phragmidium violaceum* in Chile, *FAO Plant Protect. Bull.* 25: 26-28.

Pauu, A.S. 1988, Formulation useful in applying beneficial microorganisms to seeds, *Tibetch.* 6: 276-279.

Phogat, B.S., Bhan, V.M. and Dhawan, R.S. 1990, Studies on the competing ability of sugarcane with weeds, *Indian J. Weed Sci.* 22: 37-41.

Politis, D.J. and Bruckart, W.L. 1986, Infection of musk thistle by *Puccinia carduorum* influenced by conditions of dew and plant age, *Plant Dis.* 70: 288-290.

Powell, A.K. 1992, Biocontrol product fermentation, formulation and marketing, *NATO ASI Ser. A, Life Sci.* 230: 381-387.

Rao, A.P. and Rao, A.S. 1987, New fungal diseases of some weeds, *Indian Bot. Repor.* 6: 38.

Rathee, S.S., Malik, R.K. and Punia, S.S. 1992, Effect of time of nitrogen application and weed management on pearlmillet, *Indian J. Weed Sci.* 24: 17-21.

Ridings, W.H. 1986, Biological control of stranglervine in citrus - a researcher's view, *Weed Sci.* 34: 31-32.

Ridings, W.H., Mitchell, D.J., Schoulties, C.L. and El-Gholl, N.E. 1976, Biological control of milkweed vine in Florida citrus groves with a pathotype of *Phytophthora citropthora*, In: *Proc. Int. Symp. Biol. Contr. Weeds*, ed. T.E. Freeman, University of Florida, Gainesville, Florida, USA, pp. 224-240.

Sandhu, K.S., Sandhu, B.S. and Bhatia, R.K. 1993, Studies on weed control in mungbean [*Vigna radiata* (L.) Wilzeck], *Indian J. Weed Sci.* 25: 61-65.

Sands D.C. and Miller, R.V. 1993, Evolving strategies for biological control of weeds with plant pathogens, *Pestic. Sci.* 37: 399-403.

Sankaran, S. and Rethinam, P. 1974, An evaluation of chemical and mechanical weed control methods in irrigated cotton (var. MCU 5), *Cotton Develop.* pp. 25-29.

Sastry, K.S. 1980, *Plant Virus and Mycoplasmal Diseases in India: A Bibliography,* Bharti Pub., Delhi, India, 1-270 p.

Schroeder, D. 1983, Biological control of Weeds, In: *Recent Advances in Weed Research,* ed. W.W. Fletcher, Commonwealth Agricultural Bureaux, England, USA pp. 41-78.

Schroeder, D. 1994, Potential for biological control of *Cuscuta* spp. and *Orobanche* spp., In: *Orobanche and Cuscuta Parasitic Weed Management in the Near East,* ed. R. Labrada, Report TCP/RAB/2252, Plant Protection Service, AGPP, FAO, Rome, pp. 45-69.

Schroeder, D. and Muller-Scharer, H. 1995, Biological Control of Weeds and its prospectives in Europe, *Med. Fac. Landbouww. Univ. Gent.* 60 : 2.

Simmons, E.G. 1986, *Gibbago,* A new phaeodictyoconidial genus of Hyphomycetes, *Mycotaxon* 27: 107-111.

Singh, G. and Prasad, R. 1994, Studies on the control of *Trianthema portulacastrum* L. in fodder maize, *Indian J. Weed Sci.* 26: 64-67.

Singh, S.J., Sinha, K.K., Mishra, S.S., Thakur, S.S. and Choudhary, N.K. 1992, Studies on weed management in onion (*Allium cepa* L.), *Indian J. Weed Sci.* 24: 6-10.

Taber, R.A., Mitchell, J.K. and Brown, S.M. 1988, Potential for biological control of the weed *Trianthema* with *Drechslera* (*Exserohilum*) *indica*, In: *Abstract Papers, Vth Int. Congr. of Plant Pathology,* Kyoto, Japan, pp. 130.

TeBeest, D.O. and Templeton, G.E. 1985, Mycoherbicides: progress in the biological control of weeds, *Plant Dis.* 69: 6-10.

TeBeest, D.O., Yang, X.B. and Cisar, C.R. 1992, The status of biological control of weeds with fungal pathogens, *Annu. Rev. Phytopath.* 30: 637-657.

197

Templeton, G.E. 1982, Status of weed control with plant pathogens, In: *Biological Control of Weeds with Plant Pathogens,* eds. R. Charudattan and H.L. Walker, John Wiley and Sons, New York, USA, pp. 29-44.

Templeton, G.E. 1986, Mycoherbicidal research at the University of Arkansas- past, present and future, *Weed Sci.* 34(1): 35-37.

Templeton, G.E. and Greaves, M.P. 1984, Biological control of weeds with fungal pathogens, *Trop. Pest Manage.* 30: 333-338.

Templeton, G.E., Smith, Jr. R.J., TeBeest, D.O. and Beasley, J.N. 1988, Mycoherbicides, *Arkansas Farm Res.,* pp. 7.

Tiwana, U.S. and Brar, L.S. 1990, Effect of herbicides on weed control efficiency and production potential of American cotton (*Gossypium hirsutum* L.), *Indian J. Weed Sci.* 22: 6-10.

Trujillo, E.E. 1985, Biological control of hamakua pamakani with *Cercosporella* sp. in Hawaii, In: *Proc. VI Int. Symp. Biol. Contr. Weeds,* ed. E.S. Delfosse, Agriculture Canada, Ottawa, pp. 661-671.

Trujillo, E.E., Aragaki, M. and Shoemaker, R.A. 1988, Infection, disease development, and axenic cultures of *Entyloma compositarum,* the cause of hamakua pamakani blight in Hawaii, *Plant Dis.* 72: 355-357.·

Umarani, R. and Selveraj, J.A. 1995, Seed development and maturation studies in *Trianthema portulacastrum,* *Indian J. Weed Sci.* 27: 217-218.

Walker, H.L. 1983, Control of sicklepod, showy crotalaria and coffee senna with a fungal pathogen, *U.S. Patent No.* 4,390,360.

Walker, H.L. and Boyette, C.D. 1985, Biocontrol of sicklepod (*Cassia obtusifolia*) in soybeans (*Glycine max*) with *Alternaria cassiae, Weed Sci.* 33: 212-215.

Wang, R. 1990, Biological control of weeds in China: A status report, In: *Proc. VII Int. Symp. Biol. Contr. Weeds,* ed. E.S. Delfosse, CSIRO Publications, Australia, pp. 689-693.

Wapshere, A.J. 1982, Biological control of weeds, In : *Biology and Ecology of Weeds,* eds. W. Holzner and N. Numata, Junk Publisher, The Hague, Netherlands. pp. 47-56.

Wapshere, A.J. 1985, Effectiveness of biological control agents for weeds: present quandaries, *Agric. Ecol. Environ.* 13:261-280.

Watson, A.K. 1989, Current advances in bioherbicide research, *Brighton Crop Protection Conference-Weeds,* 3: 987-996.

Watson, A.K. 1991, The classical apporach with plant pathogens, In: *Microbial Control of Weeds,* ed. D.O. TeBeest, Chapman and Hall, New York, USA, pp. 3-23.

Watson, A.K. 1994, Current status of bioherbicide development and prospects for rice in Asia, In: *Integrated Management of Paddy and Aquatic Weeds in Asia,* Book No. 45, Food and Fertilizer Technology Center for the Asian and Pacific Region, pp. 195-201.

Watson, A.K., Mabbayad, M.O., Zhang, W., Masangkay-Watson, R.F., DeLuna-Couture, L.Z., Yandoc, C.B., Paulitz, T.C. and Mortimer, A.M. 1997, Progress of a biological weed control project in rice-based cropping systems in Southeast Asia, In: *Proc.16th Asian - Pacific Weed Science Society Conference,* ed. A. Rajan, Malaysian Plant Protection Society, Malaysia, pp. 342-344.

Wymore, L.A., Watson, A.K. and Gotlieb, A.R. 1988, Interaction between *Colletotrichum coccodes* and thidiazuron for control of velvetleaf (*Abutilon theophrasti* L.), *Weed Sci.* 35: 377-382.

Zettler, F.W. and Freeman,T.E. 1972, Use of plant pathogens as biocontrols of aquatic weeds, *Annu. Rev. Phytopath.* 10: 455-470.

BIOLOGICAL CONTROL OF WEEDS WITH PLANT PATHOGENS

Rajni Gupta and K.G. Mukerji

Department of Botany
University of Delhi
Delhi-110007, INDIA

1. INTRODUCTION

Pest-weeds, diseases, insects, nematodes and rodents compete with people and plants for food and some spread disease. For many years, our main weapons against pest were pesticides (or biocides), chemical substances that can kill organisms that we consider to be undesirable. Herbicides are the chemical substances used to kill weeds i.e. unwanted plants that compete with crop plants for soil nutrients. Chemical herbicides are the most effective immediate solution to most weed problems. Use of chemicals, however can be harmful to wildlife, to the environment and to the humans. Pests develop the resistance to chemicals that's why there is move towards alternative methods of controlling the weeds which are the scourage of agriculture world wide. Fungal diseases are one of the worst threats to cultivated plants resulting in billion dollars loss annually but what the plant pathologists, had overlooked is the extent to which wild plants are vulnerable to fungal infections, presenting an opportunity to turn fungal diseases to the farmer's advantage as weed killers.

Biological weed control involves using one living organism to control the activities of another organism. The objective of biological weed control is not the eradication of weeds but the reduction and long term establishment of weed density cut to a sub-economic level. Permanent weed control has been obtained in a sufficient number of cases to establish biological weed control as valuable method of weed suppression. Biological weed control is approached from one of the two strategies depending on the pathogen discovered i.e., classical strategy and mycoherbicidal strategy.

2. CLASSICAL STRATEGY

As the success of control depends on the ability of the pathogen to self perpetuate, disperse and reduce the host population in the infested region, special care is taken to select agent which are strongly pathogenic to the plant and are ecoclimatically suitable for the target area. During subsequent field surveys, collections are made of all pathogens causing damage to the weed not only those already known, and if this can be done at different times of the season it will enable organisms attacking different plant stages to be collected. Priority should be given to pathogens attacking all stages of development, from very young seedlings to the mature plants. Those pathogens with high virulence and wide spread occurrence are normally considered as suitable for further studies for their biological control effectiveness.

Plant pathogens can only be used in biological control of weeds if there is sufficient evidence that these will not be harmful to useful plants. Most of this evidence can be obtained by glass house tests for host specificity, but field observations can also be supporting evidence. Further indications regarding host specificity of a pathogen can also be had from its taxonomic position. Plants are selected on the basis of their phylogenetic relationship to the target weed. This method is most commonly used in the current testing programmes and includes in order of relationship (i) other forms of the target species; (ii) other species of the same genus; (iii) other members of the tribe; (iv) other members of the sub family; (v) other members of the family; (vi) other members of the order. In some cases where a pathogen has never been recorded on plants outside a family. Thus *Phragmidium violaceum* from blackberry was tested only on the plants belonging to the Rosaceae, as the genus *Phragmidium* has been recorded only on plants of this family (Bruzzese *et al.*, 1986).

In classical method, weeds are inoculated with obligate parasites using inoculum gathered from naturally infected plants. Phatak *et al.* (1983) have demonstrated the feasibility of controlling canada thistle with *Puccinia obtegens*, yellow nut sedge with *P. canaliculata* and Johnson grass with *Sphacelotheca holci*. Scheepens and Vanzon (1982) have cited similar use of *Puccinia punctiformis* to control canada thistle, *Cirsium arvense*, and of cynophages to control blue green algae. Unlike the obligate parasites used as classical agents, these pathogens are native or naturalised in their respective regions and occur on native or naturalised weeds, generally causing endemic diseases. But similar to the inoculative agents, they are capable of self dissemination and of cauisng epidemic build up after an application of inoculum. The amount of inoculum utilized and the area of weed infestation treated are small in the inoculative approach they are generally large in the case of the inundative method. Pathak *et al.* (1983) reported the feasibility and potential use of the rust fungus *Puccinia chondrillina* and potential use of the rust and *P. canaliculata* for the contact of infestations of skeleton weed, *Chondrilla juncea*. In Florida, investigations were initiated to determine the feasibility of the practical use of *Phytophthora citrophthora* to control milkweed vine. *Morrenia odorata* in citrus groves (Burnett et al., 1973).

3. MYCOHERBICIDAL STRATEGY

This method consists of applications of massive doses of inoculum to the weed population to create a fast and high level of epidemic when conditions conducive for disease development and weed control are present (Charudattan, 1984b; Templeton *et al.*, 1979). Daniel *et al.* (1943) first introduced the mycoherbicide concept and demonstrated that an endemic pathogen might be rendered completely destructive to its weed host by applying a massive dose of inoculum at a particularly susceptible stage of weed growth. Mycoherbicide can be defined as "the plant pathogenic fungi developed and used in the bioherbicidal strategy to control weeds in the same way as chemical herbicides are used (Tebeest *et al.*, 1985) or as living products that control specific weeds in agriculture as
200

effectively as chemicals (Templeton *et al.,* 1986)". Mycoherbicides effect their influence either by parasitism-competition or antagonism (Charudattan, 1984).

3.1. Advantage of Mycoherbicide

Role of chemical pesticides is currently under closer scrutiny than ever before. As safety standards in developed countries increase and new data on health and environmental hazards become available, more products are being withdrawn from the market. Many of these banned pesticides find their way into tropical countries where environmental pressure groups are non-existent or less active and in which safety regulations are ill defined. In such cases, dumping would be a more accurate description of what is actually happening. For conventional chemical herbicides, the long term prospects are particularly critical, because they constitute the largest part of the pesticide market (>60%). In the short term, however most countries and all tropical ones cannot afford to consider theoretical or real environmental problems, while they are struggling to feed their populations and to increase agricultural exports. Mycoherbicides have several advantages over conventional chemical pesticides. These are (i) cost effective; (ii) could also be used as alternatives to herbicides when the weed has developed resistance; (iii) more selective than most chemicals making them a better choice to control weeds which are close relatives of the crops that they infest; (iv) not presistent in the environment unlike many chemical herbicides; (v) safe for humans, animals, other plants and the environment and (vi) can only be used when the target weed does not have chemical and mechanical controls (Charudattan, 1990).

3.2. Advantage of Fungal Pathogens used as Mycoherbicides

The organism used to control weeds should be easy to culture and yield infective mycelia or spores for inoculum and store, have a good viability and shelf life, and yield infections and disease cycle over a range of environmental conditions, expensive to use reliable at a high and predictable level of control, and safe for the user and the environment (Tebeest *et al.,* 1985). Many of these characteristics are exhibited by plant pathogenic fungi that infect weeds and most of the fungi that affect leaves and stems. This is mainly because it is easier to spray a suspension of fungal spores on the shoots than roots. Two more fundamental reasons for exploiting shoot pathogens is that they provide large number of spores and they are more specific to particular plants than root pathogens. When a mycoherbicides is applied either as a wettable powder or as a suspension after diluting it with standard spray equipment, the pathogen should return to the back ground levels by natural constraints after weed's death. However, the fungi have received maximum attention (Templeton, 1982) as they offer a wide choice of being biocontrol agents. Most importantly, unlike bacteria and viral pathogens which need insect vectors, natural openings, or wounds for entry into hosts, fungi are capable of active penetration, which makes them the desired candidates for spray applications.

3.3. Steps in the Development of a Mycoherbicide

Mycoherbicides are usually plant pathogens that affect leaves and stems and are already present in the infested region of the target weed, inundation of the environment with a massive dose of inoculum still requires them to be host specific, and to be effective they should still be sufficiently virulent. Harris (1971) has described six steps in the development of a programme of biological control of weeds with insects, (i) determine the suitability of a weed for biological control; (ii) Survey the plant's natural enemies;

(iii) study and evaluate the ecology of the several natural enemies; (iv) determine their host specificity, hence safety; (v) follow their introduction and establishment; and (vi) evaluate their effect. The steps applicable to plant pathogens are (i) extensive search for plant pathogens in the natural population of a weed; (ii) ecological studies like climate and geographical features of the native areas of the pathogen that have a major effect on the virulence disease intensity, infectivity and capacity to spread from one plant to another plant in the areas where weed create nuisance and (iii) determination of safety of the pathogen by finding out the taxonomic position of the pathogen and by performing host specificity tests.

Host specificity determination i.e. host range of a potential biocontrol agent by selecting those plant species that are potential hosts of the organism. Spurr and Van Dyke (1982) reported the four decisions which mark the research progression for any weed-fungal pathogens system are (i) select a target weed; (ii) select a fungal pathogen; (iii) select a biocontrol strategy and (iv) continue or terminate the biological control project.

4. MYCOHERBICIDAL STATUS

The usefulness of mycoherbicides depends on the ability to remove the constraints the weed with massive dose of spores formulated so that they are not dependent on a narrow range of environmental conditions. Fungal pathogens which are presently being used as mycoherbicides on a practical scale, or those that are in advanced stage of development. An extensive list of projects has been compiled by Scheepens and Van zon (1982) but to date there are many commercial available mycoherbicides preparation available.

The mycoherbicide 'Collego' has been used commercially since 1982 in Arkansas to control Northern Jointvetch (*Aeschynomene virginica*) a leguminous weed in rice (*Oryza sativa*) fields. It is marketed as a dry-formulation consisting of 15% viable, dry conidia of *Colletotrichum gloeosporioides* f. sp. *aeschynomene* and 85% inert ingredients. Each formation batch is assayed or packaged to contain 7.54×10^4 viable conidia per bag, the amount required for treatment of 4.05 hactares at the rates of 93.5 ha^{-1} (Bowers, 1986). Mycoherbicides Devine has been used commercially since 1981 in Florida cirtus groves to control milkweed vive, *Morrenia odorata* (Ridings, 1986). It is marketed as a wet formulation of chlamydospores of *Phytophthora palmivora* with a shelf life of six weeks in cold storage. It is applied at the rate of 8×10^4 chalmydospores m^{-2} to the surface of moist soil under citrus tree.

The persimmon wilt fungus, *Acremonium diospyri* is not commercially available but is routinely used as a mycoherbicide to control Persimon trees (*Diospyrous virginiae*) in range land in South Central Okhaloma. It has been used since 1960 to control trees upto 10 cm in diameter (Griffith, 1970).

'Velgo' is a potential mycoherbicide for velvet leaf (*Abutilion theophrasti*) control in corn (*Zea mays*) and soybeans (*Glycine max*) in the U.S. corn belt and Southern Ontario, Canada. It is a strain of *Colletotrichm coccoides* (Wymore *et al.*, 1988). Commercialization is anticipated in 3 to 5 years as a combination treatment with reduced rates of several velvet leaf alone. The fungus alone at the rate of 1×10^9 spores m^{-2} kills 40% of plants inoculated at the two or three leaf stage.

'Luboa' 2 is a selected strain of *Colletotrichum gloeosporiodies* f. sp. *cuscutae* that is used as a mycoherbicide in Peoples' Republic of China to control dodder (*Cuscuta chinensis* and *C. australis*) parasitic on broad cast planted soybeans. It has been used since its discovery in 1963 for practical control of this parasitic weed. Spore concentrations of 2×10^7 spores ml^{-1} are applied with a hand sprayer untill run off.

202

'Biomal' is a potential mycoherbicide for control of round leaved mallow *Malva pusilla* in wheat (*Triticum aestivum*) and lentil (*Lens culinaris*) in the provinces of Manitoba and Saskatchewan, Canada and the northern tries of wheat producing states in the United States. It is a selected strian of *Colletotrichum gloeosporioides* f. sp. *malvae*. It is applied in spore suspension containing $2x10^9$ spores at the rate of $5x10^2$ per hectare (Mortensen, 1988).

'Casst' is a potential mycoherbicide for control of sickel pod (*Cassia obtusifolia*) and coffee senna (*Cassia occidentalis*) in soybeans and peanuts (*Arachis hypogea*) in the Southern states of the United States (Walker and Riley, 1982). it is a strain of *Alternaria cassiae* and is applied at the rate of 1.1 kg ha^{-1} in 76.7 litre of water with a oil based adjuvant. The spore concentration in the spray tank is $7.5x10^4$ spores ml^{-1}.

A potential mycoherbicide for control of Bathrust burr (Spiny cockelbur or *Xanthium spinosum*) has been develoepd in New South Wales, Australia, with a strain of *Colletotrichum orbiculare* that is indigenous in the area where the weed is an economic problem (Auld, *et al.,* 1988). Small scale field tests with the pathogen indicate that it kills large weed plants in relatively dry conditions when heavily inoculated. The pathogen does not kill common cockelbur (*Xanthium pennsylvanicum*) when spray inoculated but if infected into young stems it causes death (Sutton, 1980).

Another host pathgoen interaction that has been examnined in detail for mycoherbicide potential is *Cercospora rodmanii* on waterhyacinth (*Eichhornia crassipes*) (Charudattan *et al.,*1983). Mycelium of this leaf spot inducing pathogen was grown for 3 weeks on potato dextrose broth containing 5% yeast extract blended and applied at the rate of 1.1 gm wet weight m^{-2}. After inoculation of the plants, grown in buckets of nutrients supplemented water disease stress caused a significant reduction in the net rate of leaf production. However, success of this biocontrol strategy was limited by a low rate of plant kill and the ability of the host to compensate for diseae loss by rapid leaf production.

In South Africa, Morris (1984) reported a severe leaf spot and stem blight of *Emex australis* caused by an unidentified species of *Phomopsis*. At least 60 species of *Phomopsis* are known to be pathogenic on plants (Wehmeyer, 1933) and for many other species it is not known whether the fungus is pathogenic or a secondary invader. Shivas (1992) reported the presence of *Phomopsis emicis* on the plant *Emex australis*, where it caused a severe stem blight. *Phomopsis emicis* was obtained from infected stems, petioles, leaves and fruit by culturing surface sterilized peices of *Emex australis* onto Potato Dextrose Agar medium. Morin *et al.* (1994) isolated many fungi i.e. *Phomopsis, Colletotrichum* and *Alternaria* from diseased foliage of *Xanthium*. *Cyperus rotundus* is regarded as the worlds worst weed (Holm *et al.,* 1977) and is one of the major constraints to sustainable agriculture in many tropical countries. There are some specific and damaging pathogens with seemingly restricted distributions (Evans, 1987), notably *Phytophthora cyperi* and *Entyloma cyperi* that could be considered for classical biological control. Perhaps one of the most novel approaches to biological control of weeds using pathogens has been explored in the closely related species *Cyperus esculentus* in the United States. A mycoherbicide has been formulated and is in the final stages of registration and commercialization (Phatak, 1992). In the United States, this rust strain is restricted to *C. esculnetus* and searches are contnuing for strains virulent to *C. rotundus* (Evans, 1995).

Loss of virulence in culture is a pervasive problem in plant pathology, some reported loss of virulence actually involve loss of viability or improper handling of pathogens and plant material. Reversible loss of virulence has been attribute to changes in characters determnined by cytoplasmic inheritance.

5. CONCLUSION

Biological contrtol can be used to substantially reduce our dependence upon chemical pesticides and achieve pest control without negative non-target effect to achieve significant transition to biological pest-control, however, will require new knowledge of pests and their natural enemies and realignment of research and development responsibilities for pest control among public and private sectors. New biotechnological technuiques open new opportunities for genetic improvement of biological control agents, especially in the bio-pesticide tactic and provide some incentive for private sector involvement in biological control.

REFERENCES

Auld, B.A., McRae, C.F. and Say, M.M. 1988, Possible control of *Xanthium spinosum* by a fungus, *Agri. Ecos. Environ.* 21: 219-223.

Bruzzese, E., Hasan, S. 1983, A whole leaf clearing and staining technique of specificity studies of rust fungi, *Plant Pathol.* 32: 335.

Burnett, H.C., Tucker, D.P.H., Patterson, M.E. and Ridings, W.H. 1973, Biological Control of Milkweed vine with a race of *Phytophthora citrophthora*, In : *Proc. Florida State Horticultural Soc.* 86 : 111-115.

Bowers, R.C. 1986, Commercialization of Collego an industrialist view, *Weed Sci.* 34 : 24-25.

Charudattan, R. 1984, Microbial control of plant pathogen and weeds, *J. Georgia Entomol. Soc.* 19(2): 40-62.

Charudattan, R. 1990, Release of fungi: Large scale use of fungi as biological weed control agents, In: Risk Assessment in Agricultural Biotechnology, *Proc. Internat. Conf. Tech.*, University of California, California, pp. 70-84.

Daniel, J.T., Templeton, G.E., Smith, R.J. Jr. and Fox. W.T. 1973, Biological control of northern jointvetch in rice with an endemic fungal disease, *Weed Sci.* 21: 303-307.

Evans, H.C.1987, Funal pathogens of some substropical and tropical weeds and the possibilities for biological cotrol, *Biocont. News Inf.* 8: 7-30.

Evans, H.C., Carrion, G. and Ruiz-Beliw 1995, Mycobiota of the giant sensitive plant, *Minosa pigra sensulato* in the Neotropics, *Mycol. Res.* 99: 420-428.

Griffith, C.A. 1970, Annual Report, Samuel Roberts Nble Foundation, Inc: Ardmore, Oklahoma.

Harris, P. 1971, Current approaches to biological control of weed, *Tech. Comm. CIBC* 4: 67-76.

Holm, L.G., Plucknett, D.L., Pancho, J.V. and Herberger, J.P. 1977, The World's Worst Weed, University Press of Hawaii, Honolulu, USA.

Morris, M.J. 1984, Additioal diseases of *Emex australis* in South Africa, *Phytophylactica* No. 171-175.

Mortensen, K. 1988, The potnetial of an endemic fungus, *Colletotrichum gloeosporioides* for biological control of round leaves mallow *Malva pusilla* and velvet leaf *Abutilon theophrastii*, *Weed Sci.* 36: 473-478.

Pathak, S.C. 1992, Development and commercialization of therust *Puccinia canaliculata* for biological control of yellow nut sedge (*Cyperus esculentus*), In : *Proc. First Internat. Weed Control Progress*, ed. R.G. Richardson, Weed Science Society of Victoria Inc., Melbourne, Australia, pp. 388-391.

Phatak, S.C., Sumer, D.R., Wells, H.D., Bell, D.K. and Glaze, N.C. 1983, Biological control of yellow nutsedge with the indigenous rust fungus *Puccinia canaliculata*, *Science* (Washington D.C.) 219: 1446-1447.

Ridings, W.H. 1986, Biological control of strangler vine in citrus - researcher's view, *Weed Sci.* 34(1): 31-32.

Scheepens, P.C. and Vanzon, H.C.J. 1982, Microbial herbicides, In : *Microbial and Viral Pesticides*, ed. E. Kubstack, Marcel Dekker Inc., New York, pp. 624-641.

Shivas, R.G. 1992, *Phomopsis emicis* sp. nov. on *Emex australis* in South Africa and Western Australia, *Mycol. Res.* 96(1) : 75-77.

Spurr, H.W. Jr. and Van Dyke, C.G. 1982, Biocontrol of weeds with fungal pathogens, four decisions mark the research progression, In: *Biological Control of Weeds with Plant Pathogens*, eds. R. Charudattan and H.L. Walker, John Wiley & Sons, New York, USA, pp. 238-239.

Sutton, B.C. 1980, *The Coetomycetes: Fungi Imperfecti with Pycnidia, Acervuli and stromata*, Common Wealth Mycological Instiute, Kew, Surrey, U.K.

Te-Beest, D.O. and Templeton, G.E. 1985, Mycoherbicides progress in the biological control of weeds, *Plant Dis.* 69: 6-10.

Templeton, G.E. 1982. Status of weed control with plant pathogens, In: *Biological Control of Weeds with Plant Pathogens*, eds. R. Charudattan and H.L. Walker, John Wiley & Sons, New York, USA, pp. 29-44.

Templeton, G.E., Te-Beest, D.O. and Smnith, R.J. Jr. 1979, Biological weed control with mycoherbicides, *Ann. Rev. Phytophth.* 17: 301-310.

Templeton, G.E., Smith, R.J. Jr. and Te-Beest, D.O. 1986, Progress and potential of weed control with mycoherbicides, *Rev. Weed Sci.* 2: 1-14.

Walker, H.L. and Riley, J.A. 1982, Evaluation of *Alternaria cassiae* for the biocontrol of sickel pod (*Cassia obtusifolia*), *Weed Sci.* 30 : 651-654.

Wehmeyer, L.E. 1933, The genus *Diaporthe* Nitschke and its segregates, University of Michigan Studies, Scientific Series 9, University of Michigan Press, Michigan, USA.

Wymore, L.A., Piorier, C. and Watson, A.K. 1988, *Colletotrichum coccodes* a potential bioherbicide for control of velvet leaf (*Abutition theophrasti*), *Plant Dis.* 72: 534-538.

BIOLOGICAL CONTROL OF WEEDS IN INDIA

K. P. Jayanth

Biocontrol Research Laboratories
A Division of Pest Control (India) Ltd.
Post Box No. 3228, 479, 5ᵗʰ Cross
H.M.T. Layout, R. T. Nagar
Bangalore-560 032, Karnataka, INDIA

1. INTRODUCTION

Weeds play an important role in human affairs over much of the earth. The major characteristics of weeds are their unwanted occurrence, undesirable features and ability to adopt to a disturbed environment (Combellack, 1992). Despite measures adopted for their control, weeds are estimated to reduce world food supplies by about 11.5% annually (Parker and Fryer, 1975). Many of our problem weeds are of exotic origin, having been introduced accidentally or deliberately as ornamental plants, etc. They flourish in the new environment as they have escaped from the natural enemies, which suppress their vigour and aggressiveness in their native lands.

The four basic approaches to weed control are mechanical, cultural, chemical and biological control. Until the advent of modern herbicides only the first two methods were generally available to the farmer. Much of the weed research worldwide, in the past four decades, has been oriented towards herbicide use (Watson, 1992). Cultural and chemical control practices are currently the main approaches to weed control. These methods offer short-term solutions at considerable annual expenditure of resources and energies. The increased consciousness of the quality of the environment and the realization that pesticides cannot provide long term solutions to most of our insect and weed problems have enhanced the importance of alternative methods of control. Biological control is one such method, which is environmentally sound and economical as it is self-perpetuating.

Biological control is ideally suited to situations where a few aggressive weeds, often introduced ones, dominate large areas. At relatively low cost, it can permanently reduce such weeds to scattered plants and isolated clumps. Without biological control there is usually no economical method of solving this type of weed problem (Andres, 1977). However, the specificity of biological weed control agents is a disadvantage in monocultured crops, most of which harbour a complex of weed species.

The employment of biological control does not lead to eradication. For biological control to be continuously successful small numbers of the natural enemy must always be present to attack those plants that survive and to prevent an increase in their numbers. Insects rank first among the organisms employed for biological weed control. Plant feeding fishes and mites have also been used for weed control. Recently plant pathogenic fungi are gaining importance as biocontrol agents of aquatic and terrestrial weeds.

Classical, conservation and augmentation approaches are generally employed in biological control. However, classical biological control of naturalized weeds by the introduction of exotic control organisms from the native range is the most frequently used method in biological control of weeds. The conservation approach involving environmental manipulation to enhance the effect of existing native or exotic control organisms has not received much attention in biological control of weeds. Recently augmentation approach using periodic release and/ or redistribution of native natural enemies has attained certain importance, particularly using native pathogens.

2. SCOPE OF BIOLOGICAL WEED CONTROL

Most weeds are introduced species, which occur at much greater population densities than they do in their natural habitat. Their pest status is often due to competition with crops and pastures, toxicity to livestock, alteration of the habitat in natural wilderness areas or in natural or artificial water bodies. In biological control, population densities of the weed are reduced by the introduction, from the country of origin of the weed, of specific natural enemies (Andres, 1977).

The employment of biological control does not lead to eradication. For biological control to be continuously successful small numbers of the host must always be present to attack those plants that survive and to prevent an increase in their numbers. Biological control has most often been used against alien plants that have become weedy. Native plants are also increasingly becoming targets of biological control attempts.

Insects constitute the largest group of natural enemies of weed plants. Mites have also been used occasionally in biological control. Fishes, snails, ducks and mammals (manatee) have been used in the control of aquatic weeds. Weeds may also be damaged by a variety of pathogens (fungi, viruses, bacteria) or by parasitic plants (dodder, witch weed) (Andres and Goeden, 1971).

Insects are important in biological control due to their great variety and numbers, good knowledge of the systematics, life history and plant associations and the availability of a range of natural enemies suited to particular ecological situations. Most insect species are intimately adapted to their host plants, with high degree of host specialization in certain taxa and cause conspicuous damage. Moreover they are easy to handle (Andres, 1977).

3. HISTORY OF BIOLOGICAL WEED CONTROL

The earliest record of biological control of a weed was the large scale destruction of *Opuntia vulgaris* in central and northern India by *Dactylopius ceylonicus* (Green), introduced from Brazil in the mistaken belief that it was *D. coccus* Costa, a species cultured as a source of carmine dye. The intentional introduction of the same insect during 1836-1838 into southern India is the first successful intentional use of an insect to control a weed. The subsequent introduction of *D. ceylonicus* to Sri Lanka about 1865 and the successful control of *O. vulgaris* there constituted the first international transfer of a natural enemy for biological control of weeds (Goeden, 1988).

It was only 36 years later that the first deliberate foreign exploration for identifying suitable natural enemies for a weed was carried out. Albert Koebele who was earlier involved in the successful cottony cushion scale control project of 1888, which marked the beginning of modern biological control methods, shipped 23 species of insects to Hawaii after exploring the jungles of south-central Mexico during 1902. Among these 8 species established throughout the island group by 1905 (Goeden, 1988). However, colonization of natural enemies continued up to 1970, bringing about partial to substantial control of *L. camara* over the years (Julien, 1992).

The first spectacular success in a biological weed control project was reported from Australia. *Cactoblastis cactorum* (Berg.), of Argentine origin, released in 1925, brought about virtual complete control of *Opuntia inermis* and *Opuntia stricta* from 24 million hectares of formerly infested land by 1935 and restored the same to agricultural use (Goeden, 1988). However, this project involved the study of over 150 species of insects in North and South America, shipment of 48 species to Australia, release of 18 of them and field establishment of 11 insects. Between 1925 and 1933 about 2.7 billion mass-cultured and field collected eggs of *C. cactorum* are reported to have been distributed in the first such effort ever undertaken in a biological weed control project.

The spectacular success achieved in the biological control of cactus in Australia aroused interest in utilizing this method against other weeds. Initially most weeds targeted for biological control were introduced perennial weeds of relatively undisturbed range land. However, from late 1950's projects were initiated on aquatic and semi-aquatic weeds, annuals and biennials, crop land and ruderal weeds in Australia, Canada and the USA.

The utilization of insects for the control of aquatic weeds had its inception with the introduction of the flea beetle *Agasicles hygrophila* Selman and Vogt into the USA for the biological control of the alligator weed *Alternanthera philoxeroides* (Bennett, 1984). The successful project on biological control of the annual weed *Chondrilla juncea* is an important milestone in that the targeted weed invaded crop lands. It also involved the first deliberate intercontinental transfer of a plant pathogen, viz. *Puccinia chondrillina* Bubak and Sydenham (Fungus: Uredinales) for the control of a weed (Cullen *et al.*, 1973).

Up to early 1920's introduced weed control agents were released into the field without confirming their safety to economically important plants. The concept of host-specificity tests came into vogue only during the cactus control project in Australia. The currently employed procedures in host-specificity determination of exotic biological control agents of weeds have evolved over the past 75 years, based on experience gained during the execution of biological control programmes worldwide. A series of eight international symposia on biological control of weeds starting 1969 provided opportunities for discussion at the international level for improving the procedures.

In the past 90 years, since the initiation of the *Lantana* project in Hawaii, there has been a dramatic increase in activity in the field of biological control of weeds worldwide. Outstanding results have been achieved in the control of *Cactus* spp. in Australia (1930's); *Hypericum perforatum* in California (1940's); *Emex* spp. in Hawaii (1950's); *Alternanthera philoxeroides* in Southern USA (1960's); *Chondrilla juncea* in Australia (1970's) and *Salvinia molesta* in Australia and India (1980's). Up to the year 1980 there were 174 projects worldwide to control 101 species of weeds, of which 68 (39%) were considered successful and led to some level of control of 48 weeds. In all, 171 species of insects, two mites, one nematode and four fungi were used, of which 115 (71%) became established and 51 (34%) effective in at least one situation (Julien *et al.*, 1984).

The rate of effectiveness of released biocontrol agents of weeds was exceptionally high between 1920 and 1940 due to few oligophagous insects controlling a range of cactus species in numerous countries. As the number and diversity of target weeds and control agents increased the success rates declined. This is due to the fact that all weeds are not amenable to biological control and some insect groups are less effective than others are. Thus noctuid moths and tephritid flies are less than 15% effective compared to 35% for chrysomelid beetles and 67% for dactylopids (Julien, 1989).

4. COST OF BIOLOGICAL CONTROL OF WEEDS

Classical biological control is relatively inexpensive when applied in the proper situation. As in the case of other methods the benefits derived from biological weed control is proportional to the productive capacity of the area occupied by the weed. But better benefits can be obtained for biological control even in relatively unproductive areas, as it is self perpetuating and derives its energies from the weed itself (Andres, 1977). For example *Opuntia triacantha* was controlled in island of Nevis, West Indies between 1957 and 1960 at a cost of $1,500. This achieved a reduction in weed density that would have cost between $30,000 to $60,000 annually with herbicides (Harris, 1979). Similarly it has been estimated that biological control of water fern in Kerala by release of *C. salviniae* has brought about an annual saving of Rs. 6.8 million on account of labour for paddy cultivation alone (Singh, 1989).

As biological control of weeds developed from an intuitive approach to one requiring proof of safety of agents prior to release, the cost of undertaking biocontrol projects increased enormously. Harris (1979) estimated that every insect released costs on average two scientist years whether it combats the weed or not. Countries such as Australia, USA and Canada, due to better financial support, are now in the forefront, carrying out research on all aspects of biological weed control, including foreign exploration. Biocontrol agents, identified and tested by the above countries, are imported, tested and utilized by developing countries such as India. Nevertheless, such transfer projects have ushered in a tradition of international cooperation to solve major weed problems worldwide.

5. PROCEDURES INVOLVED IN BIOLOGICAL CONTROL OF ALIEN WEEDS

The procedures involved in a classsical biological control programme are: (i) determine the suitability of the weed for biological control; (ii) conduct surveys for suitable natural enemies in the weed's native range; (iii) select the most suitable natural enemies; (iv) study the host-specificity of these organisms to acertain their saftey and (v) evaluate their effectiveness on the weed population. These can be divided broadly into (i) Pre-introduction studies; (ii) Introduction and host-specificity tests under quarantine conditions and (iii) Field releases, monitoring and evaluation.

5.1. Pre-Introduction Studies

Before embarking on a biological control programme it is advisable to ascertain whether the plant has useful attributes and to resolve any possible conflicts of interests. The three main sources of conflict are economical, ecological and aesthetic. The main obstacles to resolving this problem are inadequate information and the difficulty in

210

expressing losses and benefits associated with a weed in terms of monetary value. Losses from weeds are generally a direct result of their interference with planned productivity; benefits are usually attributed to specific characteristics of the weedy plant. A close look at these characteristics may reveal that they can be provided by other plants or that the weeds can still be beneficial at lower densities (Andres, 1981).

For biological control of weeds host-specific natural enemies are generally made use of. It is generally accepted that the greatest number of safe, effective and virulent biological control agents for weeds occur in the centre of diversification of the weed genus. If suitable organisms do not exist in the weed's native range, surveys in parts of the exotic range may be required (Schroeder, 1983).

Preliminary surveys are conducted in the area of origin of the weed, to collect natural enemies for identification and to locate the best centre for long-term studies. These surveys give an estimate about the relative significance of the more important control agents. Surveys are simultaneously carried out in the country where control is required to determine if any of the natural enemies are already present. Costly reintroduction of non-effective insects can thus be avoided.

After completion of surveys, natural enemies for importation are selected on the basis of further studies on their biologies, host range, type of damage inflicted, period of attack, reproductive potential, feeding behaviour, extrinsic mortality factors, distribution, compatibility with other control agents, etc.

The determination of potential effectiveness of biocontrol agents is a very complex problem and the question of whether an agent selected will become established and effects the density and abundance of the target weed cannot be answered in advance. However, the analysis of successful biological control projects have pointed out that effective control agents are those which attack the plant at a critical time of its life cycle, such as when carbohydrate levels are very low or when the plant is subjected to environmental stress (Schroeder, 1983).

5.2. Introduction and Host-Specificity Tests

Before any natural enemies can be imported into India, permission of the Plant Protection Adviser to Govt. of India (PPA) has to be obtained. Shipments are received by air-freight and cleared through customs and quarantine authorities at the port of entry. The parcel is opened in a quarantine laboratory. The material is screened for parasitoids, predators and pathogens, after which laboratory cultures are established.

The crucial point in a programme for the biological control of a weed is to determine whether a candidate agent can be introduced to control a weed without the danger that it may also damage desirable plants. Therefore, most countries including India require that potential biocontrol agents be screened to determine beyond reasonable doubt, that they will not damage any desirable plant, after releases. Host-specificity tests are carried out under quarantine conditions to determine the safety of the introduced insects to economically important plants in the country. Plant species which are potential hosts of the organisms in question are selected on the basis of established procedures.

The list of plants selected for these studies include:

(i)   Plants related to the target weed and other recorded hosts of the candidate agent, however dubious such records may be.

(ii) Host plants of species closely related to the candidate agent,

(iii) Unrelated plants having morphological or biochemical characteristics in common with the target weed,

(iv) Crop plants, the entomology and mycology of which is little known and those for geographical, climatic and ecological reasons have not been exposed to attack by the candidate agent.

During the screening programme, all test plants are exposed to the potential control agents in a series of tests. These determine if the insect will accept a test plant as food and if oviposition occurs on the test plant and if so do the larvae develop normally. Tests are replicated several times with a sufficiently large number of potential control agents. If screening tests demonstrate that the potential biocontrol agent feeds and develops exclusively on the target weed, the agent is considered to be safe.

5.3. Field Releases, Monitoring and Evaluation

Permission of the PPA, based on the recommendations of an Experts' Committee is a prerequisite for initiating field releases of exotic natural enemies of weeds in India. Initially permission is granted for limited filed trials, at selected locations, under close scientific supervision. Large-scale field trials are only permited if no adverse effects are noticed during these studies.

The number of individual organisms needed for successful establishment varies widely for different species and depends on the environmental resistance of the release area. Since it is not possible to obtain a large number of insects from their native range for direct field releases, it is necessary to standardize methods for multiplying the insects in the laboratory. However, from a genetic point of view continuous laboratory breeding may be detrimental, resulting in the selection of organisms adapted to laboratory conditions (Hopper et al., 1993).

The selection of appropriate release sites is of major importance for the successful establishment of otherwise suitable organisms. The initial release should be made in the most favourable habitat. It is also important to select release sites which are well protected against interference from grazing animals, farm implements, other weed control operations, etc.

The assessment of the results of biological control of weed projects is of primary importance. Once natural enemies are released, their establishment, population build up, dispersal from release areas, adaptability to local climatic conditions, effect of local natural enemies, etc., should be monitored regularly. Two types of evaluation are required before a project is completed: documentation of the economics of the project and an analysis of the scientific aspects (Schroeder, 1983).

An economic assessment of biocontrol projects requires the determination of the pre- and post-biocontrol costs and losses associated with the target weed as well as the cost of biocontrol operation, to determine the cost: benefit ratio. The losses caused by a weed are often difficult to determine and vary with the site, the weed density, value of crop losses and other parameters. It is generally easy to establish costs of alternative control methods (chemical and mechanical), but difficult to estimate the cost of biological control. This will mainly depend on the type of weed, the availability of suitable control agents, the time needed for study, the speed of successful establishment and the number of agents needed.

Sometimes, the decision to initiate a biocontrol project will not always depend on economic consideration, because the benefits of successful biological control cannot be expressed in monetary terms in all cases. This may be true where control is required of widespread alien weeds which negatively affect the native flora and fauna or in the case of poisonous and nuisance weeds. Such weeds are often unsuitable targets for conventional control and biological control may be the only available solution.

Nevertheless, it is highly desirable to determine the monetary benefits from biological weed control, where appropriate.

The scientific evaluation of biocontrol projects, successes and failures, is of prime importance in obtaining the feedback necessary to form concepts for improved success of future biocontrol projects. In cases where biocontrol agents become established without controlling the target weed, it can be evaluated in terms of stress on the weed. Biocontrol can be regarded as a matter of increasing stress load to a critical level at which point the plant declines (Harris, 1981).

6. BIOLOGICAL CONTROL OF WEEDS IN INDIA

Although several well known alien weeds occur in India, only a few sporadic biological control attempts were made until 1980. After the spectacular success achieved against *Opuntia* spp. releases of exotic natural enemies were only made against *L. camara*, *Ageratina adenophora* and *Chromolaena odorata*. With the initiation of the All India Coordinated Research Project on Biological Control of Crop Pests and Weeds concerted efforts were made from 1982 for the biological control of water hyacinth (*Eichhornia crassipes*); water fern (*Salvinia molesta*); *C. odorata* and *Parthenium hysterophorus*.

6.1. Biological Control of *Lantana camara*

Lantana camara L. (Verbenaceae), introduced into India as an ornamental plant in 1809, is now a serious weed in many parts of the country (Muniappan and Viraktamath, 1986). It is mainly a weed of plantation crops and pastures. Dense thickets are formed in pastures, which shade out and encroach upon desirable plants and form pure stands over a period of time.

Biological control trials against *L. camara* were initiated in 1921 with the introduction of the seed fly *Ophiomyia lantanae* Froggatt (Diptera: Agromyzidae) from Hawaii in 1921. Although establishment was obtained, the insect failed to control the weed. *Teleonemia scrupulosa* Stal. (Hemiptera: Tingidae) imported from Australia in 1941, but not released as it was feared that it might attack teak, is now widespread. The insect escaped from quarantine but does not attack teak in the field. *T. scrupulosa* is reported to cause defoliation of *L. camara* in some parts of the country, without bringing about control (Muniappan and Viraktamath, 1986). Recently, *Erythmelus teleonemiae* Subba Rao (Hymenoptera: Mymaridae) was reported to cause up to 80% parasitism of *T. scrupulosa* egg, which may account for the non-effectiveness of this natural enemy in suppressing *L. camara* (Jayanth and Ganga Visalakshy, 1992).

Although *Octotoma scabripennis* Guerin and *Uroplata girardi* Pic. (Coleoptera: Chrysomelidae), *Salbia haemorrhoidalis* Guenee (Lepidoptera: Pyralidae), *Diastema tigris* Guenee (Lepidoptera: Noctuidae) and *Leptobyrsa decora* Drake (Hemiptera: Tingidae) were imported and released, only the first two are reported to have established (Muniappan and Viraktamath, 1986; Sankaran, 1973). However, there are no reports in the literature on their field efficacy.

6.2. Biological Control of *Ageratina adenophora*

Ageratina adenophora (Sprengel) R. King and H. Robinson (Asteraceae) is a serious weed of grazing lands, forests and plantations in hilly areas of the country. The tephritid gall fly *Procecidochares utilis* Stone was introduced in 1963 and released for biological control of *A. adenophora*. Although the insect has established its effectiveness is hampered by attack of parasites (Sankaran, 1973). *P. utilis* is reported to have spread into Nepal, where it has become well distributed (Kapoor and Malla, 1978).

6.3. Biological Control of *Chromolaena odorata*

Chromolaena odorata (L.) R. King and H. Robinson (Asteraceae), of Neotropical origin, is a serious weed in plantations, grazing lands and open forests along the Western ghats in Karnataka, Tamil Nadu and Kerala. It is also a common weed in Assam, Maharashtra, Orissa and West Bengal (Chacko and Narasimham, 1988).

6.3.1. Importation of exotic natural enemies

In its native home *C. odorata* is attacked by about 225 species of insects (Cruttwell, 1974). *Pareuchaetes pseudoinsulata* Rego Barros (Lepidoptera: Arctiidae) was imported by the Commonwealth Institute of Biological Control (CIBC) from Trinidad and its safety to economically important plants confirmed (Sankaran and Sugathan, 1974). The flower feeding weevil *Apion brunneonigrum* Beguin-Billecocq (Coleoptera: Apionidae), was imported and released directly in the field, without attempting laboratory multiplication, in 1976 by CIBC (Chacko and Narasimham, 1988) and in 1982 by IIHR. In 1986 IIHR obtained a culture of *Mescinia parvula* Zeller (Lepidoptera: Pyralidae) but could not establish a laboratory culture due to paucity of material received.

6.3.2. Field evaluation

Field trials were initiated with *P. pseudoinsulata* in 1973, at many localities in Kodagu district in Karnataka. However, these efforts did not result in field establishment. Further attempts made during 1978-79 also did not yield positive results (Singh, 1989). *P. pseudoinsulata* was imported again by IIHR in 1982. As releases of these insects in Kodagu and Kerala also did not result in establishment, about 70 larvae of the insect collected from the field in Sri Lanka were obtained in September, 1984 through CIBC. Releases of the insect in Sri Lanka in 1973 had resulted in establishment and defoliation of the weed (Dharmadhikari *et al.*, 1977).

Releases of the Sri Lankan strain of *P. pseudoinsulata* resulted in establishment and partial control of the weed in the campus of the Kerala Agricultural University (Joy *et al.*, 1985a). Further releases resulted in establishment of the insect in a rubber plantation and clearance of the weed in an area of about two hectares. However, additional releases of the insect since then have failed to increase its spread.

Between October, 1984 and December, 1987 a total of 61,345 larvae of *P. pseudoinsulata* were multiplied in the laboratory at IIHR, Bangalore and released at different locations in Karnataka. Since the release spots, other than in Bangalore district, were located more than 250 km away, a method was developed to store young larvae at 15°C in a B.O.D. incubator for up to 45 days (Jayanth and Ganga Visalakshy, 1989a). It was thus possible to accumulate larvae for carrying out large scale releases at distant locations.

Among the release spots in Karnataka, establishment of *P. pseudoinsulata* was noticed in July 1988, about 9 months after releases, at Kamela in Suliya taluq of Dakshina Kannada district. The insect was noticed to cause large-scale defoliation of *C.*

odorata in a one hectare area of uncultivated land, fully infested by *C. odorata*, in a private estate. Although, *P. pseudoinsulata* was noticed to disperse naturally over more than 1000 sq km area by January, 1993 it was found to be distributed sporadically, defoliating the weed in pockets. But, observations carried out during October, 1995 revealed that although *P. pseudoinsulata* population continued to survive in the field, the insect was noticed to cause only minor damage to the leaves of the weed.

The results of the studies carried out till date indicate the need for initiating concerted efforts for bringing about biological control of this noxious weed. These involve redistribution *of P. pseudoinsulata*, reintroduction of *A. brunneonigrum* and *M. parvula* besides importation and releases of additional host-specific natural enemies such as *Procecidochares connexa* Macq (Diptera: Tephritidae).

6.4. Biological Control of *Parthenium hysterophorus*

Parthenium hysterophorus L. (Asteraceae) is a native of southern parts of North America, Central America, the West Indies and the central parts of South America. It has accidentally been introduced, over the past five decades, into many countries including Australia, China, India, Israel, Madagascar, Mozambique, Nepal, South America and Vietnam (Joel and Liston, 1986; Towers *et al.*, 1977). It is a serious weed of pastures, waste lands and also of agricultural fields in most parts of India (Krishnamurthy *et al.*, 1977).

P. hysterophorus is known to suppress local vegetation by release of growth inhibitors through leaching, exudation of roots, decay of residues, etc. The weed is also reported to affect the growth and yield of several plants (Mohandas, 1981). If left unchecked, it is likely to affect natural diversity and cause extinction of native flora. However, the weed has gained notoriety mainly due to the health hazards, such as nasobronchial allergy and dermatitis, it poses to humans (Towers *et al.*, 1977). It is a threat to livestock as it reduces the availability of fodder in pastures (Vartak, 1968) and affects animal health when ingested (Narasimhan *et al.*, 1980).

6.4.1. Importation of exotic natural enemies

Since *P. hysterophorus* has reportedly entered India in the seed form, along with imported food grains (Vartak, 1968), the natural enemies that keep it under check in its native home were left behind. Although a number of insects attack *P. hysterophorus* in India (Kumar *et al.*, 1979) none of these are host specific. Extensive surveys carried out in Mexico revealed that all parts of the parthenium plant are attacked by a large number of insects (McClay, 1980) and fungal pathogens like *Puccinia abrupta* var. *partheniicola* (Evans, 1987).

The leaf feeding beetle *Zygogramma bicolorata* Pallister (Coleoptera: Chrysomelidae), a flower feeding weevil *Smicronyx lutulentus* (Coleoptera: Curculionidae) and a stem galling moth *Epiblema strenuana* (Walker) (Lepidoptera: Tortricidae) were introduced during 1983-1985 for biological control trials against *P. hysterophorus*. Cultures of *Z. bicolorata* and *E. strenuana* were established under quarantine conditions, while that of *S. lutulentus* could not be multiplied, as live insects were not received from the shipment. As host-specificity tests revealed that *E. strenuana* is capable of multiplying on the oil seed crop niger (*Guizotia abyssinica* - Astaraceae), the culture of this insect was terminated (Jayanth, 1987a). Detailed host specificity tests carried out in Mexico, Australia and India (Jayanth and Nagarkatti, 1987a; McFadyen and McClay, 1981) confirmed that *Z. bicolorata* is incapable of completing its development on any cultivated crops.

6.4.2. Field evaluation

Z. bicolorata established readily under field conditions in Bangalore after releases were initiated in 1984 (Jayanth, 1987b). However, it started building up damaging population levels only from July 1988, probably after getting acclimatized to local conditions. Detailed observations revealed that an insect density of one adult per plant caused skeletonization of leaves within 4-8 weeks, provided this density is achieved early in the rainy season. *Z. bicolorata* is now dispersed over more than 200,000 sq km in Karnataka, Tamil Nadu and Andhra Pradesh, causing large scale defoliation. Reports of its establishment and spread have also come in from Jammu, Punjab, Haryana and Madhya Pradesh.

Defoliation of *P. hysterophorus* due to feeding by adults and larvae of *Z. bicolorata* was found to cause up to 98% reduction in flower production, even though the insect does not feed directly on the flowers (Jayanth and Geetha Bail, 1994). But the early stage larvae congregate and feed on the terminal and axillary buds, thus preventing the emergence of flowers. Once the flowers present on the plants, at the time of defoliation, mature and the seeds drop to the ground, no further flowers are added due to the destruction of flower buds.

During studies carried out over a period of three years, in an experimental plot at IIHR, 40 different species of plants, including eight grasses, were observed to grow in the areas formerly occupied by parthenium, due to defoliation by *Z. bicolorata* (Jayanth and Ganga Visalakshy, 1996). A change in the plant species complex is evident in Bangalore and surrounding areas, with a large number of plants growing in areas which were fully under parthenium cover. Although pure stands of parthenium are still visible at many pockets, these are generally disturbed areas.

Field studies also showed that ploughing of fallow agricultural fields, after defoliation of parthenium by *Z. bicolorata*, resulted in renewed weed growth (Jayanth and Ganga Visalakshy, 1996). This may be due to the presence of over 200,000 viable seeds of parthenium per sq m of infested soil, as reported by Joshi (1990).

The results of the studies carried out so far indicate that *Z. bicolorata* has the potential to reduce the density of *P. hysterophorus* and encourage the growth of vegetation formerly suppressed by this weed. However, this insect alone may not be sufficient to bring down the population of the weed to the desired level. Besides *Z. bicolorata* is not likely to be effective in all situations, where parthenium is prevalent. Laboratory studies revealed that the insect is capable of breeding at temperatures ranging from 15-35°C, although 20-30°C was the most suitable. It was also observed that adults diapausing within the soil cannot tolerate continuous exposure to 40 and 45°C for more than 10 days and 21 hrs, respectively, suggesting that the beetle is unlikely to survive in a state of diapause in parts of the country, where the summer temperatures exceed 45°C (Jayanth and Geetha Bali, 1993).

It may be desirable to import additional natural enemies such as the leaf mining moth *Bucculatrix parthenica* Bradley (Lepidoptera: Bucculatricidae) and the seed feeding weevil *S. lutulentus*. Similarly, importation of host-specific plant pathogens such as *P. abrupta* var. *partheniicola*, which has recently been introduced into Australia (Tomley, 1990), may also contribute to the successful biological control of this noxious weed throughout India.

6.5. Biological Control of Water Hyacinth

Eichhornia crassipes (Martius) Solms-Laubach (Pontederiaceae) (water hyacinth) is considered to be the most serious aquatic weed worldwide (Holms *et al.*, 1977). It is a

216

sturdy plant, uses solar energy efficiently and is enormously productive. It propagates by vegetative and sexual methods and adapts to changing climate and water quality. Under ideal conditions water hyacinth plants can double their biomass in 10 days (Vietmeyer, 1975). The seeds of the weed sink to the bottom mud, where they can remain viable for as long as 20 years (Gopal and Sharma, 1981).

6.5.1. Importation of exotic natural enemies

Two weevils *Neochetina eichhorniae* Warner and *N. bruchi* Hustache (Coleoptera: Curculionidae) and a mite *Orthogalumna terebrantis* Wallwork (Acarina: Galumnidae), of Argentinian origin, were imported from the USA in March, 1982 for biological control trials. Detailed host-specificity tests involving 76 species of plants belonging to 42 families with the weevils (Jayanth and Nagarkatti, 1987b; Nagarkatti and Jayanth, 1984) and 88 species of plants belonging to 42 families with the mite (Jayanth and Nagarkatti, 1988) confirmed their safety to cultivated plants in the country.

6.5.2. Field evaluation

N. eichhorniae and *N. bruchi* established readily under field conditions in Bangalore. Both *N. eichhorniae* and *N. bruchi*, separately and in combination, were found to be effective in suppressing water hycinth. The rate of suppression was found to be slower in a tank which was fully sedimented, due to silt coverage of root hairs, on which the insects pupate (Jayanth, 1987c, 1988a,b).

Fluctuations in weed coverage, ranging from 2-30% were noticed in all the tanks, for 2-3 years after the initial colapse. However, the insects were found capable of bringing down weed populations every time an increase was noticed, without additional releases. Studies also indicated that *N. eichhorniae* and *N. bruchi* can be released even in tanks that dry up during summer as they are capable of tolerating drought and surviving without food for as long as 48 days provided some moisture is available (Jayanth and Ganga Visalakshy, 1990).

Spectacular biological control of water hyacinth was achieved in the 286 sq km Loktak Lake in Manipur, 75% of which was covered by the weed, within 3 years after releases of about 18,500 adults of the weevils during 1987-88 (Jayanth and Ganga Visalakshy, 1989b). Adults of the weevils have also been released at different locations in 15 states within the country.

Although *O. terebrantis* has also established under field conditions in Bangalore, after releases were intiated in 1986, the mite does not appear to be capable of suppressing water hyacinth on its own. Heavy population build up was noticed in all the tanks, where releases had been carried out, leading to browning of leaf laminae (Jayanth and Ganga Visalakshy, 1989c). *O. terebrantis* has established well in Kerala also, but promising results are yet to be achieved.

The results of the studies carried out to date clearly show that *N. eichhorniae* and *N. bruchi* are effective biological control agents of water hyacinth and also have potential to bring about permanent solution to the problems posed by this weed, throughout the country. An operation of such magnitude would, in the normal course, involve a lot of funds for setting up infrastructural facilities, recruiting and training technical manpower for mass multiplication of insects, etc. This can be overcome by carrying releases of the weevils in two or three undisturbed weed infested tanks in different regions in the country, that are well connected by air, rail and road. After one or two years, adults collected from the above areas can be redistributed throughout the region. This is based on the observation that releases of around 2000 adults are sufficient to get a good breeding population established, irrespective of the area of weed coverage and the development of a simple and inexpensive technique for dispatching large numbers of weevils for direct field releases (Jayanth, 1988c).

6.6. Biological Control of *Salvinia molesta*

Salvinia molesta D.S. Mitchell (Salviniaceae) is a non-flowering plant native to south-eastern Brazil (Forno and Harley, 1979), which has been spread into many countries around the world by man, probably as an ornamental. The weed probably reached India in about 1955 from Sri Lanka. It spread extensively in Kerala, especially in the southern districts, affecting the daily life of about 5 million people. The weed was reported to cover 75,000 acres of paddy fields in Kuttanad, which is considered to be the rice bowl of Kerala, making paddy cultivation highly expensive due to the cost involved in the removal of the weed mat before starting cultivation (Joy, 1978).

6.6.1. Importation of natural enemies

For biological control of water fern *Cyrtobagous salviniae* Calder and Sands (Coleoptera: Curculionidae), native to Brazil, was imported from Australia in 1982. The host-specificity of this insect was confirmed by testing 75 plants belonging to 41 families (Jayanth and Nagarkatti, 1987c).

6.6.2. Field evaluation

The efficacy of *C. salviniae* in suppressing water fern was demonstrated in a lily pond in Bangalore, where complete control was obtained in 14 months (Jayanth, 1987d). The insect was supplied by IIHR to Kerala Agricultural University, Trichur for multiplication and releases in Kerala. The insect established readily and brought about 99% weed suppression within 12-16 months at most release spots (Joy *et al.*, 1985b). The weevils are reported to have cleared over 1000 sq km of water surface infested by the weed in Kerala, within three years after releases were initiated (Joy, 1986). Although fluctuations were noticed in the weed coverage the insect has been capable of keeping water fern populations below the economic injury level. Most of the water bodies in southern Kerala, which were choked by water fern are now free of the weed.

7. CONCLUSIONS

Biological control of weeds has an excellent track record and none of the organisms introduced for weed control have become pests of crops. Nevertheless, there have been numerous instances where introduced biological weed control agents fed to a limited extend on non-target plants during the initial population explosion phase. Feeding by *T. scrupulosa* on *Sesamum indicum* in Uganda (Greathead, 1971) and *Cactoblastis cactorum* (Bergroth) on tomato in Australia (Williams, 1954) are two such examples. Normally such instances of feeding on non-hosts do no recur and are not given much importance.

Although the risk involved in biological weed control agents temporarily feeding on non-hosts is negligible compared to those posed by chemicals, such instances get magnified due to high visibility. If not handled properly they become emotional issues and can threaten further progress in biological weed control. Thus, the recently reported feeding by the parthenium beetle *Z. biolorata* on another weed *Xanthium strumarium* and also on sunflower (*Helianthus annuus*) were considered as pointers towards an expanding host range (Kumar, 1992). However, laboratory and field studies suggested that accumulation of deposits of parthenium pollen, containing the phagostimulant parthenin, on sunflower plants growing adjacent to the weed stand, during periods of scanty rainfall, may have caused feeding by adults of the beetle (Jayanth *et al.*, 1993). It is therefore high time that biocontrol workers recognize that such feeding on non-hosts during the initial population explosion phase, is not unusual and investigate the causes, so as to convince the public and the funding agencies.

The science of biological control of weeds has made significant contribution to the control of alien weeds in many countries around the globe. However, not all weeds are amenable to biological control and unfortunately we still cannot predict the impact of introduced biological weed control agents. Thus about 65% of weed insects released up to 1985 established, whereas only 25% of them were effective (Julien, 1989). There is thus, an urgent need to standardize methods for selecting weed control agents suited to particular ecoclimatic regions.

Detailed studies on the ecology and population dynamics of weeds and their natural enemies in their native home may provide clues on selecting efficient biocontrol agents. But, such studies are expensive and may increase pre-introduction costs of agents. However, costs may be reduced through international cooperation. Alien weeds do not respect state boundaries and can be seen infesting several countries in a particular region. Therefore the cost of surveys, pre-introduction studies, host-specificity tests, etc., can be minimized if all those countries cooperate in jointly sponsoring a weed control programme.

The success rate in the utilization of host-specific plant pathogens for weed control has been 66.6% compared to 25% for insects (Julien, 1989). Australia has been in the forefront in this field and there is an urgent need for other countries including India to recognize the importance of pathogens in the control of alien weeds.

REFERENCES

Andres, L.A. 1977, The biological control of weeds, In: *Integrated Control of Weeds*, eds. J.D. Fryer and S. Matsunaka, University of Tokyo Press, Tokyo, Japan, pp. 153-176.

Andres, L.A. 1981, *Biological Control in Crop Production*, ed. G.C. Papavizas, BARC Symposium No.5, Allenheld, Osmun, Totowa, Japan, pp. 341-349.

Andres, L.A. and Goeden, R.D. 1971, *Biological Control*, ed. C.B. Huffaker, Plenum Publishing Corporation, New York, USA, pp. 143-164.

Bennett, F.D. 1984, Biological control of aquatic weeds, In: *Water Hyacinth*, ed. G. Thyagarajan, *UNEP : Rep. Proc.*, Series 7, pp. 14-40.

Chacko, M.J. and Narasimham, A.U. 1988, Biocontrol attempts against *Chromolaena odorata* in India - a review, In: *Proc. First Internat. Workshop Biol. Cont. Chromolaena odorata*, pp. 65-79.

Combellack, J. H. 1992, The importance of weeds and ways of reducing concerns about methods for their control, In: *Proc. First Internat. Weed Cont. Cong.*, eds. J.H. Combellak, K.J. Levick, J. Parsons and R.G. Richardson, Vol. I, Weed Control Society of Victoria Inc., Melbourne, Australia, pp. 43-63.

Cruttwell, R.E. 1974, Insects attacking *Eupatorium odoratum* in the Neotropics, 4, An annotated list of the insects and mites recorded from *Eupatorium odoratum* L., with a key to the types of damage found in Trinidad, *Technical Bulletin, Commonwealth Institute of Biological Control*, 17: 87-125.

Cullen, J. M., Kable, P. F. and Catt, M. 1973, Epidemic spread of a rust imported for biological control, *Nature* 244: 462-404.

Dharmadhikari, P. R., Perera, P. A. C. R. and Hassan, T. M. F. 1977, The introduction of *Ammalo insulata* for the control of *Eupatorium odoratum* in Sri Lanka, *Technical Bulletin, Commonwealth Institute of Biological Control* 18: 129-135.

Evans, H. C. 1987, Life cycle of *Puccinia abrupta* var. *partheniicola*, a potential biocontrol agent of *Parthenium hysterophorus*, *Trans. Brit. Mycol. Soc.* 88: 105-111.

Forno, I.W. and Harley, K.L.S. 1979, The occurrence of *Salvinia molesta* in Brazil, *Aqu. Bot.* 6: 185-187.

Goeden, R.D. 1988, A capsule history of biological control of weeds, *Biocon. News Infor.* 9: 55-61.

Gopal, B. and Sharma, K.P. 1981, *Water Hyacinth (Eichhornia crassipes): The Most Troublesome Weed of the World*, Hindasia Publishers, New Delhi, India, 129 + 91 pp.

Greathead, D. J. 1971, Progress in the biological control of *Lantana camara* in East Africa and discussion of problems raised by the unexpected reaction of some of the more promising insects to *Sesamum indicum*, In: *Proc. II Internat. Symp. Biol. Cont. Weeds, Commonwealth Agricultural Bureau*, pp. 89-94.

Harris, P. 1979, Cost of biological control of weeds in Canada, *Weed Sci.* 27: 242-250.

Harris, P. 1981, *Biological Control in Crop Production*, ed. G.C. Papavizas, BARC Symposium No. 5, Allenheld, Osmun, Totowa, pp. 33-340

Holms, L.G., Plucknet, D.L., Pancho, J.V. and Herberger, J.P. 1977, *The World's Worst Weeds: Distribution and Biology*, University Press, Hawaii, Honolulu, USA, 609 pp.

Hopper, K. R., Roush, R.T. and Powell, W. 1993, Management of genetics of biological control introductions, *Ann. Rev. Entomol.* 38: 27-51.

Jayanth, K.P. 1987a, Investigations on the host-specificity of *Epiblema strenuana* (Walker) (Lepidoptera: Tortricidae) introduced for biological control trials against *Parthenium hysterophorus* in India, *J. Biol. Cont.* 1: 133-137.

Jayanth, K.P. 1987b, Introduction and establishment of *Zygogramma bicolorata* on *Parthenium hysterophorus* in Bangalore, India, *Curr. Scie.* 56: 310-311

Jayanth, K.P. 1987c, Biological control of water hyacinth in India, *Technical Bulletin* No.3, Indian Institute of Horticultural Research, Bangalore, India, 28 pp.

Jayanth, K.P. 1987d, Biological control of water fern *Salvinia molesta* infesting a lily pond in Bangalore (India) by *Cyrtobagous salviniae*, *Entomophaga* 32: 163-165.

Jayanth, K.P. 1988a, Successful biological control of water hyacinth (*Eichhornia crassipes*) by *Neochetina eichhorniae* (Coleoptera: Curculionidae) in Bangalore, India, *Trop. Pest Manag.* 34: 263-266.

Jayanth, K.P. 1988b, Biological control of water hyacinth in India by release of the exotic weevil *Neochetina bruchi*, *Curr. Scie.* 57: 968-970.

Jayanth, K.P. 1988c, A simple packing method for large-scale shipment of water hyacinth weevils, *J. Biol. Con.* 2: 139-140.

Jayanth, K.P. and Ganga Visalakshy, P.N. 1989a, A method to store larvae of *Pareuchaetes pseudoinsulata* Rego Barros (Lepidoptera: Arctiidae), a potential biocontrol agent of *Chromolaena odorata* (Compositae) under low temperature, *J. Biol. Cont.* 3: 137-138.

Jayanth, K.P. and Ganga Visalakshy, P.N. 1989b, Introduction and establishment of *Neochetina eichhorniae* and *N. bruchi* on water hyacinth in Loktak lake, Manipur, *Sci. Cult.* 55: 505-506.

Jayanth, K.P. and Ganga Visalakshy, P.N. 1989c, Establishment of the exotic mite *Orthogalumna terebrantis* on water hyacinth in Bangalore, India, *J. Biol. Cont.* 3 :75-76.

Jayanth, K.P. and Ganga Visalakshy, P.N. 1990, Studies on drought tolerance in the water hyacinth weevils *Neochetina eichhorniae* and *N. bruchi* (Coleoptera: Curculionidae). *J. Biol. Con.* 4: 116-119.

Jayanth, K.P. and Ganga Visalakshy, P.N. 1992, Suppression of the lantana bug *Teleonemia scrupulosa* by *Erythmelus teleonemiae* in Bangalore, India, *FAO Pl. Prot. Bull.* 40: 164.

Jayanth, K.P. and Ganga Visalakshy, P.N. 1996, Succession of vegetation after suppression of parthenium weed by *Zygogramma bicolorata* in Bangalore, India, *Biol. Agri. Horticul.* 12: 303-309.

Jayanth, K.P. and Geetha Bali 1993, Diapause behaviour of *Zygogramma bicolorata* (Coleoptera: Chrysomelidae), a biological control agent for *Parthenium hysterophorus* (Asteraceae) in Bangalore, India, *Bull. Entomol. Res.* 83: 383-388.

Jayanth, K.P. and Geetha Bali 1994, Biological control of *Parthenium hysterophorus* by the beetle *Zygogramma bicolorata* in India, *FAO Pl. Prot. Bull.* 42: 207-213.

Jayanth, K.P. and Nagarkatti, S. 1987a, Investigations on the host specificity and damage potential of *Zygogramma bicolorata* Pallister (Coleoptera: Chrysomelidae) introduced into India for the biological control of *Parthenium hysterophorus*, *Entomon* 12: 141-145.

Jayanth, K.P. and Nagarkatti, S. 1987b, Host-specificity of *Neochetina bruchi* Hustache (Coleoptera: Curculionidae) introduced into India for biological control of water hyacinth, *Entomon* 12: 385-390.

Jayanth, K.P. and Nagarkatti, S. 1987c, Host-specificity of *Cyrtobagous salviniae* Calder and Sands (Coleoptera: Curculionidae) introduced into India for the control of *Salvinia molesta*, *Entomon* 12: 1-6.

Jayanth, K.P. and Nagarkatti, S. 1988, Host-specificity of *Orthogalumna terebrantis* Wallwork (Acarina: Galuminidae) introduced for biological control of water hyacinth in India, *J. Biol. Cont.* 2: 46-49.

Jayanth, K.P., Mohandas, S., Asokan, R. and Ganga Visalakshy, P.N. 1993, Parthenium pollen induced feeding by *Zygogramma bicolorata* (Coleoptera: Chrysomelidae) on sunflower, *Bull. Entom. Res.* 83: 595-598.

Joel, D.M. and Liston, A. 1986, New adventive weeds in Israel, *Isr. J. Bot.* 35: 215-223.

Joshi, S. 1990, Parthenium - its biological control, In: *Karnataka State of Environment Report* IV, ed. C.J. Saldanha, Center for Taxonomic Studies, St. Joseph's College, Bangalore, India, pp. 61-73.

Joy, P.J. 1978, Ecology and control of *Salvinia* (African payal) the molesting weed of Kerala, *Technical Bulletin* No.2, Kerala Agricultural University, 40 pp.

Joy, P.J. 1986, *Salvinia* control in India, *Biocont. News Infor.* 7: 142.

Joy P.J., Sathesan, N.V. and Lyla K.R. 1985a, Biological control of weeds in Kerala, In: *Proc. Nat. Symp. Entom. Ins.,* Calicut, pp. 247-251.

Joy, P.J., Sathesan, N.V., Lyla, K.R. and Joseph, D. 1985b, Successful biological control of the floating weed *Salvinia molesta* Mitchell using the weevil *Cyrtobagous salviniae* Calder and Sands in Kerala (India), In: *Proc. 10th Conf. Asian Pacific Weed Sci. Soci.* pp. 622-626.

Julien, M.H. 1989, Biological control of weeds worldwide: trends, rates of success and future, *Biocon. News Infor.* 10: 299-306.

Julien, M.H. 1992, *Biological Control of Weeds: A World Catalogue of Insects and their Target Weeds,* CAB International, UK and Australian Centre for International Agricultural Research, Canberra, 186 pp.

Julien, M.H., Kerr, J.D. and Chan, R.R. 1984, Biological control of weeds; an evaluation, *Prot. Ecol.* 7: 3-25.

Kapoor, V. C. and Malla, V. K. 1978, Infestation of the gall fruit fly, *Procecidochares utilis* (Stone) on crofton weed, *Eupatorium adenophorum* Sprengel in Katmandu, *Indian J. Entom.* 40: 337-339.

Krishnamurthy, K., Ramachandra Prasad, T.V., Muniyappa, T.V. and Venkata Rao, B.V. 1977, Parthenium, a new pernicious weed in India, *Technical Series* No. 17, University of Agricultural Sciences, Bangalore, 46 pp.

Kumar, A.R.V. 1992, Is the Mexican beetle *Zygogramma bicolorata* (Coleoptera: Chrysomelidae) expanding its host range? *Curr. Scie.* 63: 729-730.

Kumar, S., Jayaraj, S. and Muthukrishnan, T.S. 1979, Natural enemies of *Parthenium hysterophorus* Linn, *J. Entomol. Res.* 3: 32-35.

McClay, A.S. 1980, Studies on some potential biocontrol agents for *Parthenium hysterophorus* in Mexico, In: *Proc. Fifth Internat. Symp. Biol. Cont. Weeds,* Brisbane, Australia, pp. 471-482.

McFadyen, R.E. and McClay, A.S. 1981, Two new insects for the biological control of parthenium weed in Queensland, In: *Proc. Sixth Australian Weed Conf.* 1: 145-149.

Mohandas, S. 1981, Agrophysiological impact of the allelopathic effects of *Parthenium hysterophorus* L., Science Academy Medals for Young Scientists Lectures, pp. 275-285.

Muniappan, R. and Viraktamath, C.A. 1986, Status of biological control of the weed *Lantana camara* in India, *Trop. Pest Manag.* 32: 40-42.

Nagarkatti, S. and Jayanth, K.P. 1984, Screening biological control agents of water hyacinth for their safety to economically important plants in India, I, *Neochetina eichhorniae* Warner, In : *Water Hyacinth,* ed. G. Thyagarajan, *UNEP Rep. Proc.* Series 7, 1005 pp.

Narasimhan, T.R., Ananth, M., Narayana Swamy, M., Rajendra Babu, M., Mangala, A. and Subba Rao, P.V. 1980, Toxicity of *Parthenium hysterophorus* L.: Parthenosis in cattle and buffaloes, *Indian J. Anim. Sci.* 50: 173-178.

Parker, C. and Fryer, J.D. 1975, Weed control problems causing major reductions in world food supplies, *FAO Pl. Prot. Bull.* 23: 83-93.

Sankaran, T. 1973, Biological control of weeds in India, a review of introduction and current investigations of natural enemies, In: *Proc. Second Internat. Symp. Biol. Cont. Weeds,* Commonwealth Agricultural Bureau, pp. 82-88.

Sankaran, T. and Sugathan, G. 1974, Host specificity tests and field trials with *Ammalo insulata* (Wlk.) (Lep: Arctiidae) in India, Mimeographed Report, Commonwealth Institute of Biological Control, 11 pp.

Schroeder, D. 1983, *Recent Advances in Weed Research,* ed. W.W. Fletcher, Commonwealth Agricultural Bureaugh, Slough, pp. 41-78.

Singh, S.P. 1989, Biological Suppression of Weeds, *Technical Bulletin* No.1, Biological Control Centre, National Centre for Integrateed Pest Management, Bangalore, 27 pp.

Tomley, A.J. 1990, Parthenium rust *Puccinia abrupta* var. *partheniicola,* In: *Proc. 9th Australian Weeds Conf.* pp. 511-512.

Towers, G.H.N., Mitchell, J.C., Rodriguez, E. and Bennett, F.D. 1977, Biology and chemistry of *Parthenium hysterophorus* L., a problem weed in India, *J. Scie. Indus. Res.* 12: 672-684.

Vartak, V.D. 1968, Weed that threatens crops and grass lands in Maharashtra, *Indian Farming* 18: 23-43.

Vietmeyer, N.D. 1975, The beautiful blue devil, An IPPC papers reprint from Natural History Magazine, November 1975, The American Museum of Natural History, 5 pp.

Watson, A.K. 1992, Biological and other alternate control measures, In: *Proc. First Internat. Weed Cont. Congr.,* ed. J.H. Combellack, K.J. Levick, J. Parsons and R.G. Richardson, Vol. I, Weed Control Society of Victoria Inc., Melbourne, Australia, pp. 43-63.

Williams, J.R. 1954, The biological control of weeds, In: *Rep. Sixth Comm. Entom. Conf.,* Commonwealth Institute of Entomology, London, pp. 85-98.

BIOLOGICAL CONTROL OF ROT DISEASES OF SMALL CARDAMOM

Joseph Thomas

Indian Cardamom Research Institute
(Spices Board), Myladumpara
P.O. Kailasanadu – 685 553, Kerala, INDIA

1. INTRODUCTION

Biological control of diseases of crop plants has become an integral part of disease management systems in recent years. Biological control offers promising solutions to pesticidal hazards, environmental pollution, pesticide residues in crops and to ever increasing cost of plant protection operations. The term biological control is used to denote the use of one species of organism to control or eliminate another organism which is harmful to the crop plants or animals. Various scientists have attempted to define biological control (Baker and Cook, 1974; Cook, 1988; Cook and Baker ,1983; Lupton, 1984). Cook (1988) defined biological control as the use of organisms, genes or gene products to regulate a pathogen and can be used with strategies intended to keep (i) inoculum density below an economic threshold level (ii) retard or exclude infection and (iii) maximise the plants' system for self defence. Thus biological control in its broadest sense envisages the use of microorganisms which are antagonistic to pathogens or any bioagent that restricts the activity of a pest or the use of specific genes which are potential suppressors of pathogenic effects so as to ultimately partially or fully relieve the host from the deleterious effects of the pathogen or the pest. Biological control of plant pathogens is brought about by reducing the population levels of pathogens by the bioagents. Bioagents used to control plant diseases are potential inhibitors or suppressors of pathogens and they bring about effective disease control. A number of biocontrol agents belonging to fungal, bacterial and viral groups are known to possess antagonistic activity against specific pathogens. Biocontrol has become one of the major components in modern concept of organic farming which greatly restricts the use of chemicals in agriculture, and encourages the exclusive use of only natural products and beneficial microorganisms in farming systems. Microorganisms which act as bioagents bring about biological control through their interaction with pathogens in one or more mechanisms such as hyperparasitism, antagonism, immunization,or cross-protection and hypovirulance which ultimately resulting in the reduction of pathogens' activity on crop plants. Besides reducing the cost of chemical pesticides, the bio-agents contribute in great deal to maintain the ecological balance in the biosphere. A number of antagonistic fungi and bacteria are being developed in recent years as biocontrol agents. Among these, the well known and the most widely used bioagent is the genus *Trichoderma* thanks to its high degree of antagonism to a wide variety of pathogenic fungi. The effectiveness of *Trichoderma* and other bioagents in combating diseases of cardamom is reviewed in this chapter.

2. CARDAMOM AND ITS MAJOR DISEASES

Small cardamom *(Elettaria cardamomum* Maton) the well known "Queen of Spices" is a herbaceous spice crop of the family Zingiberaceae. It is cultivated in the evergreen rainforests of the western ghats of southern India. About 80,000 ha are under cardamom cultivation and its distribution is 60, 30 and 10 percent area in the states of Kerala, Karnataka and Tamilnadu respectively. The crop thrives well at altitudes ranging from 600 to 1200 meters above msl and requires a cool humid climate with an annual rainfall of about 200 to 2500 mm distributed over a period of 190 to 200 days. Cardamom requires red lateritic to black soils highly enriched with organic matter or humous and low pH ranging from 4.5 to 5.8. A forest canopy with about 40-50 percent shade is most ideal for the crop. India's annual prodction of cardamom is about 6500 to 7000 metric tonnes. Until a decade back, India had world monopoly in the production and export of cardamom. However, due to stiff competition, India was pushed down to second position as Gautemala came up in the first position. Due to the attack of various diseases, pests, etc the country is facing a lot of constraints in the profitable production of the crop. Effective management of pests and diseases using safe and low cost methods has an important role in making the crop production more economical.

The crop is susceptible to the attack of a number of diseases caused by fungal and viral pathogens. Although about 22 different types of diseases at various growth stages of cardamom were reported (Naidu and Thomas, 1992), only a few are alarming such as the viral diseases and rot diseases. The bio control system of disease management described in this chapter is for controlling the soil borne fungal diseases only which cause rotting and damage to the crop. These are 'Azhukal', rhizome rot and seedling rot diseases. Minor diseases affecting foliage and capsules are air borne and their occurrence is only sporadic. These generally do not become severe problems.

2.1. Azhukal or Capsule Rot

Capsule rot also called in Malayalam vernacular as 'Azhukal' is a major rot disease affecting the capsules and leaves (Menon *et al.*, 1972). The disease appears during south west monsoon season in plantations of Kerala and in some parts of Tamilnadu where high rainfall is received. Although capsules and leaves are affected, the whole plant rots and perishes when the disease appears in a severe form. During years of heavy disease incidence, crop loss as high as 40 percent has been reported (Anon, 1989). The causal organism is *Phytophthora nicotianae* Breda de Hann var. *nicotianae* Waterhouse (Thankamma and Pillai, 1973); *P.palmivora* Butler (Radha and Joseph, 1974) and *P. meadii* Mc Rae (Anon, 1986). However, the predominant species causing azhukal is *P. meadii*. The pathogen survives in the soil and on infected plant debris in the form of chlamydospores and oospores. During favourable conditions such as continuous rainfall,

224

high soil moisture, low temperature and high relative humidity, the propagules of pathogen become active and initiate infection.

2.2. Rhizome Rot

Rhizome rot commonly known as clump rot is a wide spread disease in Kerala , Karnataka and in heavy rainfall areas of Tamilnadu as such as the Anamalai hills. This disease is characterised by development of rotting symptoms on young shoots and, rhizomes, and flaccidity and yellowing of leaves. Disease becomes severe during July-August months and infected plants often completely decay and perish within a short period. The disease is caused by soil-borne fungal pathogens such as *Rhizoctonia solani* Kuhn (Subba Rao, 1938); *Pythium vexans* de Barry (Ramakrishnan,1949) and *Fusarium oxysporum* (Thomas and Vijayan,1994). Overcrowding of plants, excess shade, high soil moisture and humid atmosphere are congenial conditions for disease development.

2.3. Nursery Rots

Cardamom is propagated mainly through seedlings raised in nurseries.Often two sets of nurseries are maintained;the primary nursery or germination beds, and the secondary nursery where six month old seedling are transplanted from the former. Diseases such as seed rot, damping off clump rot and root rots are commonly observed in the nurseries.

2.3.1. Damping off

Often seeds sown in germination beds as well as young seedlings are affected by damping off disease. Seedlings wilt and suddenly collapse in patches following development of rotting symptoms. Over crowding and excess soil moisture are congenial conditions for disease development. The disease is caused by *Pythium vexans* (Middleton, 1943); *Rhizoctonia solani* (Subba Rao,1938; Wilson *et al.,*1979) and *Fusarium oxysporum* (Siddaramaiah, 1988). Damping off incidence in unprotected nurseries goes upto 78 percent (Siddaramaiah, 1988a). Another seedling rot caused by *Sclerotium rolfsii* characaterised by rotting of leaves, leaf sheaths and psuedostems of seedlings was also reported (Siddaramaiah, 1988 b).

2.3.2. Seedling rot

This type of seedling rot is often found in the secondary nursery where the seedlings are of about 6 to 18 months old. Symptoms appear in the form of flaccidity of leaves followed by leaf yellowing and wilting. The rhizomes and roots decay and seedlings collapse. The pathogenic fungi involved are *P. vexans* and *R. solani* as in the case of damping off disease. Ali and Venugopal (1993) had observed that root knot nematode *Meloidogyne incognita* is also associated with fungal pathoges.

2.4. Nematode Diseases

Nematode infection in cardamom nurseries as well as in plantations is a serious problem. Ali (1983) reported about 20 genera of various nematodes; however only the root knot nematode *Meloidogyne incognita* and the burrowing nematode *Radopholus similis* (Ramana and Eapen,1992) are the major disease causing types. The symptoms produced are stunting of plants with reduced tillering, rosetting and narrowing of leaves.

The flowering and fruit setting are highly reduced. Often knots and swellings are observed in the roots of affected plants.

3. ANTAGONISTS AS BIOCONTROL AGENTS

The various diseases of cardamom have been brought under control with chemical fungicides. As fungicides being highly expensive and often they leave toxic residues in crops, attempts have been initiated by the Indian Cardamom Research Institute from 1990 onwards to search for suitable bioagents. A large number of fungal and bacterial antagonists were known as bioagents active against various pathogens. However in the pilot trials a few bioagents such as *Trichoderma* spp., *Laetisaria arvalis, Bacillus subtilis, Aspergillus* spp. and a few actenomycetes were tested against major rot pathogens. Among these, *T.viride* and *T.harzianum* are two exotic isolates obtained from Tamilnadu Agricultural University, Coimbatore and a number of local isolates of *T. viride* and *T. harzianum* collected from cardamom soils were tested (Vijayan *et al.,* 1994) for their interaction with rot pathogens both in the laboratory and in the field.

3.1. *In vitro* Screening of Bioagents

It is important that the antagonists which are to be used as biocontrol agents should possess certain basic requirements such as fastness of growth and multiplication, high rate of sporulation or production of propagules and inhibitory properties over the pathoges. These are generally ascertained through *in vitro* screening against target pathogens. The growth rates of the pathogens and the antagonists have been determined in monocultures and their mutual growth rates or inhibition were studied in dual culture plates. (Thomas *et al.,* 1993). The fungal antagonists *Trichoderma* sp. and *Laetisaria* sp. showed almost double the growth rates as that of *Phytophthora* while these showed almost identical growth rates with *P. vexans* and *R.solani* when cultured on PDA. It required only four days growth for the antagonist to attain a colony diameter of 100 mm while for *Phytophthora*,8 days were required and for *P.vexans* and *R.solani* only three days were required to attain equal colony size. The various local isolates of *Trichoderma* spp. exibited varying degrees of growth rates but theses were always at much higher levels than *Phytophthora.* The fast growing antagonists *Trichoderma* sp. and *Laetisaria* sp. inhibited the pathogens to a higher degree by mostly preventing the growth and advancement of the pathogen. The antagonists induced nearly 60 to 70 percent inhibition of *Pythium, Rhizoctonia,* and *Phytophthora* within 5 to 6 days when grown in paired culture (Thomas *et al.,* 1993).The fast growing antagonists often over grew the pathogens (Mukherjee *et al.,*1989; Thomas *et al.,* 1993). Dhanapal and Thomas (1996) studied the growth and inhibition rates of several local isolates of *T. viride* and *T.harzianum* and listed out a few strains which are very effective bioagents. Other antagoinists such as *Aspergillus* sp., Actenomycetes and *Bacillus subtilis* also showed antagonostic property against pathogenic fungi but to a lesser extent as compared to *Trichoderma.* Interaction of antagonists and pathogens studied by dual culturing on agar coated cellophane as well as by inverted plate method also showed that the rot pathogens were inhibited by *Trichoderma.* Growth inhibition of pathogen was mainly through the production of volatile compounds (Dennis and Webster,1971) and through degradation of hyphae of pathogen caused by the intertwining hyphae of *Trichoderma* as reported by Elad *et al.* (1982). All isolates of *Trichoderma* spp. and *L. arvalis* tested showed promosing results on growth inhibition of pathogens under *in vitro* conditions.

226

3.2. Biocontrol of Seedling Rot Diseases

As it is seen that rotting of cardamom seedlings can be caused by one or more pathogens viz., *Pythium vexans, Rhizoctonia solani, Fusarium oxysporum,* etc. the bioagents which showed antagonism with *in vitro* studies should also be equally effective under *in vivo* conditions. The bioefficacy of antagonists were tested in cardamom seedlings under green house conditions. Thomas *et al.* (1994) studied the comparative infection and mortality rates of cardamom seedlings infected by *P. vexans, R.solani* and their combinations as against their infection in seedlings under the influence of the bioagents such as *T. viride, T.harzianum, L. arvalis* and *Bacillus subtilis*. When young cardamom seedlings were inoculated with either or both of the pathogens alone, the infection rates ranged from 28 to 58 percent while the seedling mortality ranged from 15 to 38 percent. The infection and mortality rates of seedlings were reduced to 15 to 38 percent and 4 to 10 percent respectively when the seedling were pre-treated with the antagonists. In double infections of seedlings by *P. vexans* and *R.solani, T.harzianum* was found to be the most suited bioagent as there was only less than 5 percent mortality of seedlings as compared to unprotected seedlings which showed motrality upto 38 percent. Both *L.arvalis* and *B.subtilis* also reduced disease incidence in the case of infection caused by *P.vexans*. However, when the pathogen was *R.solani* or *R.solani* together with *P. vexans* these bioagents were not as effective as the *Trichoderma* spp. As seedling rot is often caused by both these pathogens, it is important that a suitable antagonist which shows a wide spectrum of suppressiveness should be judiciously selected for all practical purposes for effective disease control. In these studies *T.harzianum* consistently showed antagonistic effects to both the pathogens. The efficacy of *T.harzianum* in suppressing *R.solani* induced diseases has been well demonstrated (Marshall, 1982; Smith and Webner 1987) and that on *Pythium* spp. has been shown by Liftshitz *et al.* (1984). Pre-treatment of nursery beds and potting mixtures with *T. harzianum* fifteen days prior to transplanting of cardamom seedlings has been found to offer better protection from natural infection of seedlings caused by soil inhabiting *Pythium* spp. or *Rhizoctonia* spp. Infection of seedlings by *Fusarium oxysporum,* the common root rot pathogen of cardamom seedlings was also found to be greatly reduced in nursery beds pre-treated with *T.harzanium* and *T.viride*.

Seedling rot caused by *Phytophthora* spp. is not generally observed as a major problem. However, occassionally seedling rot is observed in *Phytophthora* sick soils. Bhai *et al.* (1993) have studied the effect of bioagents in seedlings inoculated with *P.meadii*. When seedlings were planted in *P. meadii* incoporated soil,the seedling mortality was as high as 53 percent. When *T.viride* or *T.harzanium* were incorporated in nursery beds one week prior to *P. meadii* inoculation, the seedling mortality was only 24 and 33 percent respectively. Soil application of antagonists in sick soils has been reported as an effective method of biocontrol of soil borne *Phytophthora* spp. (Tsao *et al.,* 1988).

3.3. Biocontrol in Plantations

The biocontrol efficacy of *Trichoderma* spp.and *L.arvalis* in managing capsule rot and rhizome rot diseases in cardamom plantations was studied in detail (Thomas *et al.,* 1997). Bhai *et al.* (1993) incorporated *T. viride* and *T. harzianum* in *Phytophthora* sick cardamom plantations at a population level of 5×10^5 CFU prior to the appearance of disease symptoms and repeated the applications twice at monthly intervals and noted that azhukal or capsule rot incidence was greatly reduced with a decrease in the disease potential index (DPI) in the soil (Table-1). Observations recorded on rainfall and DPI revealed that during high and continuous rainfall periods, the soil population level of *Phytophthora* increases drastically with a corresponding increase in disease severity in untreated soils. On the other hand, regular soil application of bioagents reduces the population level of *Phytophthora* obviously due to increased soil abtibiosis, resulting in drastic reduction in disease severity. Continued presence of *Trichoderma* spp. in sick soils seems to be essential requirement for bringing down the *Phytophthora* population under control.

Table 1. Effect of soil application of *Trichoderma* on incidence of 'azhukal' disease and population levels *Phytophthora meadii.*

Antagonist	Seedling/ Plantation	% Disease incidence	DPI *
T. viride	Seedlings	22	2
T.harzianum	Seedlings	33	2
Control	(No antagonist)	45	16
T.viride	Plantation	14	1
	Sick plot I		
T. harzianum	,,	18	2
Control	(No antagonist)	32	16
T.viride	Sick plot II	23	4
T.harzianum	,,	40	4
Control	(No antagonist)	50	32

* Disease Potential Index

The bio-control efficacy of *Trichoderma* spp. has been exploited to control the rhizome rot or clump rot disease in cardamom plantations. Vijayan *et al.* (1994) evaluated a number of isolates of *T.viride* and *T. harzanium* for testing their comparative efficacy in bringing about field control of the disease. Among the nine isolates tested (Table 2) in

Table 2. Rhizome rot incidence in field plots treated with isolates of *Trichoderma* spp.

Code No	Isolates	Disease incidence (%)	Disease control (%)
T1	*T.viride*	6.88(15.18)	42.63
T6	,,	12.50(20.02)	24.34
T7	,,	6.86(14.07)	46.82
T8	,,	6.61(14.07)	45.11
T13	,,	9.35(16.52)	39.46
T14	,,	4.80(12.63)	52.29
T2	*T.harzianum*	2.65(8.80)	69.08
T4	,,	5.13(12.35)	53.34
T12	,,	4.23(11.85)	55.21
Control	(no antagonist)	19.98(26.46)	--
		L.S.D.(1%)	(12.96)
		L.S.D(5%)	(9.46)

T1 & T2 are exotic isolates.

Figures in paranthesis are transformed values

the field eight isolates were found to be effective in reducing the disease, although their comparative suppressive efficacies varied. Those isolates which showed disease control in the field were also found to be highly inhibitory under *in vitro* conditions. On the contrary there were a few native isolates which showed high degree of inhibition under *in vitro* conditions but these were not effective in the field control of rhizome rot disease. Field survival of *Trichoderma* in inoculated soils showed that *Trichoderma* could be recovered from inoculated soils five months after incorporation in the field (Dhanapal and Thomas, 1996). Such prolonged survival in the soil often depends on the availability of food materials (Papavizas, 1981).

3.4. Biocontrol of Nematode Diseases

Nematode diseases caused by species of *Meloidogyne* and *Radopholus* pose considerable damage in cardamom nurseries and plantations. Biocontrol of these nematodes with the use of antagonistic fungi and VA Mycorrhizae has been achieved to some extent. *Paecilomyces lilacinus* is a soil inhabiting fungal antagonist often found in cardamom soils. Nursery beds treated with *P. lilacunus* at the time of seed sowing, rhizome formation and tillering stages reduced root knot nematode population upto 74.2 percent and incidence of rhizome rot disease upto 19.7 percent (Eapen and Venugopal, 1995). *P. lilacinus* multiplied on coffee husk and neem oil cake was reported to be an excellent formulation for reducing the population of nematodes (Eapen, 1995). The bio-control effects of *Trichoderma* spp. on root knot nematode of cardamom has been studied in detail. Suppression of root knot nematodes resulting in improved growth of cardamom seedlings in nurseries has been reported with *T. harzianum*, (NRCS, 1991,1994,IISR, 1995) and the suppression was more pronounced when native soils were used for growing the seedlings. Eapen and Venugopal (1995) have shown that isolates of *Trichoderma* spp. have a broad spectrum of biocontrol activity against a number of pathogenic fungi and nematodes. *Trichoderma* isolates when incorporated to cardamom nursery beds affected with root knot nematodes and seedling rot diseases have resulted in remarkable decrease in the population level of nematodes as well as reduction of seedling rot disease. Works carried out on the interaction of antagonistic fungi and nematodes at Indian Institute of Spices Research, Calicut (IISR, 1995) have shown that these fungi colonised the egg masses of root knot nematodes resulting in distortion of nematode eggs. The culture filtrates of these antagonistic fungi also showed nematicidal properties.

Vesicular Arbuscular Mycorrhizal (VAM) fungi also sometimes act as effective bio-control agents in addition to their contribution to plant nourishment. Smith (1987) has shown that VAM association enables the host plant to attain more tolerance to nematode infection due to improved phosphorous status of the host or by competition or antagonism between the nematode and the VAM fungus. In cardamom,the VAM fungi such as *Gigaspora margarita* and *Glomus fasciculatum* suppressed *Meloidogyne incognita* and improved the growth of cardamom seedlings (Thomas *et al.,* 1989).

3.5. Mechanism of Biocontrol

Trichoderma spp. is the most widely used biocontrol agent which possess all the required ideal qualities such as easiness of isolation and culturing, ability to rapidly grow and colonize on a wide variety of food bases, ability to act as a mycoparasite and attacking a wide variety of fungal pathogens. The biocontrol of plant diseases is brought about in one or more ways such as mycoparasitism, antibiosis, competition, etc.

3.5.1. Hyperparasitism

In *in vitro* studies with dual cultures of *Trichoderma* and pathogenic fungi namely *R.solani* and *P. meadii*, the hyperparasitic activity of the former was noticed on the hyphae of the pathogens in the form of hyphal coiling, penetration and hyphal lysis. Such type of hyphal interactions with pathogens are commonly observed characteristic feature of *Trichoderma*. Weinding (1932) observed that *Trichoderma* hyphae growing adjacent to *Rhizoctonia* hyphae coiled around the latter and later found growing within it.

In experiments on *in vitro* studies, cases of *Trichoderma* penetrating into the hyphae of *R.solani* and later causing the lysis of the hyphae of *R.solani* were observed. Intermingling areas of hyphae of *T. viride* and *P.meadii* also showed lysis and erosion of hyphae of *P.meadii* . Elad *et al.* (1982, 1983) correlated such lysis with the secretion of wall dissolving enzymes by the antagonists. The hyper-parasitic relationship of *Trichoderma* to *R.solani* was described in detail (Wu, *et al.*, 1986).

3.5.2. Antibiosis

The growth inhibition observed in *in vitro* studies is due to the production of antibiotic compounds by *Trichoderma* in the medium . Godtfredson and Vangedal (1965) described Trichodermin a sesquiterpene as an antibiotic compound produced by *Trichoderma*. Dennis and Webster (1971) reported that acetaldehyde was the major volatile antibiotic produced by *Trichoderma*. In the dual culture plates *Trichoderma* inhibited the growth of *P. meadii* almost completely. With *P. vexans* and *R. solani* , the inhibition was not complete as the growth rate of *R. solani* and *P. vexans* were higher than that of *P. meadii*. Several workers have reported a number of *Trichoderma* exudates having antibiotic properties. Dermadine, a product of *Trichoderma* is an unsaturated monobasic acid active against a number of pathogenic fungi (Pyke and Dietz , 1977).

3.5.3. Competition

The protection *Trichoderma* provides to the crops from infection with soil borne pathogens depends on its ability to compete with the latter for food and site. Field experiments with *T.viride* and *T.harzianum* to control rhizome rot disease have shown that a population level of 10^5 units per gram of soil is required for effective disease reduction.Lower population levels of the antagonist or its decline in soil over a long period often resulted in failure of suppression of pathogens. This shows that a certain level of *Trichoderma* population is required to compete with soil pathogens .Wells (1988) considers that the omnipresence of *Trichoderma* in agricultural and natural soils through out the world is a first hand evidence that it is an excellent competitor for space and nutritional resources.

4. DEVELOPEMENT OF BIOFUNGICIDE

In the history of biocontrol, no agents had ever been well studied and widely used as the genus *Trichoderma*. As Wells (1988) rightly points out, *Trichoderma* is an excellent model bioagent since it is easy to isolate and culture, its universal presence, its rapid growth on a wide variety of substances, attacks a large number of pathogens, acting as a mycoparasite by competing for food and space. That *Trichoderma* is harmless to crop plants or humans and an active destroyer of many pathogenic fungi, it has succeeded in

230

being elevated to the status of a biofungicide. Cates (1990) visualized the large scale use of *Trichoderma* as a biofungicide for the future for controlling plant diseases. Wells *et al.* (1982) and Baker and Cook (1983) have demonstrated under field conditions that *T. harzianum* is an effective biocontrol agent when mixed with a food base. Since then several attempts had been made to develop a suitable formulation of *Trichoderma* that can easily be used as a biofungicide for all practical purposes.

4.1. Application Technique

Application of the bioagent to the specific site is needed for effective utilization of the biocontrol agent. In the pilot field trials for control of rhizome rot disease of cardamom *Trichoderma* was applied in the field as agar culture mascerates having a CFU value of 6 x10^5 per gram (Thomas *et al.*, 1993). A good delivery system is essential for the bioagent for its field application. Several workers have used various methods for application of the bioagent depending upon the nature of the diseases. For the control of soil-borne fungal diseases, the bioagent needs to be applied to the rhizosphere of the plant and this is usually done after incorporating the bioagent in a suitable carrier medium. Lewis and Papavizas (1980) used sterilized sand and corn meal mixture for growing the antagonist and it was broadcasted in the field to control belly rot of cucumber caused by *Rhizoctonia solani*. For biological control of root pathogens Cook and Baker (1983) used seed treatment with the bioagent. As cardamom rot pathogens are soil borne and being invaders of the rhizome and roots, soil application of *Trichoderma* in the rhizosphere zone is more appropriate mode of application.

4.2. Seed Treatment

As cardamom is mainly propagated through seedlings which are grown in nurseries , application of the bioagent as seed coating was found ideal. A simplified method was developed for coating cardamom seeds with spores of *T.viride* or *T. harzianum*. An aqueous suspension of *Trichoderma* spores is prepared from well sporulated agar cultures and is mixed with an equal volume of warm rice water or gelatin as an adhesive. Washed cardamom seeds are thoroughly mixed with this spore suspension and allowed to dry for a day. *Trichoderma* coated seeds always exhibited high germination rates and the seedlings raised from such seeds were free of damping off disease. This method was found to be an excellent application technique of the bioagent to control seed rot and damping off disease in cardamom nurseries. It also gives the antagonists to express its competition to the pathogens in its vicinity. Coating of cardamom seeds with *Trichoderma* spores also results in the long term preservation and protection from storage diseases.

4.3. Carrier Media

It is important that the bioagent needs to be incorporated in a suitable carrier medium for its storage in a viable state and for field application at a later stage. Various types of crop residues, agricultural wastes, talc powder and a number of inert materials have been reported to act as carrier media for *Trichoderma* (Kousalya and Jeyarajan, 1990; Mukopadhyaya, 1987). Materials selected as carrier media are sometimes inert materials such as the talc powder, vermiculite, perilite, sand, etc. or agricultural wastes or bye-products. Bhai *et al.* (1994) have evaluated a number of agricultural wastes which could be used as carrier and multiplication media at the same time. They have reported that sterilized tea waste, coffee husk or a mixture of coffee husk and cattle manure were ideal combinations for the fast growth and multiplication of both *T. viride* and *T.harzianum* (Table 3), which produced maximum sporulation in these carrier media. Wheat bran and saw dust mixture have been used as carrier media for the mass multiplication of *T.harzianum* (Elad *et al.,* 1980; Mukhopadhyaya *et al.,* 1986). In many of these cases, the materials used for carrier cum multiplication medium were either expensive or locally not available in cardamom plantations. As cardamom is grown in the plantation belt where other plantation crops such as coffee, tea, etc. are also grown, coffee husk or tea waste or farm yard manure are locally availabe and are least expensive. The best suited carrier medium for *T.viride* and *T.harzianum* was found to be sterilized coffee husk, tea waste or their mixture with dried farm yard manure which not only supported the antagonists, but also act as an excellent growth and storage medium with abundant sporulation of the antagonists. The shelf life of such formulations was found to be nearly 12 to 18 months.

Table.3. Comparative sporulation of *Trichoderma* sp. in different carrier media.

Carrier media		*T.harzianum*	*T.viride*
		X 10^6cfu/g	X 10^6cfu/g
1.	Farm Yard Manure (FYM)	0.19	83.33
2.	Neem Cake (NC)	0.13	0.67
3.	NC+FYM 1:1	0.11	0.67
4.	NC + FYM 1:3	0.06	0.32
5.	Coffee husk (fresh)	5.00	86.67
6.	Coffee husk (decomposed)	29.00	107.33
7.	Coffee husk + FYM 1:1 (fresh)	16.00	146.67
8.	Coffee husk + FYM 1:1 (decomposed)	5.67	122.67
9.	Tea waste	18.67	168.67
10.	Tea waste + FYM 1:1	30.00	53.33
11.	Tea waste + FYM 3:1	17.33	105.33
12.	Soil (Control)	3.00	3.00
	GM	10.43	73.22
	SE	1.77	22.82
	CD	5.20	66.93

4.4. Commercial Formulations

Commercial manufacturing and distribution of bioagents is now a days increasing to a great extent, but the market coverage is only very little as compared to the well established chemical pesticides. Though at present a number of commercial formulations of *Trichoderma* bioagent are available in the market, most of them are not popular and widely marketed (Table 4). The commercial formulations of bioagent will be easily accepted by the agriculturist provided that the formulations should be of high quality, efficacious, and economical. Though presently, a number of commercial formulations of *Trichoderma* spp. are available, most of them are in inert materials which act only as carrier media but do not support the growth, sporulation and storage of the antagonist. More often their shelf life period also is very short. As the formulation medium is inert in

Table 4. Commercial formulations of *Trichoderma* spp.

Sl No.	Name of Product	Name of bioagent	Carrier media	Country
1.	F stop	*T. harzianum*	Not known	USA
2.	Binab.T.	*T. viride*	"	Sweeden
3.	Trichodermin	"	"	Bulgaria
4.	Antagon T.V	"	Vermiculite	India
5.	Bio-care F.	"	Sand based	"
6.	Triderma-Lupin	*T. harzianum*	Talc powder	"
7.	Margo-Ecoderma	*T. viride*	"	"
8.	Trichosan	*T. viride*	"	"
9.	Gees T. viride	*T. viride*	Mica mixture	"
10.	Green - Fertderma	*T. viride*	Neem cake	"
11.	Trinjavika	*T. harzianum*	Coffee husk	"
12.	ICRI, Trichoderma	*T. harzianum* Farm yard - manure mixture	Coffee husk	"

most of the cases, the population level of the antagonist in the formulation is often found at very low concentrations when the product reaches the hands of the consumer. The inert meterial formulation when directly applied to the soil becomes unable to multiply in the absence of a suitable growth promoting food bases. Formulations in agricultural waste products, oil cakes, bran of grains, farm yard manure, etc. are found to be excellent growth carrier and storage media. However, unfortunately few industrialists have thought of developing their products in this line.

5. BIOAGENTS IN INTEGRATED SYSTEMS.

Although bioagents have now become an effective substitute for chemical fungicides, the best method to control the diseases is through an integrated approach giving a significant role to the bioagents. *Trichoderma* spp. is being used as one of the major component in the integrated management of several diseases (Chet and Henis, 1985; Papavizas, 1985). As *Trichoderma* dominates in the soil treated with fungicides, it can easily proliferate and produce antibiotics or compete for nutrients.

In the present biocontrol practices of cardamom rot diseses, *Trichoderma* is used as a major component along with plant sanitation, soil amendments, fungicidal applications and cultural practices. Integrated management of capsule rot of cardamom has been reported (Bhai *et al.*, 1997). The rhizome rot disease incidence and its spread in plantations were brought under effective control using the IDM strategies with *T. oside T.harzianum* (Thomas *et al.*, 1997). For integration of the bioagent with fungicides, it is essential that their compatibility should be worked out. Fungicides such as Bordeaux

233

mixture and Emisan (methoxy ethyl mercuric chloride) completely inhibited the growth of *Trichoderma* under *in vitro* conditions while with other tested fungicides the percentage inhibition varied from 41.7 to 89 (Table 5). The field survival of *Trichoderma* population in the soil treated with fungicides showed a gradual decline in the population level of the antagonist over a period of 100 days as evidenced by their decreasing cfu values. However, a similar trend in decrease of *Trichoderma* population in soil was also observed in the case of soils where *Trichoderma* was inoculated and no fungicides were applied. Results of compatibility tests with *Trichoderma* and fungicides in the field have shown that the latter did not significantly affect the bioagent's population. If fungicides are drenched over the antagonists, there is always a reduction in the rate of colonisation of the antagonist but during later period, the antagonist establishes in the soil.

Table 5 Survival of *Trichoderma harzianum* in the soil after soil application of fungicides

Sl. No.	Fungicides		% *in vitro* inhibition	cfu x 10^2 in the soil		
				20 days	50 days	100 days
1.	B.M.	1%	100.0	21.0	3.0	7.5
2.	Bavistin	0.2%	80.5	18.5	2.5	1.5
3.	Emisan	0.2%	100.0	15.5	4.5	4.5
4.	Dithane-M45	0.2%	61.7	21.5	2.0	8.0
5.	Ridomil Mz	0.2%	79.0	22.5	4.5	6.5
6.	P.N.C.B.	0.2%	61.1	18.5	8.5	10.5
7.	Thiride	0.2%	89.0	28.5	4.5	2.5
8.	Ziride	0.2%	89.0	19.5	4.0	5.0
9.	C.O.C	0.2%	41.7	18.0	2.5	9.5
10.	Control	--	--	18.5	2.0	6.0

In the experiments on integrated management of capsule rot disease of cardamom, Bhai *et al.* (1997) have observed that a combination of plant sanitation, application of fungicides such as foliar spray with Bordeaux mixture or Akomin, or soil drenching with copper oxychloride have to be given as prophylactic treatments before commencement of the disease. The bioagent is mass multiplied and applied to the rhizosphere of the plants fifteen to twenty days after the fungicide treatment, for two to three rounds at one month interval. Soil amendments with neem cake was also found to reduce the inoculum level in the field.

6. CONCLUSIONS

The biocontrol strategy of plant disease management has a lot of practical implications. First of all the bioagent should be easily available, have utmost efficacy and easiness of application with nominal costs. Though biocontrol is superior to the conventional method of plant disease control, its integration with other components is always better in achieving the goal in a sustainable agriculture system.Cardamom being a

234

spice crop of high value and flavour qualities it becomes inevitable choice to safe guard the crop from hazardous chemical pesticides. The development and use of biocontrol agents open new horizons in the field of disease management with utmost safety and effectiveness. The *Trichoderma* spp, need to be tested against other diseases also for making the biocontrol a successful persuit in the long march to fight against disease problems of this crop It is essential to continue and extend the search to newer fields such as the development of highly efficient *Trichoderma* strains that can withstand varying ecological conditions such as the drought, floods, frosts, acidity etc.to fight against the ever changing virulent isolates of the pathogens.

REFERENCES

Ali, S.S.1983, Nematode problems in cardamom and their control measures, In : *Proc* VIIth *Workshop of All India Coordinated Spices and Cashewnut Imporvement Project,* Calicut, India

Ali,S.S. and Venugopal, M.N. 1993, Prevalance of damping off and rhizome rot diseases in nematode infested cardamom nurseries in Karnataka, *Curr. Nematol.* 4 (1): 19-24.

Anonymous, 1986, *Annual Report,* 1986, ICRI, Myladumpara, pp. 51-53.

Anonymous 1989, *Biannual report,* 1987-89, ICRI, Myladumpara, pp. 41-47.

Baker, K.F. and Cook, R.J. 1974, *Biological Control of Plant Pathaogens,* ed. W. H.Freeman, San Fransisco, USA, 433 p.

Bhai, S.R., Thomas, J. and Naidu, R. 1993, Biological control of azhukal disese of small cardamom caused by *Phytophthora meadii* Mc Rae, *J.Plantn. Crops* 21 (Suppl) : 134-139.

Bhai, S.R., Thomas J. and Naidu, R. 1994, Evaluation of carrier media for field application of *Trichoderma* spp. in cardamom growing soils, *J.Plantn.Crops* 22 (1):50-52.

Bhai, S.R., Thomas J. and Sarma Y.R. 1997, Biocontrol of capsule rot of cardamom, Paper presented in *International Congress on Integrated Management of Plant Diseases for Sustainable Agriculture,* New Delhi, India.

Cates, D. 1990, Biological fungicide closer to market, *Ag. Consultant,* August, 11.

Chet, I. and Henis,Y. 1985, *Trichoderma* as a biocontrol agent against soil-borne root pathogens, In : *Ecology and Management of Soil-borne Plant Pathogens,* ed. C.A. Parker, American Phytopathological Society, St. Paul, Minnesota, USA, 110p.

Cook, R.J. 1988, Biological control of soil-borne plant pathogens, In : *Proc. (Abs.) Fifth Internat. Cong. Pl. Pathol.,* Kyoto, Japan, p. 19.

Cook, R.J. and Baker, K.F. 1983, The nature and practical of biological control of plant pathogens Amer. Phytopathol. Soc., St. Paul, Minnesota, USA 53 pp.

Dennis, C. and Wbster, J. 1971, Antagonistic proporties of species groups of *Trichoderma* II, Production of volatile antibiotic, *Trans. Brit. Mycol. Soc.* 57 : 25.

Dhanapal, K. and Thomas, J. 1996, Evaluation of *Trichoderma* isolates against rot pathogens of cardamom, In : *Recent Advances in Biocontrol of Plant Pathogens,* eds. K.Manibhushan Rao and A.Mahadevan, Today and Tomorrow Printers and Publishers, New Delhi, India, pp. 67-75.

Eapen, S.J. 1995, *Final Project Report,* Investigations on plant parasitic nematodes associated with cardamom, Indian institute of spices Research, Calicut, India, 39 pp.

Eapen,S.J. and Venugopal,M.N. 1995, Field evaluation of *Trichoderma* spp. and *Paecilomuces lilacinus* for control of root knot nematodes and fungal diseases of cardamom nurseries (Abs.), *Indian J. Nematol.* 25:15-16.

Elad,Y., Chet, I. and Katan T, 1980. *Trichoderma harzianum,* a biological agent effective against *Sclerotium rolfsii* and *Rhizoctonia solani., Phytopath.* 70: 119-121.

Elad,Y., Chet, I. and Hennis,Y. 1982, Degradation of plant pathogenic fungi by *Trichoderma harzianum, Can. J. Microbiol.* 28:719-725.

Elad, Y., Chet, I, Boyle, P. and Hennis,Y. 1983, Parasitism of *Trichoderma* spp.on *Rhizoctonia solani* and *Sclerotium rolfsii,* Scanning election microscopy and flourscence microscopy, *Phytopath.* 73:85-88.

Godtfredsen, W.O. and Vangedal, S. 1965, Trichodermin a new sesquiterpene antibiotic, *Acta. Chem. Scand.* 19 : 1088.

Indian Institute of Spices Research 1995, Annual Report for 1993-94, Calicut, India, pp. 89.

Kousalya, G. and Jeyarajan 1990, Mass multiplication of *Trichoderma* spp., *J. Biol. Cont.* 4 (1) 17:1-10.

Lifshitz, R., Sneh, B. and Baker, R. 1984, Soil suppressiveness to a plant pathogenic *Pythium* sp., *Phytopath.* 74:9:1054-1061.

Lewis, J.A. and Papuavizas, G.C. 1980, Integrated control of *Rhizoctonia* fruit rot of cucumber, *Phytoopath.* 70 : 85.

Liptton, F.G.H. 1984, Biological control, The plant breeding objectives, *Ann. Appl.Biol.* 104.

Marshall, D.S. 1982, Effect of *Trichoderma harzianum* seed treatment and *Rhizoctonia solani* inoculum concentration on damping off of snap been, *Plant Dis.* 66 : 778.

Middleton, J.T. 1943, The Taxonomy Host range and Geographic distribution of the genus *Pythium*, *Mem. Torrey Bot. Club* 20:171.

Meno, M.R., Sajoo, B.V., Ramakrishnsn,C.K. and Remadevi, L. 1972, A new *Phytophthora* disease of cardamom, *Curr. Sci.* 41 (6) : 231.

Mukherjee, P.K., Upodhyaya, J.P. and Mukopadhyaya, A.N. 1989, Biological control of *Pythium* damping off of Cauliflower by *Trichoderma harzianum*, *J.Biocont.* 3 (2):119.124.

Mukhopadhyay, A.N., Brahamabhatt, A. and Patel, G.J.1986, *Trichoderma harzianum* a potential biocontrol agent of tobacco damping off, *Tob. Res.* 12 : 26-35.

Mukhopadhyay, A.N. 1987, Biological control of soil-borne plant pathogens by *Trichoderma* spp., *Indian J. Mycol. Pl. Pathol.* 17 : 1-10

Naidu, R. and Thomas J., 1992, Diseases of cardamom an approach for integrated management, In : *Proc. Nat. Sem. Black Pepper and Cardamom*, Calicut, India.

National Research Centre for Spices, 1991, *Annual Report* for 1993-94, Calicut, India, pp.75.

National Research Centre for Spices, 1994, *Annual Report* for 1993-94, Calicut, India, pp.65.

Papavizas,G.C. 1981, *Biological Control in Crop Production*, Allen held. Totowa, Nengaland, 305 p.

Papavizas, G.C. and Lewis, J.A. 1981, Introduction and augmentation of biological agents for the control of soil-borne plant pathogens, In : *Biological Control in Crop Production*.

Papavizas, G.C. 1985, *Trichoderma* and *Gliocladium*, biology, ecology and potentialfor biocontrol, *Ann. Rev. Phytopath.* 23-23.

Pyke, T.R. and Dietz, A. U-21.963. a new abtibiotic I , Discovery and bilogical activity, *Appl. Microbiol.* 14 : 506-10.

Radha, K. and Thomas, J. 1974, Investigations on the bud rot disease (*Phytophthora palmivora* Butl) of coconut PL-480, *Final Report*, 1968-1973, C.P.C.R.I., Kayamkulam, pp.30.

Ramana,K.V. and Eapen, S.J. 1992, Plant parasitic nematodes of black pepper and cardamom and their management, In : *Proc. Nati. Sem. on Black Pepper and Cardamom*, Calicut, India pp. 43-47.

Ramakrishnan, T.S. 1949, The occurrence of *Pythium vexans* de Barry in South India, *Indian Phytopath.* 2: 27-30.

Siddaramaiah A. L. 1988, Seed rot and seedling wilt a new disease of cardamom, *Curr. Res.* 17:34-35.

Siddaramaiah, A.L., Khan, M.M. and Nataraju, S.P. 1988a, Incidence and management of damping off and clump rot of cardamom, In : *Proc. Workshop on Strategies of the Managementof Rot Disease Incidence in Plantation Crops*.

Siddaramaiah,A.L. 1988b, Stem, sheath and leaf rot diseases of cardamom caused by *Sclerotium rolfsii* from India, *Curr. Res.* 17:51.

Smith, E.M. and Webner, E.C. 1987, Biological and chemical measures integrated with deep soil cultivation against crater disease of wheat, *Phytophylactica* 19 : 87-90.

Subba Rao, M.K. 1938, Report of the mycologist 1937-1938, *Admn. Rept. Tea Sci. Dept. Unit plant. Assoc. S. India*, 1937-38, pp.28-42.

Thankamma, L. and Pillai, P.N.R. 1973, Furit rot and leaf roi disesae of cardamom in India, *FAO Plant Prot. Bull.* 21:83-84.

Thomas, J., Bhai, R.S., Vijayan, A.K. and Naidu, R. 1993, Evaluation of antagonists and their efficacy in managing rot diseases of small cardamom, *J. Biocont.* 7 : 29-36.

Thomas, J. and Vijayan, A.K. 1994, Occurrence severity, etiology and control of rhizome rot diseaase of Cardamom, PLACROSYM (Abstr.), Calicut, India.

Thomas, J., Bhai, S. R., Dhanapal, K. and Vijayan,A.K.1997, Integrated management of rot diseases of cardamom, Paper presented in *International Congress on Integrated Disease Management for Sustainable Agriculture*, New Delhi, India.

Thomas,G.V., Sundararaju,P.,Ali,S.S. and Ghai, S K. 1989, Individual and interactive effects of VA Mycorrhizal fungi and root knote nematode *Meloidogyne incognita* in cardamom, *Trop. Agric.* 66 : 21-24.

Tsao, P.H., Sztejnberg, Shou, R. and Fanag. G. 1988, Biological control of *Phytophthora* root rots with antagonistic fungi, In : *Proc. 5th Intr. Congress, on Plant Pathology*, Kyoto, Japan.

236

Vijayan, A.K., Thomas, J. Dhanapal K. and Naidu, R. 1994, Field evaluation of *Trichoderma* isolates in the biocontrol of rhizome rot diseases of small cardamom, *J.Biocont.* 8(2):111-114.

Weindling, R. 1932, *Trichoderma lignorum* as a parasite of other fungi, *Phytopath.* 22: 837.

Wells,H.D. 1988, *Trichoderma* as a biocontrol agent, In : *Biocontrol of Plant Diseases*, Vol.I, eds. K.G. Mukerji and K.L.Garg, CBS Publishers and Distributors, Delhi, India.

Wells, H.D., Bell,D.K. and Jaworski,C.K. 1982, Efficacy of *Trichoderma harzianum* as a biocontrol for *Sclerotium rolfsii*, *Phytopath.* 62 : 442.

Wilson,K.I.,Sasi,P.S. and Rahagopalan, B. 1979, Damping off of cardamom caused by *Rhizoctonia solani* Kuhn, *Curr. Sci.* 48-364.

Wu, W.S., Lin., S.D., Chang, Y. and Tschen, J. 1986, Hyperparasitic relationships between antagonists and *Rhizoctonia solani*, *Plant Prot. Bull.* 28 : 91-100.

BIOCONTROL OF PULSE DISEASES

M.V. Reddy[1], B. Srinivasulu[2] and T. Pramila Devi[2]

[1]Regional Agricultural Research Station
Lam, Guntur - 522 034, Andhra Pradesh, INDIA
[2]Department of Plant Pathology
National Centre for Integrated Pest Management (ICAR)
Pusa Campus, New Delhi-110 012, INDIA

1. INTRODUCTION

In Indian agriculture, pulses play an important role in maintaining soil fertility and supplying portein to the large vegetarian population of the country. India is a major pulse growing country in the world, sharing 35-36 per cent and 27 - 28 per cent of the area and production of these crops, respectively. The commonly grown pulse crops in India are chickpea (*Cicer arietinum*), pigeonpea (*Cajanus cajan*), urdbean (*Vigna mungo*), mothbean (*Vigna radiata*), horsegram (*Macrotyloma biflorus*), mothbean (*Vigna aconitifolia*), lathyrus (*Lathyrus sativus*), lentil (*Lens culinaris*), cowpea (*Vigna unguiculata*), drybean (*Phaseolus vulgaris*) and peas (*Pisum sativum*). Diseases and insect pests are the major constraints to their production. Most of the pulse crops are grown under rainfed conditions, hence less attention is paid for the management of these crops and protection against diseases gets least priority.

Important diseases of pulse crops in India are as follows : Pigeonpea (i) wilt (*Fusarium udum* Butler); (ii) sterility mosaic (Causal agent unknown) (Amin *et al.*, 1993a,c; Kannaiyan *et al.*, 1984); chickpea (i) wilt (*F. oxysporum*, Schlecht. emend. Snyd. and Hans f.sp. *ciceri* (Padwick). Synd. and Hans), (ii) dry root rot (*Rhizoctonia bataticola* (Taub) Butler, (iii) Ascochyta blight (*Ascochyta rabiei* (Pass.) Labrousse), (iv) stunt (pea leaf roll virus) (Reddy and Vishwa Dhar, 1997), Mungbean and Urdbean (i) Yellow Mosaic Virus (MYMV) and *Cercospora* leaf spot (*Cercospora canescens* Ell. and Martin and *C. cruenta* Sacc), (ii) powdery mildew (*Erysiphe polygoni* D.C.) (Poehlman, 1991; Satyanarayana, 1989; Singh *et al.*,1989); Lentil (i) wilt (*F. oxysporum* Schlecht. f. sp. *lentis* Snyd. and Hans.) (ii) rust (*Uromyces fabae* (Pers.) de Bary), (iii) ascochyta blight (*Ascochyta fabae* Speg. f. sp., *lentis* Cossen), (iv) root rots (*R. solani* Kuhn, *M. phaseolina* (Tassi) Goid, *F. solani* (Mart.) App. and Wollen) (Khare, 1981); Fieldpea (i) powdery mildew (*E. polygoni*), (ii) rust (*U. pisi* (Pers.) Wint.); (iii) wilt (*F. oxysporum* Schlecht. f. sp. *pisi* Snyd. and Hans.) (Reddy and Vishwadhar, 1997).

2. CHICKPEA

2.1. Wilt (*Fusarium oxysporum* f. sp. *ciceri*)

Seed treatment with spore suspension of biocontrol agent *Gliocladium virens* and Foltaf at 2 g/kg seed reduced *Fusarium* wilt (Singh *et al.*, 1993). Coating chickpea seeds with biocontrol agent *Bacillus subtaillis, Gliocladium virens, Trichoderma harzianum* and *T. viride* significantly controlled *Fusarium oxysporum f.* sp. *ciceri* wilt by 30 - 45.8 per cent and integration of bicontrol agent and carboxin increased the seed yield by 25.4 - 42.6 per cent (De *et al.*, 1996). Chickepa and lentil seeds treated with *Gliocladium virens* (107 conidia/ml) and then with 0.1 per cent carboxin effectively reduced soil borne population of *Fusarium oxysporum, Rhizoctonia solani* and *Sclerotium rolfsii* (Mukhopadhyay *et al.*, 1992). Five bacterial isolates (2 of *Bacillus subtilis* and 3 of other *Bacillus* spp.), 2 actinomycets (*Streptomyces griseus* and *Sreptomyces* sp.), and 5 Fungi (*Aspergillus fumigatus, Pencillium funiculosum, P. pinophilum, Trichoderma harzianum* and *T. viride*) were antagonistic *in vitro* to *F. oxysporum* f. sp. *ciceri* causing vascular wilt of *Cicer arietinum* (Dhedhi *et al.*, 1990).

Integration of soil solarization (for 6 weeks), VA mycorrhizal fungus (VAM), *Glomus fasciculatum* inoculation (12 g/hill) and seed treatment with carbosulfan (3% w/w) was highly effective in reducing population levels of wilt pathogen *F. oxysporum* f. sp. *ciceri.*

Soil application of *Trichoderma harzianum* gave 53.5 - 85.7 per cent disease control in the glass house. In the field, integrated use of *T. harzianum* with fungicidal seed treatments significantly reduced the incidence of chickpea wilt complex and increased crop yield. Seed treatment with vitavax and ziram resulted in 29.5 per cent disease control, this control increased to 63.3 per cent when *T. harzianum* was also applied (Kaur and Mukhopadhyay, 1992).

Pseudomonas fluorescens strains showed inhibitory action against chickpea wilt pathogen. When the treated seeds were sown in soil, the antagonist moved to the rhizosphere and survived well in it. Biopriming of seeds increased rhizosphere population. When seed treatment was followed by root zone application, the efficacy of *P. fluorescens* formulations increased. *P. fluorescens* did not inhibit the beneficial N-fixing *Rhizobium* and *Azospirillum in vitro* (Vidhyasekaran and Muthamilan, 1995). Integration of soil solarization (for 6 weeks), VA mycorrhizal fungus (VAM), *Glomus fasciculatum* inoculation (12 g/hill) and seed treatment with carbosulfan (3 per cent w/w) was highly effective in reducing population levels of both pathogens root knot disease and wilt incidence (Rao and Krishnappa, 1995).

Seed bacterization of *Bacillus* strain SR 2 reduced the number of wilted chickepeas in wilt sick soil. (Kumar, 1996).

Bacterization of chickpea seeds with a siderophore - producing fluorescent pseudomonad RBT 13 reduced the number of chickpea wilted plants in wilt sick soil by 52 per cent (Kumar and Dube, 1992).

A siderophore producing *Pseudomonas* strain isolated from rhizosphere of tea showed *in vitro* antibiotic activity towards *F. oxysporum* f. sp. *ciceri* of chickpea and *Fusarium udum* of red gram. Addition of iron into the culture media decreased the inhibition (Kumar *et al.*, 1996).

2.2. Dry Root Rot (*Rhizoctonia bataticola*)

Seed treatment by fungal filtrates of *Paecilomyces lilacinus* significantly controlled root rot disease complex of chickpea caused by *Macrophomina phaseolina* and *Meloidogyne incognita* race 3. In integrated control, combination of green manuring of *Cymbopogon citratus* and *P. lilacinus* was more effective (Siddiqui and Mahmood, 1993). *Paecilomyces lilacinus* and *Bacillus subtilis* were used for the biocontrol of the root rot disease complex of chickpea caused by *Meloidogyne incognita* race 3 and *Macrophomina phaseolina.*

Individually, *P. lilacinus* treatemnt was better against *M. incognita* while *B. subtilis* against *M. phaseolina* (Siddiqui and Mahmood, 1993). Seed treatment with *Bacillus subtilis* in chickpea reduced the pathogenicity of *M. phaseolina* (Siddiqui and Mahmood, 1995). Culture filtrates of *Paecilomyces lilacinus, Aspergillus niger* and *A. flavus* were effective agaisnt *M. phaseolina* (Siddiqui and Husnin, 1991). *Paecilomyces lilacinus, Bacillus lichoniformis* and *Alcaligenes faecalis* were found to be best biocontrol agents against *M. phaseolina* (Siddiqui and Mahmood, 1992). The growth promoting bacterium *Pseudomonas aeruginosa* (Strains Pa 6 and Pa 12) significantly reduced infection by *Macrophomina phaseolina* and *Rhizoctonia solani* on chickpeas. Combined use of *Bradyrhizobium* sp. (TAL 480) and *P. aeruginosa* completely controlled infection (Ishrat - Izhar *et al.*, 1995).

2.3. Wet Root Root (*Rhizoctonia solani*)

Media containing composted grape marc (CGM) or composted separated cattle manure (CSM) reduced disease caused by *Rhizoctonia solani.* The mechanism of suppression is due to the presence of antagonistic microorganisms in composts, as gamma irradiation of the composts elminiated the suppressive effect (Gorodecki and Hadar, 1990).

The growth promoting bacterium *Pseudomonas aeruginosa* (Strains Pa 6 and Pa 12) significantly reduced infection by *Rhizoctonia solani.* Combined use of *Bradyrhizobium* sp. (TAL 480) and *P. aeruginosa* completely controlled infection by *R. solani* more nodules/plant were produced where Bradyrhizobia was used with strains of *P. aeruginosa* compared with using Bradyrhizobia alone (Ishrat - Izhar, 1995).

2.4. *Phytophthora* Root Rot (*Phytophthora megasperma* f. sp. *medicaginis*)

Pseudomanas lepacia (7 strains) and *P. fluorescens* (2 strains) were found to be potential biocontrol agents against *P. megasperma* f. sp. *medicaginis in vitro* (Myatt *et al.,* 1993).

2.5. *Botrytis* Gray Mold (*Botrytis cinerea*)

Vinclozolin - tolerant isolates of *Trichoderma viride* were effective in controlling Botrytis mould of chickepea. The action of *Trichoderma* was both preventive and curative. When applied prior to pathogen application, it prevented the pathogen from invading and colonizing the plants. When applied after the pathogen, *Trichoderma* restricted growth of the lesions and prevented sporulation of the pathogen by rapid saprophytic colonization of the dead tissues (Mukherjee *et al.,* 1995). *Trichoderma* isolate T 15 reduced gray mould on incoulated susceptible pot grown plants, when sprayed as a spore suspension 1day before incoulation with the pathogen (Mukherjee and Haware, 1993).

2.6. Seed Mycoflora

Extracts from the leaves of *Diplaziam esculentum* and *Microsporum punctatum* were effective in inhibiting the growth of *Rhizoctonia solani* and *Fusarium oxysporum* on chickpea seeds (Yasmeen and Saxena, 1992).

2.7. Seed Rots and Damping - off

Seed treatment with 2 *Pseudomonas fluorescens* strs, Q 29 Z - 80 and M 82 - 80, increased emergence and yields of chickpeas. These seed treatments were equivalent to captan, metalaxyl or *Penicillium oxalicum* treatments (Kaiser and Hannan, 1989).

3. PIGEONPEA

3.1. Wilt (*Fusarium udum*)

Three *Trichoderma* species, *T. viride, T. harzianum, T. koningii* showed maximum antagonistic potential agaisnt *F. udum* at pH 6.5. The antagonistic potential of *Trichoderma* spp. was not much affected by changing the C/N ratio. However, *T. koningio* was comparatively more susceptible to pH changes (Himani and Bhatnagar, 1996). *Trichoderma harzianum* and *Bacillus subtilis* produced a wide zone of inhibition against *Fusarium udum* and inhibited spore germination completely at 8.42 × 107 spores ml and 8 × 107 cells/ml, respectively. Seeds coated with the antagonists germinated better than untreated seeds and produced longer roots and shoots when stown in either wilt infested or sterilized soil (Sumitha and Gaikwad, 1995). A vesicular arbuscular fungus *Gigaspora margarita* and two biocontrol agents for nematodes, *Paecilomyces lilacinus* and *Verticillium chlamydosporium,* alone or in combination, increased shoot dry weight, number of nodules, phosphorous content and reduced nematode multiplication and wilting index. Simultaneous use of the biocontrol agents and the vesicular arbuscular fungus gave better control of the disese complex than did their individual applications (Siddiqui and Mohamood, 1995). Addition of the soil bacterium *Bacillus subtilis* resulted in an increse in chlamydospore production by *Fusarium udum* (Chakraborthy and Gupta, 1995).

Trichoderma viride present in the rhizosphere soil resistant NP 15 as well as susceptible T 21 varieties of pigeonpea was the most effective in controlling the *Fusarium* wilt followed by *Aspergillus niger, Streptomyces* sp., *Penicillium* sp. and *Bacillus* sp. Pre-inoculation of pigeonpea seedlings with soil borne fungi non-pathogenic to pigeonpea (*Fusarium oxysporum* f. sp. *niveum, F. oxysporum* f. sp. *ciceri, F. solani* f. sp. *pisi* and *Cepholosporium sacchari*) before challenge inoculation with the pathogen *F. udum* gave good disease control (81.6 per cent protection). A direct correlation was found between the spore concentration of the non-pathogens and degree of protection and an indirect correlation between the period (2-7 d) between non-pathogen inoculation and challenge inoculation and degree of protection (Chakraborty and Gupta, 1995).

3.2. Phytophthora Blight : (*Phytophthora drechleri* f. sp. *cajani*)

An Indian isolate (AFI) and a Canadian isolate of *Bacillus subtilis* inhibited the growth of *Phytophthora drechleri* f. sp. *cajani in vitro. P. drechleri* f. sp. *cajani* failed to grow in a 10-fold concentrated cell - free culture filtrate of AFI. Increasing the concentration of cell-free culture filtrate of AFI in Richard's solution decreased the dry weight of the fungus (Podile and Dube, 1987).

3.3. Seed Mycoflora

The pigeonpea mycoflora, *Alternaria alternata, Curvularia lunata* and *Cladosporium* sp. were inhibited by *Trichoderma viride. T. harzianum* was not antagonistic to any of the fungi (Lokesh and Hiremath, 1988).

4. BLACKGRAM OR URDBEAN

4.1. Root Rot (*Macrophomina phaseolina*)

Seed treatment with microbial antagonists, *Trichoderma hamatum, T. harzianum, Aspergillus candidus* and *Paecilomyces lilacinus* significantly reduced *Macrophomina phaseolina* infection (Shahzad *et al.,* 1991). The best control of *Macrophomina phaseolina* in blackgram was given by carbendazim (92.2 per cent plant stand), followed by *Trichoderma harzianum* z. + decomposed coirpith (90.6 per cent) (Mani and Marimuthu, 1994).

4.2. *Cercospora* Lesf Spot (*Cercospora canescens*)

Germination of conidia of *C. canescens* was reduced when incubated with spores of *Penicillium oxalicum, Fusarium semitectum, Aspergillus niger* and *A. luchuensis.* Leaf spot severity was reduced by spraying with *P. oxalicum, F. semitectum, A. niger* 2 days before inoculation of *C. canescens* (Rao and Mllaiah, 1988).

4.3. Root Knot (*Meloidogyne incognita*)

Inoculation of blackgram with *Glomus fasciculatum* 15 and 20 days earlier than inoculation with root - knot nematode (*Meloidogyne incognita*) controlled nematode population and increased biomass production (Sankaranayana and Sundara Babu, 1994).

5. GREENGRAM (MUNGBEAN)

5.1. Root Rots (*Macrophomina phaseolina, Rhizoctonia solani, Fusarium* spp.)

The efficacy of *Trichoderma harzianum, Gliocladium virens, Streptomyces* sp., and *Rhizobium meliloti* against *M. phaseolina* on *Vigna radiata* was increased when the antagonists were applied with fertilizers. Maximum reduction (72 per cent) in the population of sclerotia in the soil was obtained where Hi-gro was used with *T. harzianum* or *G. virens.* Infection by *M. phaseolina* was least where soil was amended with a combination of Hi-gro + *T. harzianum* or where plant - food was used with either *T. harzianum* or *G. verens* (84· per cent) (Zaki and Ghaffar, 1995). *Bradyrhizobium* sp. and *Paecilomyces lilacinus* either alone or mixed with cotton cake or neem cake significantly reduced infection by *Macrophomina phaseolina, Rhizoctonia solani* and *Fusarium* spp. on roots of greengram. The combined use of *Bradyrhizobium* sp. or *P. lilacinus* with a low amount of neem cake or cotton cake is uggested for the control of root rot disease of greengram (Ehteshamul - Haque *et al.,* 1995).

Seed pelleting of microbial antagonists, *Stachybotris atra,* combined with *Rhizobium meliloti* was more effective in suppressing colonization by *M. phaseolina, Rhizoctonia solani* and *Fusarium solani* (Dawar *et al.,* 1993). Dry root rot of redgram caused by *M. phaseolina* was reduced by applications of *Trichoderma viride* as row treatments 2 days before sowing, under acid soil conditions (Raghuchander *et al.,* 1993). *Epicoccum purpurascens, Trichoderma harzianum* and *T. viride* were antagonistic to *M. phaseolina. E. purpurascens* and *T. viride* showed antibiosis while *T. harzianum* exhibited hyperparasitism as the mechanism of pathogen inhibition (Singh *et al.,* 1993). *Rhizobium meliloti* (2 strains), *R. leguminosarum* and *Bradyrhizobium japonicum* were antagonistic to *Macrophomina phaseolina, Rhizoctonia solani* and *Fusarium* spp. when applied as seed or soil treatments (Ghaffar, 1993). In *in vitro* tests, *Rhizobium meliloti* inhibited growth of the soil borne root infecting fungi, *Macrophomina phaseolina, Rhizoctonia solani* and *Fusarium solani,* while *Bradyrhizobium japonicum* inhibited *M. phaseolina* and *R. solani.*

In field tests, *R. meliloti, R. leguminosarum* and *B. japonicum* used either as a seed dressing or as a soil drench reduced infection by *M. phaseolina, R. solani* and *Fusarium* spp.

(Etheshamul - Haque and Ghaffar, 1993). *Trichoderma viride* applied as a seed coating reduced mortality of greengram due to *M. phaseolina* from 19 to 8 per cent, in unsterilized soil under green house conditions (Kehri *et al.*, 1991). Seed treatment or soil drenching with *Trichoderma harzianum, Paecilomyces lilacinus, Gliocladium virens* controlled root rot caused by *Macrophomina phaseolina, Rhizoctonia solani* and *Fusarium* spp. on greengram under field conditions (Etheshamul - Haque *et al.*, 1990). Tests on petri agar plates and culture filtrate tests indicate antagonism by *Epicoccum purpurascens* (Singh and Sekhon, 1988). Infection by *M. phaseolina* was significantly reduced following treatment of mungbean seeds with *Trichoderma harzianum, Gliocladium virens, Paecilomyces lilacinus* or *Streptomyces* sp. and *Rhizobium meliloti* (Hussain *et al.*, 1990) on potato dextrose agar *Bacillus subtilis* str. F-29-3 induced *Rhizoctonia solani* to form bulging hyphae and to accumulate chitin like substances in the hyphae, and the culture filtrate of *B. subtilis* stimulated sclerotium formation of *Rhizoctonia solani*.

When *Vigna radiata* seeds coated with *B. subtilis* were grwon in originally amended soils, infection by *R. solani* was reduced. Chromatographic studies of the antibiotic substance indicate that it belongs to an antibiotic complex containing bacilysin and fengymycin (Fegycin). It is concluded that fengymycin produced by *B. subtilis* F-29-3 is involved in the protection of plants from infection by *R. solani* (Tschen, 1987). Seed dressings with antagonists *T. harzianum, T. hamatum* and *Paecilomyces lilacinus* using gum arabic as a sticker, reduced infection by the root rot fungi, *R. solani, M. phaseolina* and *Fusarium* spp. Combined use of antagonists and organic fertilizers gave better results than their separate use (Ghaffar, 1986). *P. lilacinus*, a soil borne fungus and a pathogen of *Meloidogyne* eggs was found to inhibit growth of *M. phaseolina* and *R. solani in vitro*. *P. lilacinus* reduced colonization of roots by *M. phaseolina* by 33 per cent, where as infection by *R. solani* was reduced by 67 per cent on mungbean (Shahzad and Ghaffar, 1989).

5.2. Bacterial Leaf Spot (*Xanthomonas compestris* pv. vignaeradiatae)

Seven bacteria and 6 fungi from the phylloplane of *Vigna radiata* inhibited growth of *X. campestris* pv. *vignaeradiatae* in culture. *Flavobacterium* sp. and *Penicillium oxalicum* were the most inhibitory. A ratio of 1:49 of the pathogen with *P. oxalicum* or *Flavobacterium* sp. completely controlled the disease. Application of the antagonist before inoculation with the pathogen gave better control than post inoculation treatment (Jindal and Thind, 1994). Several phylloplane bacteria *(Erwinia herbicola, Flavobacterium* sp., *Micrococcus luteus, Serratia marcescens)* and fungi (*Penicillium oxalicum*) protected leaves of *Vigna radiata* from bacterial leaf spot, when applied 48 h before inoculation with the pathogen (Bora *et al.*, 1993).

5.3. Mungbean Yellow Mosaic Virus

The incidence of mungbean yellow mosaic virus (MBYMV) was lower in mungbean crops raised from *Rhizobium* peat culture treated seeds. The yield loss due to MBYMV infection was 81.66 per cent in plants raised from untreated seeds, whereas it was only 64.2 per cent in plants raised from *Rhizobium* treated seeds (Fugro and Mishra, 1995).

5.4. Cucumber Mosaic Virus (CMV)

Satellite RNAs associated with *Vigna radiata* and cucumber isolates were obtained with an ability to ameliorate the symptoms of CMV infection (Hsu *et al.*, 1994).

5.5. Root Knot (*Meloidogyne incongnita*)

Application of *Pasteuria penetrans* and *Paecilomycis lilacitus* significantly reduced root knot indices on mungbean when these organisms were used individually or in combination with each other (Zaki and Maqbool, 1990). In a pot experiment, complete control of *Meloidogyne incognita* infection in mungbean was observed in plants grown in 3 per cent w/w cotton cake mixed with 25 ml of *Bradyrhizobium* sp. (3.5 × 109 cells/ml) (Abid *et al.*, 1992).

6. PEAS

6.1. Root Rot (*Rhizoctonia solani*)

Streptomyces lydicus WYEC 108 showed *in vitro* antagonism against *Rhizoctonia solani* and *Pythium ultimum* in liquid medium, by inhibiting fungal growth (Yuan and Crawford, 1995). *Pseudomonas fluorescens* (Strain PRA 25) and *P. cepacia* (Strain AMMD) increased seedling emergence and decreased disease incidence and severity by *Rhizoctonia solani* (Xi *et al.*, 1996). Dressing of pea seeds with *Trichoderma koningii, T. viride, Gliocladium catenulatum* and *G. roseum* protected from *Ascochyta pisi, Fusarium culmorum, F. oxysporum* f. sp. *pisi, F. solani, Rhizoctonia solani* and *Sclerotinia sclerotiorum* (Lacicowa and Pieta, 1994). A reverse correlation between hyperparasitic and antibiotic activities was noticed with *Trichoderma aureoviride* isolates 2288 and 1518, *T. harzianum* 1325, *T. viride* 1426 and *Gliocladium virens* 32 tested against *Fusarium oxysporum, F. solani, Pythium* sp., *Rhizoctonia solani* and *Sclerotinia sclerotiorum* causing root rots of pea (Velikanov *et al.*, 1994). The antagonistic activity of *B. subtilis* against *R. solani* on peas was greater where freshly introduced spores were present in the soil (Bochow *et al.*, 1995). Various isolates of *Fusarium, Gliocladium* and *Penicillium* spp. were able to reduce root rot caused by *Fusarium solani* f. sp. *pisi* (Castejon - Munoz and Oyarzun, 1995).

6.2. Seed Rots

Pseudomonas fluorescens, strain 63-28 R inoculated seeds revealed that the bacteria multiplied abundantly at the root surface and colonized a small number of epidermal × cortical cells of peas and protected from *Pythium ultimum* infection (Benhamou *et al.*, 1996).

6.3. Root Knot

Meloidogyne incognita multiplication was reduced in the presence of *Rhizobium* on *Pisum sativum* (Siddiqui *et al.*, 1995).

7. LENTIL

7.1. Root Rot (*Macrophomina phaseolina*)

Trichoderma harzianum and *T. lignorum* (*T. viride*) were used for biological control of *Fusarium oxysporum, F. culmorum, F. solani* and *Rhizoctonia solani* on lentils (Sadowski and Kryziak, 1991). *Gliocladium virens, Trichoderma harzianum* and *Paecilomyces lilacinus* significantly reduced root rot infection of lentil caused by *Macrophomina phaseolina, Rhizoctonia solani* and *Fusarium* spp. The antagonists were more effective when applied as soil drench or inoculum multiplied on wheat bran and rice grain than as a seed dressing (Ehteshamul

- Haque *et al.*, 1992). Seed treatment with *Bacillus subtilis* increased the percentage lentil seedling survival over untreated seeds sown in *Fusarium avenaceum* infested soil and reduced root rot (Hwang, 1994).

7.2. Collor Rot

Trichoderma harzianum, T. konangii, T. viride, Gliocladium virens, Aspergillus candidus, Paecilomyces lilacinus and *Bacillus* sp. significantly inhibited the mycelial growth of *Sclerotium rolfsii* (Iqbal *et al.*, 1995).

8. CONCLUSIONS

Release of a large number of high yielding and short duration varieties, availability of appropriate technology for production and assured remunerative prices for the produce have infused new interest in the farmers in cultivation of pulses. Because of increasing concern over the misuse or over use of fungicides coupled with their high cost, harmful residues and non effectiveness against soil borne pathogens, renewed interest has been aroused in biological control of diseases in pulse crops.

REFERENCES

Abid, M., Haque, S.E., Maqbool, M.A., Ghaffar, A. 1992, Effects of oilcakes, *Bradyrhizobium* sp., *Paecilomyces lilacinus* and Furadan on root nodulation and root knot nematdoe in mungbean, *Pakistan J. Nematol.* 10 (2) : 145 - 150.

Amin, K.S., Reddy, M.V., Nene, Y.L., Raju, T.N., Shukla Prathibha, Zote, K.K., Arujunan, G., Bendre, N.J., Rathi, Y.P.S., Sinha, B.K., Gupta, R.P., Anilkumar, T.B., Chauhan, V.B., Bidari, V.B., Gurdeep Singh, Jha, D.K. and Kausalya Gangadharan 1993, Multilocation evaluation of pigeonpea (*Cajanus cajan*) for broad - based resistance to sterility moasic disease in India, *Indian J. Agric. Scie.* 63 : 542-546.

Benhamou, N., Belanger, R.R. and Paulitz, T.C. 1996, Pre-inoculation of Ri T-DNA - transformed roots with *Pseudomonas fluorescens* inhibits colonization by *Pythium ultimum* Troan ultra strucutral and cytochemical study, *Planta* 199 (1) : 105 - 117.

Bochow, H., Gantcheva, K. and Vanchter, A. 1995, Soil introduction of *Bacillus subtilis* as biocontrol agent and its population and activity dynamic, Fourth international symposium on soil and substrate infestation and disinfestation, Leuven, Belgium, *Acta - Horti.* 382 : 164 - 172.

Bora, L.C., Gangopadhyay, S. and Chand, J.N. 1993, Biolgocial control of bacterial leaf spot (*Xanthomonas compestris* pv. *vignaeradiatae* Dye) of mungbean with phylloplane antagonists, *Indian J. Mycol. Pl. Pathol.* 23 (2) : 162 - 168.

Castejon, Munoz, M. and Oyarzun, P.J. 1995, Soil receptivity to *Fusarium solani* f. sp. *pisi* and biological control of root rot of pea, *Euroepan J. Pl. Pathol.* 101 (1) : 35 - 49.

Chakraborty, A. and Gupta, P.K.S. 1995, Factors affectng cross protection of *Fusarium* wilt of pigeonpea by soil borne non-pathogenic fungi, *Phytoparasitica* 23 (4) : 323 - 334.

Chakraborty, A. and Gupta, P.K.S. 1995, Factors affecting chlamydospore formation by *Fusarium udum*, the causal organism of wilt of pigeonpea, *J. Mycopath. Res.* 33 (2) : 71 - 75.

Dawar, S., Shahzad, S., Iqbal, R. and Ghaffar, A. 1993, Effect of seed pelleting with biological antagonists in the control of root infecting fungi of cowpea and mungbean, *Pakistan J. Bot.* 25 (2) : 219 - 224.

De, R.K., Chaudhary, R.G. and Naimuddin, 1996, Comparative effecacy of biocontrol agents and fungicides for controlling chickpea wilt caused by *Fusarium oxysporum* f. sp. *ciceri, Indian J. Agric. Sci.* 66 (6) : 370 -373.

Dhedhi, B.M., Gupta, Om, Patel, V.A. and Gupta, O. 1990, Antagonistic effect of microorganisms to *Fusarium oxysporum* f. sp. *ciceri*, Indian *Paecilomyces lilacinus* with oil cakes in the control of root rot of mungbean, *Trop. Scie.* 35 (3) : 294 - 299.

Etheshamul Haque, S. and Ghaffar, A. 1993, Use of rhizobia in the control of root rot diseases of sunflower, orka, soybean and mungbean, *J. Phytopath.* 138 (2) : 157 - 163.

Ehteshamul Haque, S., Ghaffar, A. and Zaki, M.J. 1990, Biological control of root rot diseases of okra, sunflower, soybean and mungbean, *Pakistan J. Bot.* 22 (2) : 121 - 124.

Etheshamul Haque, S., Hashmi, R.Y. and Ghaffar, A. 1992, Biological control of root rot disease of lentil, *Lens* 19 (2) : 43 - 45.

Fugro, P.A. and Mishra, M.D. 1995, Effect of *Rhizobium phaseoli* on mungbean yellow mosaic virus and nodulation in mungbean, *Indian J. Mysol. Pl. Pathol.* 25 (3) : 198 - 203.

Ghaffar, A. 1986, Soil borne diseases research centre, Final research report, 110 pp.

Ghaffar, A. 1993, Rhizobia as biocontrol organisms, *BNF Bulletin* 12 (2) : 6.

Gorodecki, B. and Hadar, Y. 1990, Suppression of *Rhizoctonia solani* and *Sclerotium rolfsii* diseases in container media containing composted separated cattle manure and composted grape marc, *Crop Protec.* 9 (4) : 271 - 274.

Himani and Bhatnagar, H. 1996, Influence of environmental conditions on antagonistic activity of *Trichoderma* spp. against *Fusarium udum, Indian J. Mycol. Pl. Pathol.* 26 (1) : 58 - 63.

Hsu, Y.H., Wu, C.W., Lee, C.W., Hu, C.C. and Lin, F.Z. 1994, Satellite defense as a control strategy for cucumber mosaic virus, *Pl. Pathol. Bull.* 3 (1) : 72 - 77.

Hussain, S., Ghaffar, A. and Aslam, M. 1990, Biological control of *Macrophomina phaseolina* charcoal rot of sunflowr and mungbean, *J. Phytopath.* 130 (2) : 157 - 160.

Hwang, S.F. 1994, Potential for integrated biological and chemical control of seedling rot and pre-emergence damping off caused by *Fusarium avenaceum* in lentil with *Bacillus subtilis* and vitaflo R-280, *Zeits chrift - fur - pflanzenkrankheiten - und - pflanzenschutz* 101 (2) : 188 - 199.

Iqbal, S.M., Bakhsh, A., Hussain, S. and Malik B.A. 1995, Microbial antagonism agaisnt *Sclerotium rolfsii,* the cause of colar rot of lentil, *Lens - Newsl.* 22 (1-2) : 48 - 49.

Ishrat, Izhar, Ehteshamul Haque, S., Javeed, M., Ghaffar, A. and Izhar, I. 1995, Efficacy of *Pseudomonas aeruginosa* and *Bradyrhizobium* sp., in the control of root rot disease in chickpea, *Pakistan J. Bot.* 27 (2) : 451 - 455.

Jindal, K.K. and Thind, B.S. 1994, Evaluation of greengram microflora for the control of *Xanthomonas compestris* pv. *vignaeradiatae,* the incitant of bacterial leaft spot, *Pl. Dis. Res.* 9 (1) : 10 -19.

Kaiser, W.J. and Hannan, R.M. 1989, Biolgocial control of seed rot and pre-emergence damping off of chickepa with fluorescent pseudomonads, *Soil Biol. Biochem.* 21 (2) : 269 - 273.

Kannaiyan, J., Nene, Y.L., Reddy, M.V., Rayan, J.G. and Raju, T.N. 1984, Prevalence of pigeonpea diseases and associated crop losses in Asia, Africa and the Americas, *Trop. Pest Manag.* 30 (1) : 62 - 71.

Kaur, N.P. and Mukhopadhyay, A.N. 1992, Integrated control of chickpea wilt complex by *Trichoderma* and chemical methods in India, *Trop. Pest Manag.* 38 (4) : 372 - 375.

Kehri, H.K. and Chandra, S. 1991, Antagonism of *Trichoderma viride* to *Macrophomina phaseolina* and its application in the control of dry root rot of mung, *Indian Phytopath.* 44 (1) : 60 - 63.

Khare, M.N. 1981, *Diseases of Lentil,* eds. C. Webb and G. Hawtin, Common Wealth Agricultural Bureaux, Slough, U.K.

Kumar, B.S.D. 1996, Crop improvement and disease suppression by a *Bacillus* sp. SR 2 from peanut rhizosphere, *Indian J. Exper. Biol.* 34 (8) : 794 - 798.

Kumar, B.S.D., Balamani Bezbaruah and Bezbaruah, B. 1996, Antibiosis and plant growth promotion by a *Pseudomonas* strain isolated from soil under tea cultivation, *Indian J. Microbiol.* 36 (1) : 45 - 48.

Kumar, B.S.D. and Dube, H.C. 1992, Seed bacterization with a fluorescent *Pseudomonas* for enhanced plant growth, yield and disease control, *Soil Biol. Biochem.* 24 (6) : 539 - 542.

Lacicowa, B. and Pieta, D. 1994, Protective effect of microbiological dressing of pea seeds (*Pisum sativum* L.) against pathogenic fungi living in soil, *Annals - Universitatis - Mariae - Curie - Sklodowska - Sectio - EEE - Horticultural* 2 : 165 - 171.

Lokesh, M.S. and Hiremath, R.V. 1988, Antagonism of *Trichoderma* spp. against seed mycoflora of redgram (*Cajanus cajan*), *Pl. Pathol. Newsl.* 6 (1-2) : 31 - 32.

Mani, M.T. and Marimuthu, T. 1994, Effect of decomposed coconut coir pith, fungicides and biocontrol agents on damping off of chillies and dry root rot of blackgram, *Indian J. Mycol. Pl. Pathol.* 24 (1) : 20 - 23.

Mukherjee, P.K. and Haware, M.P. 1993, Biological control of *Botrytis* gray mold of chickpea, *Internat. Chickpea Newsl.* 28 : 14 - 15.

Mukherjee, P.K., Haware, M.P. and Jayanthi, S. 1995, Preliminary investigations in integrated biocontrol of *Botrytis* gray mold of chickpea, *Indian Phytopath.* 48 (2) : 141 - 149.

Mukhopadhyay, A.N., Shrestha, S.M. and Mukherjee, P.K. 1992, Biological seed treatment for control of soil borne plant pathogens, *FAO Pl. Prot. Bull.* 40 (1-2) : 21-30.

Myatt, P.M., Dart, P.J. and Hayward, A.C. 1993, Potential for biological control of *Phytophthora* root rot of chickpea by antagonistic root - associated bacteria, *Australian J. Agric. Res.* 44 (4) : 773 - 784.

Podile, A.R. and Dube, H.C. 1987, Antagonism of *Bacillus subtilis* to *Phytophthora drechsleri* f. sp. *cajani, Indian Phytopath.* 40 (4) : 503 - 506.

Poehlman, J.M. 1991, *The Mungbean*, Oxford and IBH publishing Co. Pvt. Ltd., New Delhi, India, 357 pp.

Raghuchander, T.,Samiappan, R. and Arjunan, G. 1993, Biocontrol of *Macrophomina* root rot of mungbean, *Indian Phytopath.* 46 (4) : 379 - 382.

Rao, P.B. and Malliah, K.V. 1988, Effect of phylloplane fungi on the leaf spot pathogen *Cercospora canesceus, Indian J. Microbiol.* 28 (1-2) : 103 - 107.

Rao, V.K. and Krishnappa, K. 1995, Integrated management of *Meloidogyne incognita - Fusarium oxysporum* f. sp. *ciceri* wilt disease complex in chickpea, *International J. Pest Manag.* 41 (4) : 234 - 237.

Reddy, M.V. and Vishwadhar 1997, Disease resistance in major pulse crops, In : *Recent Advances in Pulses Research*, eds. A.N. Asthana and Masood Ali, Indian Society of Pulses Research and Development, Kanpur, India, pp. 281 - 299.

Sadowski, S. and Kryziak, A. 1991, Trial using *Trichoderma harzianum* and *Trichoderma lingnorum* to control root rot diseases of lentil (*Lens esculenta* Moench.) No. 14 : 83 - 87.

Saleem, S., Hussain, S., Dawar, S., Ghaffar, A., Malik, B.A., Shahzad, S., Hussain, S., Dawar, S. and Ghaffar, A. 1991, Biological control of *Macrophomina phaseolina* infection on mashbean [*Vigna mungo* (L) Hepper], *Pakistan J. Bot.* 23 (1) : 131 - 134.

Satyanarayana, A. 1989, Pulses in rice fallows, varietal important in chickpea, pigeonpea and other upland crops in rice based and other cropping systems, In : *Proc. International Work Shop*, 19 - 22 March, 1989, Kathmandu, Nepal, pp. 252-354.

Shahzad, S. and Ghaffar, A. 1989, Use of *Paecilomyces lilacinus* in the control of root rot and root knot disease complex of okra and mungbean, *Pakistan J. Nematol.* 7 (1) : 47 - 53.

Siddiqui, Z.A. and Hussain, S.I. 1991, Control of *Meloidogyne incognita* and *Macrophomina phaseolina* on chickpea by fungal filtrates, *Pakistan J. Nematol.* 9 (2) : 131 - 137.

Siddiqui, Z.A. and Mahmood, I. 1992, Biolgocial control of root rot disease complex of chickpea caused by *Meloidogyne incognita* race 3 and *Macrophomina phaseolina, nematoliza Mediterranea* 20 (2) : 199 - 202.

Siddiqi, Z.A. and Mahmood, I. 1993, Biological control of *Meloidogyne incognita* race 3 and *Macrophomina phaseolina* by *Paecilomyces lilacinus* and *Bacillus subtilis* alone and in combination on chickpea, *Fund. Appl. Nematol.* 16 (3) : 215 - 218.

Siddiqi, Z.A. and Mahmood, I. 1993, Integrated control of a root rot disease complex of chickpea by fungal filtrates and green manuring, *Nematologia Mediterranea* 21 (2) : 161 - 164.

Siddiqui, Z.A. and Mahmood, I. 1995, Management of *Meloidogyne incognita* race 3 and *Macrophomina phaseolina* by fungus culture filtrates and *Bacillus subtilis* in chickpea, *Fund. Appl. Nematol.* 18 (1) : 71 - 76.

Siddiqui, Z.A. and Mahmood, I. 1995, Some observations on the management of the wilt disease complex of pigeonpea by treatment with a vesicular arbuscular fungus and biocontrol agents for nematodes, *Biores. Technol.* 54 (3) : 227 - 230.

Siddiqui, Z.A. Mahmood, I. and Ansari, M.A. 1995, Effect of differnt inoculum levels of *Meloidogyne incognita* on the growth of pea in the presence and absence of *Rhizobium, Nematologia Mediterranea* 23 (2) : 249 - 251.

Singh, R.A. Amin, K.S. and Gurha, S.N. 1989, Incidence of yellow mosaic virus on mungbean in different seasons, In : *New Frontiers in Pulses Research and Development : Proc. National Symposium*, 10 - 12 Nov., 1990, Directorate of Pulses Research, Kanpur, India, Indian Society of Pulses Research and Development, pp. 122.

Singh, R.N., Upadhyay, J.P. and Ojha, K.L. 1993, Management of chickpea wilt by fungicides and *Gliodadium, J. Appl. Biol.* 3(1-2) : 46 - 51.

Singh, R.S., Singh, N., Kang, M.S. and Singh, N. 1993, Rhizosphere mycoflora of mungbean and their interaction with *Macrophomina phaseolina, Pl. Dis. Res.* 8 (1) : 25-28.

Singh, R.S. and Sekhon, P.S. 1988, Effect of *Epicoccum purpurascens* on *Macrophomina phaseolina* causing seelding blight and charcoal rot of mungbean, *Indian Phytopath.* 41 (2) : 248 - 249.

Sumitha, R. and Gaikwad, S.J. 1995, Checking *Fusarium* wilt of pigeonpea by biological means, *J. Soils Crops* 5 (2) : 163 - 165.

Tschen, J.S.M. 1987, Control of *Rhizoctonia solani* by *Bacillus subtilis, Trans. Mycol. Soc. Japan* 28 (4) : 483 - 493.

Velikanov, L.L., Cukhonosenko, E., Yu., Nikolaeva, S.I. and Zavelishko, I.A. 1994, Comparison of hyperparasitic and antibiotic activity of the genus *Trichoderma* Pers Fr. and *Gliocladium virens* Miller, Giddens et Foster isolates towards the pathogens causing root rot of pea, *Mikologiya*, i, *Fitopatologiya* 28 (6) : 52 - 56.

Vidhyasekaran, P. and Muthamilan, M. 1965, Development of formulations of *Pseudomonas fluorescens* for control of chickpea wilt, *Plant Dis.* 79 (8) : 782 - 786.

248

Xi, K., Stephens, J.H.G. and Verma, P.R. 1996, Application of formulated rhizobacteria against root rot of field pea, *Pl. Pathol.* 45 (6) : 1150 - 1158.

Yasmeen and Saxena, S.K. 1992, Effect of fern extracts on seed mycoflora of chickpea, *Seed Res.* 20 (2) : 170 - 171.

Yuan, W.M. and Crawford, D.L. 1995, Characterization of *Streptomyces lydicus* WYEC 108 as a potential biocontrol agent against fungal root and seed rots, *Appl. Environ. Microbiol.* 61 (8) : 3119 - 3128.

Zaki, M.J. and Ghaffar, A. 1995, Combined effects of microbial antagonists and nursery fertilizers on infection of mungbean by *Macrophomina phaseolina* (Tassi) Goid, *Pakistan J. Phytopath.* 7 (1) : 17 - 20.

Zaki, M.J. and Maqbool, M.A. 1990, Effect of *Pasteuria penetrans* and *Paecilomyces lilacinus* on the control of root knot nematodes of brinjal and mung, *Pakistan J. Phytopath.* 2 (1-2) : 37 - 42.

BIOLOGICAL CONTROL OF PEARL MILLET DOWNY MILDEW : PRESENT STATUS AND FUTURE PROSPECTS

H.S. Shetty and Vasanthi U. Kumar

Downy Mildew Research Laboratory
Department of Applied Botany and Biotechnology
University of Mysore
Mysore - 570 006 Karnataka, INDIA

1. INTRODUCTION

Modern agriculture is increasingly dependant on mass monoculture of plants with an emphasis on uniformity, yield and quality often at the expense of genetic diversity. Though such practices supply staple food for world's ever increasing populations, it is an ecologically unstable situation and leads to invasion of crops by diseases, pests and weeds. Other problems associated with modern agriculture are pollution, negative health effects, overuse and overdependence on chemical pesticides, interruption of natural ecological nutrient cycling and destruction of biological communities that otherwise support crop production.

In plant protection, the world wise objective is to select and use procedures for minimising losses caused by pests and diseases in a balanced way. This is referred to as 'Integrated Pest and Disease Management' and a desirable component of such management is 'Biological Control'. It involves the use of biological processes to reduce inoculum density of a pathogen and maintain soil populations below disease threshold levels. Biological control reduces crop losses by minimally interfering with an ecosystem and damaging an environment. Thus, this method of protection is applied microbial ecology for plant disease management wherein emphasis is on the behaviour and activity of microbes in supporting plant diseases and represents an attractive approach for plant disease management since it is economical, self perpetuating and usually free from residual effects. (Mukerji and Garg, 1988a,b; Mukerji *et al.,* 1999).

2. BIOLOGICAL CONTROL

2.1. The Phenomenon

Biocontrol agent as defined by Cook and Baker (1983) in their book ' Nature and Practice of Biocontrol Agent of Plant Pathogens' is " The reduction in amount of inoculum or disease producing activity of a pathogen accompanied by or through one or more organisms other than man". Taking into account of agronomic practices, the indirect effect of practices on disease management and the necessary role of human intervention within biological control, Maloney (1995) offers a revised definition considering biological control as "the stimulation and enhancement for biological activity in order to reduce the amount of pathogens".

Biological control may be accomplished through Agronomic Practices and/or Microbial Antagonism. Agronomic practices are based on 'exploitation or manipulation of natural communities for pathogen control' (Cook 1993). It consists of cultural practices like crop rotation, sanitation, flooding before planting, soil aeration or solarisation, tilling strategies and addition of organic amendments such as compost etc. which are utilized for creating an environment favourable to antagonists, host plant resistance or both. Microbial antagonism includes the deliberate use of specific antagonist organisms for prevention and management of specific diseases. An ideal biocontrol agent should have the following characters: easy to isolate, grow rapidly on substrates, affect wide range of pathogens, rarely pathogenic on host species, produces antibiotics and competes for food and site. It may be developed through any one of the following three approaches:

(i) Isolation of naturally occurring organisms with disease suppressing capabilities. The best example is *Agrobacterium radiobacter* strain 84 which is an effective biocontrol agent against crown gall.

(ii) Isolation of strains with some protective capabilities and improvement by genetic engineering. Deletion of the ice-nucleation gene from *Pseudomonas syringae* developed it as a control agent against bacteria which cause frost damage to crops (Lindeman and Suslow, 1987).

(iii) Designing a biocontrol agent strain *ab initio* by inserting into the agents that colonize host plants harmlessly a gene that codes for a toxin active against a pest or pathogen. The agents can be artificially developed using either live strains or non viable ones (Twombly 1990).

2.2. Mechanism of Action of Biocontrol Agents

In recent years, the mechanism of antibiosis have received the bulk of attention and interdisciplinary approaches have contributed to the understanding of the biological activities that are involved in plant protection and yield enhancement. A number of reports and reviews on related topics are available (Adams, 1990; Baker, 1986; Cook *et al.*, 1996; Easwaramoorthy *et al.*, 1998; Lemanceau and Alabouvette, 1993; Madi *et al.*, 1997; Satyaprasad *et al.*, 1998; Sutton and Peng, 1993; Weller, 1988). Biological activities postulated to be involved in successful disease control include; the abilities to colonize appropriate host plant and to produce antagonistic compounds such as antibiotics, siderophores (compounds that chelate biologically available iron), ammonia, cyanide, hydrolytic enzymes and substances that promote growth (Loper 1988; Schipper *et al.*, 1987; Thomashow *et al.*, 1998; Weller and Cook 1985). The biocontrol agents may offer protection also by induction of host defense mechanisms and systemic resistance, elimination of plant signals that trigger pathogen development or competition of nutrients. Research over the past years has demonstrated that induced systemic resistance can be an alternative mechanism to antagonism for achieving biological control of plant disease (Benhaumou *et al.*, 1997; Schippers *et al.*, 1993; Tuzun and Kuc, 1991; Wei *et al.*, 1996). It also been demonstrated that systemic protection achieved through plant growth

252

promoting rhizobacteria (PGPR) is the result of induced resistance and not due to possible translocation of bacterial metabolites. Also, protection achieved against a particular disease is effective against other diseases as well (Liu *et al.*, 1992). Thus, PGPR mediated induced systemic resistance is fundamentally similar to the classical systemic acquired resistance (Hoffland *et al.*, 1996; Kloepper *et al.*, 1997; Liu *et al* 1995; Wei *et al.*, 1991).

2.3. Monitoring of Biocontrol Agent in the Environment

Detection of biocontrol agent is essential (i) to study the ecology of the introduced organisms in the environment, (ii) to monitor a biocontrol product and (iii) for risk assessment. The number of methods for detection of introduced organisms have increased dramatically in recent years and there is a variety of alternative possibilities for tracking organisms which have arisen out of advances in recombinant DNA technology, microbial potential for isolate-specific nucleic acid probes like probes to rRNA sequences. Apart from detecting the population present, these methods can be used for detecting the location and level of metabolic activity of the introduced microorganisms. It will be advantageous to combine selective and diagnostic markers and also to combine markers with other detection methods. Combining methods have the advantages of (i) greater certainity about the unambiguous detection of target strain, (ii) greater ease of detection, and iii) greater sensitivity. However, traditional methods can still be used very effectively in conjugation with newer methods. A potential way to detect the introduced organism is to use of more than one method as this allows the provision of cross checking the methods leading to increased confidence in the results obtained by different methods.

2.4. Biological Control of Pearl Millet Downy Mildew

Pearl millet (*Pennisetum glaucum* (L.) R. Br) the fifth important cereal crop of India is grown in an area of 10.5 million hectares, contributing to Indian production system by ~ 7 million tonnes. The crop is extensively grown in Haryana, Rajastan, Gujarat, Maharastra, Uttar Pradesh, Tamil Nadu, Karnataka and forms the staple food and fodder for millions. Popularisation of pearl millet cultivars in the SAT regions of the country is a national problem due to 'downy mildew' or 'green ear' disease caused by the biotic, oomycetous fungus *Sclerospora graminicola* (Sacc.) Schroet. The disease is a major biotic constraint in the cultivation of pearl millet and losses upto 30% amounting to 260 million US $ due to the downy mildew disease is reported (Shetty *et al.*, 1994). The disease is widespread in regions where pearl millet is grown. The typical symptoms of pearl millet downy mildew are yellowing or reddish brown coloration, stunted growth, sporulation and 'green ear.' The half-leaf symptom shown by a distint margin between the diseased (basal) portion and the nondiseased areas toward the tip is a characteristic symptom of the disease (Kumar and Shetty, 1998). In epidemics of pearl millet downy mildew, the pathogen population starts from a low level of initial inoculum and increases through successive cycles during growing seasons of the host. The obligate nature of the pathogen rules out the possibility of it perennating on the straw, and the fragile nature of the sporangium makes it unsuitable for survival for more than a few hours. Hence, wild host, the seed, and soilborne oospores are the sources of inoculum for the recurrence of the disease year after year. Large number of oospores are produced in downy-mildew infected plants which can easily get incorporated in the soil during natural shredding of the crop or during harvesting. Oospore survival in soil is from 8 months to 10 years and the primary source of inoculum is the soil and seedborne oospores. The secondary spread in the same season is through the asexually produced sporangia (Shetty 1987; Subramanya *et al.*, 1982). Once the host becomes infected, it produces several crops of sporangia. Under

253

favourable conditions, production of 35,000 sporangia cm^{-2} on the infected leaf and as many as 11 crops of sporangia on successive nights are recorded (Safeeulla, 1976). In nature, sporangia are produced during midnight and early hours of morning and highly influenced by the environmental factors like temperature and relative humidity. They are ephemeral. They germinate directly by germtubes and by releasing 1-10 zoospores.

The traditional methods of downy mildew management include use of cultural practices, resistant cultivars and systemic fungicide like metalaxyl (Shetty, 1990). However, cultural practices are not effective as pearl millet is a monoculture crop and hence subject to diseases. Management of pearl millet downy mildew with metalaxyl is not economical for a cheap crop like pearl millet and also there are reports available for the development of resistance by some of the oomycetous pathogens to this chemical. In addition, there is growing concern about environmental pollution also. Introduction of resistant cultivars as the management strategy is time consuming and sources of durable resistance are not available.Moreover, due to pathogenic variability in *S. graminicola*, it is difficult to develop new pearl millet lines resistant to all pathotypes of *S. graminicola*. Hence, an integrated complex management measure is required that is environmentally friendly, cost effective and complements traditional methods, and biological control offers one of the promising alternative strategies for pearl millet downy mildew management.

Studies from our department has convincingly established the availability of potential antagonistic microorganisms for deployment in pearl millet downy mildew disease management programme. We have evaluated the efficiency of some antagonists to prevent primary infection in pearl millet seedlings from the germinating oospores and secondary infection through the asexual products - sporangia and zoospores by use of metabolites from the antagonists. The strategies for potential use of antagonists in pearl millet downy mildew disease management are summarised in Figure 1.

Organisms like *Trichoderma harzianum* Rifai, *T. viride* Pers. ex Gray, *Aspergillus niger* van Tieghem and *Chaetomium globosum* Kunze ex Steud. were isolated from the rhizosphere of pearl millet. The two *Trichoderma* species used in our study were shown to inhibit asexual sporulation of the downy mildew pathogen in infected leaves when applied directly. The antagonists used are known to reduce oospore infection of pearl millet seedlings in the soil (Shishupala and Shetty, 1989). Antifungal compounds present in these antagonists may act as inhibitory agents against zoospores of *S. graminicola* and this suggests the possibility of using these organisms directly as foliar spray to reduce inoculum production in pearl millet downy mildew disease management strategy. The culture of the test fungi isolated and that of *Bacillus subtilis* were tested as foliar sprays to pearl millet seedlings of different age in reducing downy mildew disease (Shishupala, 1994). The results indicated that efficiency of culture filtrates of all test organisms to protect pearl millet from downy mildew but to varying degree depending on the day of their application (Fig 2). *T. harzianum* culture filtrate was most effeivie in reducing downy mildew incidence when applied two days before inoculation with the pathogen. In the treatment of culture filtrates with five day-old-seedlings, *T. viride*, *A. niger* and *B. subtilis* did not vary in their effectiveness significantly. When the culture filtrate treatment was given simultaneously with the inoculum on the seventh day, there was no significant difference among the antagonists in reducing disease incidence. In the treatment where culture filtrates were sprayed after two days of inoculation, *T. viride* was most effective followed by *A. niger*, *T. harzianum* and *B. subtilis*. Culture filtrate of *C. globosum* did not reduce the disease significantly when applied two days after the inoculation with the pathogen. The culture filtrates were effective in restricting the earlier infection and colonization by the pathogen as evidenced by the results obtained. Antifungal compounds present in these antagonists may also act as inhibitory agents against zoospores of *S. graminicola*. These organisms have been reported to produced secondary metabolites
254

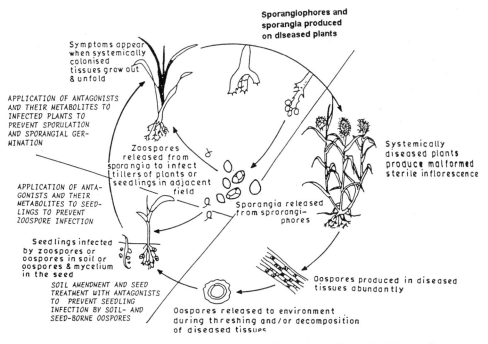

Figure 1. Strategies to break the disease cycle of pearl millet downy mildew using biocontrol agents

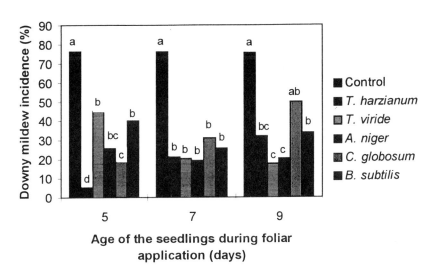

Figure 2. Bioefficacy of antagonists culture filtrate, foliar spray schedule on pearl millet downy mildew development. Bar having same letter(s) are not significantly different from each other according to Duncan's Multiple Range test (P = 0.05)

255

which were effective in preventing sporangial germination of downy mildew pathogen (Shishupala, 1989).

Field trials were also conducted to study the efficiency of antagonists applied singly or in combination with metalaxyl to reduce downy mildew. The seeds were sown in experimental plots where the soil was provided with wheat bran cultures of fungal species for some of the experiments. The culture filtrate of the organisms were sprayed separately on the 7 th and 18 th day of sowing to the seedlings of the soil amended with the respective antagonists. The effectiveness of metalaxyl formulation in combinations with antagonist treatment is shown in Table 1. Formulations along with antagonists reduced disease incidence significantly under artificial epiphytotic conditions. Soil amendment of antagonists separately was taken up singly or in combination with foliar spray of culture filtrate. Though soil amendment of antagonists alone reduced mildew incidence, the effectiveness was less when compared with their combination with metalaxyl. Results indicated that soil amendment of antagonists in combination with foliar spray of Ridomil MZ 72 WP treatment was found to be more effective for *T. harzianum* and *T. viride*. The combination of seed treatment with metalaxyl and soil amendment of antagonists resulted in little increase in the disease incidence when compared to metalaxyl seed treatment alone.

Table 1. Combination of antagonists and metalaxyl formulations to control downy mildew of pearl millet under field conditions

Treatments	Mean Downy Mildew Incidence (%)	
	30 days	60 days
Untreated control	59.9 [a]	74.4 [a]
Soil amendment		
Trichoderma harzianum	58.7 [a]	65.8 [b]
T. viride	40.8 [c]	52.3 [c]
Aspergillus niger	48.5 [b]	50.5 [c]
Chaetomium globosum	52.4 [ab]	60.9 [b]
Soil amendment + Apron 35 SD		
Seed treatment @ 2.85 g Kg-1 of seed		
T. harzianum	13.0 [e]	20.9 [f]
T. viride	18.5 [d]	26.1 [e]
A. niger	20.1 [d]	27.9 [de]
C. globosum	11.6 [e]	23.1 [de]
Soil amendment + Foliar spray of		
Radomil MZ 72 WP @ 4 g l-1 after		
20 days of sowing		
T. harzianum	6.0 [f]	11.6 [g]
T. viride	5.0 [f]	9.0 [gh]
A. niger	22.5 [d]	29.6 [de]
C. globosum	23.1 [d]	34.1 [d]
Seed treatment Apron 35 SD @ 2.85 g Kg-1 of seed	12.3 [e]	17.5 [f]
Seed treatment Apron 35 SD @ 5.7 g Kg-1 of seed	7.2 [f]	11.3 [g]
Seed treatment Apron 35 SD 5.7 g Kg-1 pf seed + Radomil MZ 72 WP @ 4g l-1 after 20 days of sowing	4.3 [g]	6.9 [h]

Numbers in the values followed by the same letter(s) are not significantly (P = 0.05) different according to Duncan's Multiple Range Test.

Combination of antagonists treatment with metalaxyl formulation was proved to be more efficient than biocontrol agent alone, even when half of the recommended dose of fungicide was used (Table 2). Combination of both the treatments was found to be better than the individual treatment. In this treatment, *C. globosum* failed to reduce the disease incidence in pearl millet significantly when compared to control.

Table 2. Efficacy of antagonists and their culture filtrates on downy mildew disease development in pearl millet under field conditions

Treatments	Mean Downy Mildew Incidence (%)	
	30 days	60 days
Untreated control	59.9 [a]	74.4 [a]
Soil amendment + Foliar spray of culture filtrate on 7th and 18th day of sowing		
Trichoderma harzianum	28.9[d]	47.4[d]
T. viride	39.7 [c]	49.6 [cd]
Aspergillus niger	45.2 [bc]	52.3 [cd]
Chaetomium globosum	66.3 [a]	69.4 [ab]
Foliar spray of culture filtrate on 7th and 18th day of sowing		
T harzianum	63.9 [a]	69.4[ab]
T. viride	51.2 [b]	56.8 [d]
A. niger	38.1 [c]	55.6 [c]
C. globosum	61.2 [a]	67.8 [b]

Numbers in the values followed by the same letter(s) are not significantly ($P = 0.05$) different according to Duncan's Multiple Range Test.

Plant growth promoting pseudomonads have been reported to control a number of plant diseases (Leeman *et al.*, 1995; M'piga *et al.*, 1997; Naik and Sen, 1992; Parke *et al.*, 1991; Vidyasekaran and Muthamilan, 1995; Wei *et al.*, 1996). In our recent study, *Pseudomonas fluorescens* isolated from the rhizosphere of pearl millet has been developed as a biocontrol agent against pearl millet downy mildew. Pure culture and a talc powder-based formulation of *P. flourescens* was found to increase the seedling vigour (Table 3, 4) and reduce the percentage of downy mildew disease incidence. Seed treatment and/or foliar application of the bacterium reduced downy mildew incidence considerably both under green house (Fig.3) and field conditions (Fig.4). Under green house conditions, the disease incidence was reduced from 86% in the control to 30% and 14% when seeds were treated with 6g and 10 g/Kg of the formulation respectively. Thus, reduction in the disease incidence observed was dependent on the dosage of the antagonist. The bacterial formulation followed by foliar application reduced downy mildew disease incidence significantly from 84% in the control to 25 and 9% when applied at 6 and 10g/Kg seed. Similar effects on management of downy mildew was achieved with the antagonist under field conditions as well. The synergistic effect of seed treatment and foliar application was found to be better reducing the disease incidence from 90% to 8% than the individual methods of treatment. The result achieved may be due to the fact that both the primary

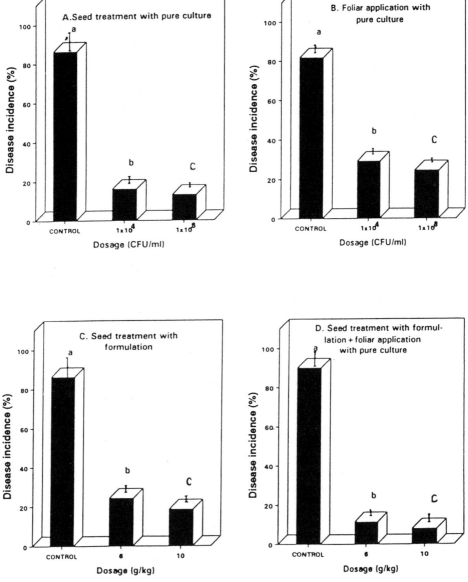

Figure 3. Effect of *Pseudomonas fluorescens* seed treatment and foliar spray on the incidence of downy mildew disease under green house conditions. The columns represented by same letters do not differ significantly (P < 0.05) according to Z-test. Vertical lines in the column represent standard error.

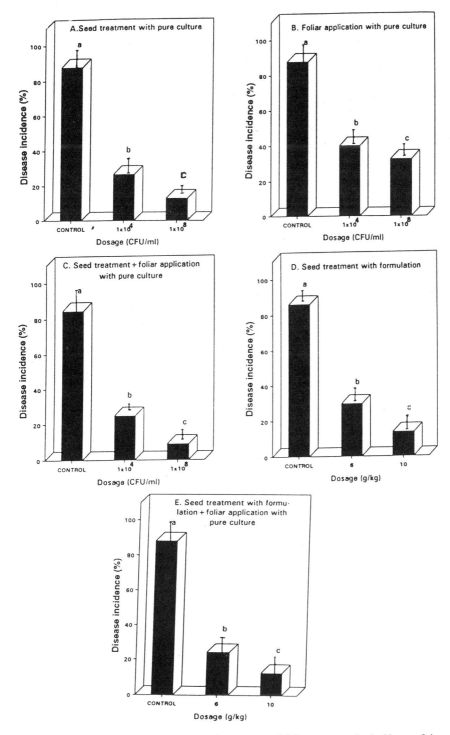

Figure 4. Effect of *Pseudomonas fluorescens* seed treatment and foliar spray on the incidence of downy mildew disease under field conditions. The columns represented by same letters do not differ significantly (P < 0.05) according to Z-test. Vertical lines in the column represent standard error.

and secondary sources of downy mildew are controlled during the combination of seed treatment and foliar spray. The protection thus achieved offer the advantages not obtainable with chemical seed protectants especially the ability of the biocontrol agent to colonize and protect pearl millet roots thus benefiting the crop. *P. fluorescence* being a bacterium is able to multiply and colonize both in the rhizosphere of seedlings and internal host tissues. Thus, they are present at or near the pathogen infection site. Also, they colonize the host tissues quickly and thus are able to prevent/reduce secondary infection through zoospores. It may be either by inhibition of the pathogen or by producing volatile compounds that affect sporulation. *P. fluorescens* apart from being a biocontrol agent for managing pearl millet downy mildew, also has growth promoting activity as evidenced by the increased root and shoot growths exhibited by pearl millet seeds treated with the antagonist. This growth promotion confirms the reports on several other crops (Parke *et al.,* 1991; Trapero-Casa *et al.,* 1990). Thus, seed bacterzation with PGPR strains is a practical way to deliver benefits of induction of resistance while hand inoculation of a pathogen inducer to each plant is much more labour intensive. Such beneficial effects of PGPR through their application as biological control agents have been documented in many of the plant pathogen interactions (Baker *et al.,* 1986; Klopper, 1991; Klopper *et al.,* 1993; Schippers *et al.,* 1995; Schroth and Hancock, 1981).

Table 3. Effect of *P. fluorescens* pure culture on seed germination and seedling vigour

Dosage CFU/ml	Percent germination	MRL (cm)	MSL (cm)	VI
Untreated	82±1.7[b]	1.9±0.12	2.8±0.12	385
1×10[4]	89±0.4[a]	3.0±0.65	4.4±0.28	658
1x10[8]	88±1.6[a]	3.3±0.35	4.7±0.23	704

Values are the means ± S.E. of 4 replicates of 100 seeds each.

MRL   - Mean root length
MSL   - Mean shoot length
VI   - Vigour index
CFU   - Colony forming units
The values in the column followed by same letter(s) are not significantly different according DMRT (P=0.05)

Table 4. Effect of *P. fluorescens* talc powder based formulation on seed germination and seedling vigour

Dosage CFU/ml	Percent germination	MRL (cm)	MSL (cm)	VI
Untreated	89±0.8[b]	2.5±0.33	4.5±0.69	623
6	90±0.9[b]	3.0±0.52	5.5±0.93	765
10	94±2.3[a]	4.3±0.57	7.4±0.80	1100

Values are the means ± S.E. of 4 replicates of 100 seeds each.

MRL   - Mean root length
MSL   - Mean shoot length
VI   - Vigour index
CFU   - Colony forming units
The values in the column followed by same letter(s) are not significantly different according DMRT (P=0.05)

The data suggest that the seed treatment system we have developed with *Pseudomonas* is an alternative to augment the existing strategies of pearl millet downy mildew management. Trials now underway will test the ability of seed treatment with the bacterium on pearl millet in diverse geogrgraphical locations.

Induced systemic resistance, the phenomenon by which a plant exhibits an enhanced level of resistance to pathogen attack after suitable stimulation, can be induced by necritizing pathogens, extracts of plants and microorganisms, and/or certain chemicals (Hammerschmidt and Kuc, 1995). Substances extracted from plants on application to other plants induce resistance in them against infection by microorganisms either by directly acting on them or acting indirectly through an effect on the host. Reports are available on the approach of using plant extracts and compost tea as an alternative to fungicides for controlling diseases and induction of systemic resistance to diseases caused by various microorganisms (Daayf *et al.,* 1995; Rajeswari and Mariappan, 1992; Tuzun and Kloepper, 1995; Verma *et al.,* 1981). These studies, although limited, indicate the biological diversity of inducers which can have practical applications. Studies in this direction to manage pearl millet downy mildew using extracts of a no of plants are in progress. Our studies have indicated that pearl millet downy mildew can be managed by using extracts of *Clerodendron inerme*. The extract was shown to inhibit sporulation of the downy mildew pathogen when applied on the directly on pearl millet leaves that were incubated for fresh crop of sporangia and release of zoospores from the sporangia. Reduction upto 60% in pearl millet downy mildew on application of *Clerodendron* plant extract was recorded.

3. FUTURE PROSPECTS

Biocontrol agent is a pragmatic and holistic approach to disease management with indirect effects on disease by antagonists and agronomic practices. It has an important application in integrated disease management. However, direct field experimentation with promising individual strains has resulted in only a very few consistently effective and economically antagonistic formulations mainly because mycoparasites are effective at high dosages of inoculum and colonise effectively in sterile, soil less mix in green house conditions than under field conditions. This is due to the inactivation of antibiotics (Williams, 1982) and the inability of biocontrol agents to establish and occupy a new ecological niche in the field and displace the pathogens. Other obstacles under field conditions include the *ad hoc* nature of the screening procedures with undue emphasis on *in vitro* methods (Gerhardsson and Larsson, 1991) and the necessity for a large quantity of the agent when applied directly to soil. Hence, methods should be developed to reduce the quantity of the agent needed. The ideal strategy is seed treatment as it is less laborious, cheap and preferred by farmers. Antagonistic microorganisms applied to seeds prior to planting colonise the rhizosphere of seedlings and thus are present at or near the pathogen's infection court where they act by producing antifungal compounds, through hyperparatism, or by competitively colonising rhizosphere substrate.. Intensive efforts should be made as formulation of antagonists to optimise their performance and to make their storage and packing convenient. Research efforts and funding in biological control needs a dramatic increase as the available sources are relatively smaller than that for chemical pesticides. Prolonged research into the ecological mechanisms that underline biocontrol agent is vital to the eventual ability of biocontrol agent strategies.

In pearl millet downy mildew management, application of biological control is still in its infancy. Continuos attempts should be made to identify new, efficient bioagents. The possibility of improving the available biocontrol agent following genetic engineering techniques like deletion and cloning of genes, protoplast fusion, mutagenesis should also be explored. Advances in the field of genetic engineering have offered significant potential to modify and improve agricultural plants at molecular level. By incorporating downy mildew resistance genes in pearl millet, biological control may be strengthened further (Mukerji, *et al.,* 1999)

Combining conventional microbial techniques with the use of latest biotechnological advancements like selected markers, sophisticated statistical approaches and computer simulations, it is possible and also essential to study the composition and microevolution of pearl millet rhizosphere and phylloplane microbial communities as it focuses the role of direct and indirect effects of microbial antagonists and agronomic practices in pathogen suppression.

Promising results reported for managing various other diseases may be exploited to improve pearl millet downy mildew management. There are reports to state that the endophytes protect host plants from biotic and abiotic environmental stress (Bush *et al.,* 1998). Their unique abilities to effectively colonise the host tissue with little or no microbial competition, survive within cortical or vascular tissue and easy movement of their products through out the plant make them potential candidates for biocontrol agents. They can also be constructed to carry genes for antibiotics/insecticides or any other desired character. Endophytes have intimate harmonious relationship with host plants. Attempts should be made to isolate suitable endophytes from pearl millet and use them for management of downy mildew disease can be explored. Among the biocontrol agents tested on pearl millet to induce resistance, *P. fluorescence* has been proved to be effective. Hence, research should be focused to achieve endophytic nature of the organism by various techniques like seed bacterization. vacuum infiltration, osmoconditioning, foliar spray talcum powder based formulation amended with methyl cellulose etc.

Solarization, a method of disinfestation wherein the moist soil is covered by plastic sheets, may be attempted to control pearl millet downy mildew. Such nonpolluting, cost effective method have been reported to be effective in controlling a number of soil borne diseases especially in tropical countries (Arora and Pandey, 1989; Gameliel and Katan 1993; Stransbaugh and Foster 1992).

The significance of ecology in the development of biocontrol agents against soilborne plant pathogens has been outlined by Deacon (1991). The impact of naturalling occurring biocontrol agent and the determination of dosage in relation to weather for the different agronomic zones can be assessed by studying the species complexity, seasonal abundance, influence of abiotic factors etc. Research should be focused to meet the criteria as effective as fungicides, rapid in action, cost or labour efficiency, ease of application and broad effectiveness.

4. CONCLUSIONS

Pearl millet, the fifth important cereal crop of India is grown in the semi-arid tropics of India in an area of 10.5 million hectares contributing to 7 million tones of annual production. The major biotic constraint in pearl millet cultivation is the 'downy mildew' disease caused by the oomycetous, biotrophic fungus *Sclerospora graminicola.* Losses upto 30% amounting to 270 million US $ have been reported. The downy mildew pathogen can serve in the soil as oospores for a long time and management of soil-borne disease is difficult. The available strategies of pearl millet downy mildew management viz., use of chemicals, resistant cultivars have their limitations of development of fungal strain resistant to the fungicide, causing pollution and health hazards and nonavailability of durable resistance. Hence, there is a need for an ecofriendly strategy of disease control and biological control is a desirable component of such management strategy.

Both fungi like *Trichoderma* and *Chaetomium*, and bacterial species like *Bacillus subtilis* and *Pseudomonas fluorescense* have been attempted as potential biocontrol agents. The antagonists tried suppressed disease incidence and promoted pearl millet growth.

REFERENCES

Adams, P. B. 1990, The potential of mycoparasites for biological control of plant diseases, *Ann. Rev. Phytopath.* 248 : 59-72.

Arora, D. K. and Pandey, A. K. 1989, Soil solarization for the control of soil-borne diseases, Theory and application, In: *Perspectives in Phytopathology,* eds. V. P. Agnihotri, N. Singh, H. S. Chaube, U. S. Singh and T. S. Dwivedi, pp. 429-438.

Baker, R. 1986, Biological control: An overview, *Can. J. Pl. Pathol.* 8 : 218-221.

Benhamou, N., Rey, P., Chérif, M., Hockenhull, J. and Tirilly, Y. 1997, Treatment with the mycoparasite, *Pythium oligandrum,* triggers the induction of defense-related reactions in tomato roots upon challenge with *Fusarium oxysporum* f. sp. *radici lycxopersici, Phytopath.* 87 : 108-122.

Bush, L. P., Wilkinson, H. H. and Schardl, L. 1997, Bioprotective alkaloids of grass-fungal endophyte symbioses, *Pl. Physiol.* 114: 1-7.

Cook, R. J. 1993, Making greater use of introduced microoragnisms for biological control of plant pathogens, *Ann. Rev. Phytopath.* 31 : 53-80.

Cook, R. J. and Baker, K. F. 1983, *The Nature and Practice of Biological Control of Plant Pathogens,* American Phytopathological Society Press, St Paul, Minnesota, USA.

Cook, R. J., Bruckart, W. L., Coulson, J. R., Humber, R. A., Lumsden, R. D., Maddox, J. V., Mc Manus, M. L., Moore, L., Meyer, S. P., Quimbay, P. C., Stack, J. P. and Vaughn, J. 1996, Safety of microorganisms intended for pest and plant disease control: A framework for scientific evaluation, *Biol. Cont.* 7: 333-351.

Daayf, F., Schmitt, A. and Belanger, R. R. 1995, The effects of plant extracts of *Reynoutria sachalinensis* on powdery mildew development and leaf physiology of long english cucumber, *Plant Dis.* 79 : 577-580.

Deacon, J. W. 1991, Significance ecology in the development of biocontrol agents against soil-borne plant pathogens, *Biocon. Scie. Technol.* 1: 5-20.

Easwaramoorthy, S. 1998, Biological control of sugarcane pests: Present status and future prospects, In: *Integrated Pest and Disease Management,* eds. R. K. Upadhyay, K. G., Mukerji, B. P. Chamola and O. P. Dubey, A. P. H. Publishing Corporation, New Delhi, India, pp. 361-380.

Gameliel, A. and Katan, J. 1993, Suppression of major and minor pathogens by fluorescent pseudomonads in solarized and nonsolarized soils, *Phytopath.* 83 :68-75.

Gerhardsson, B. and Larsson, M. 1991, Effects of *Trichoderma* and other fungal antagonists on the incidence of fungal pathogens, In: *Biotic Interactions and Soil-borne Diseases,* eds. A.B. R.Beemster, G.J. Bollen, M. Gerlagh, M. A., Ruissen, B. Schippers and A. Tempel, Elsevier Scientific Publishers B V , Amsterdam, The Netherland, pp. 121-128.

Hammerschmidt, R. and Kuc, J. 1985, *Induced Resistance to Disease in Plants,* Kluwer Academic Publishers, Dordrecht, Boston, London.

Hofzland, E., Hakulinen, J. and van Pelt, J. A. 1996, Comparison of systemic resistance induced by avirulent and nonpathogenic *Pseudomonas* species, *Phytopath.* 86: 757-762.

Lemanceau, P. and Alabouvette, C. 1993, Supression of *Fusarium* wilts by fluorescent pseudomonads: Mechanisms and application, *Biocont. Scie. Technol.* 3: 219-234.

Kloepper, J. W. 1991, Plant growth-promoting rhizobacteria as biological control agents of soil borne diseases, In : *The Biological Control of Plant Diseases,* ed. J. Peterson, Food and Fertilizer Technology Centre, Taiwan, pp. 142-152.

Kloepper, J. W., Tuzun, S., Liu, L. and Wei, G. 1993, Plant growth promoting rhizobacteria as inducers of systemic disease resistance, In : *Pest Management: Biologically Based Technologies,* eds. R. D. Lumsden and J. L. Vaughen, American Chemical Society Books, Washington, DC, USA, pp. 156-165.

Kloepper, J. W., Tuzun, S., Zehnder, W. and Wei, G. 1997, Multiple disease protection by rhizobacteria that induce systemic resistance, *Phytopath.* 87 : 136-137.

Knudsen, I. M. B., Hockenhull, J., Jensen, D. F., Gerhardson, B., Hokenberg, M., Tahvonen, R., Teperi, E., Sundheim, L. and Henriksen, B. 1997, Selection of biological control for controlling soil and seed-borne diseases in the field, *Europ. J. Pl. Pathol.* 103 : 345-355.

263

Kumar, V. U. and Shetty, H. S. 1998, Present status of cereal downy mildews and their management, In: *Integrated Pest and Disease Management,* eds. R. K. Upadhyay, K. G., Mukerji, B., P. Chamola and O., P. Dubey, A. P. H. Publishing Corporation, New Delhi, India, pp. 639-680.

Leeman, M., van Pelt, J. A., den Ouden, F. M., Heinsbroek, M., Bakker, P. A. H. M. and Schippers, B. 1995, Induction of systemic resistance by *Pseudomonas fluorescens* in radish cultivars differing in susceptibility to *Fusarium* wilt, using a novel bioassay, *Europ. J. Pl. Pathol.* 101 : 655-664.

Lindemann, J. and Suslow, T. V. 1987, Competition between ice nucleation-active wild type and ice nucleation deficient deletion strains of *Pseudomonas syringae* and *Pseudomonas flourescens* biovar I and biological control of froast injury on straberry blossoms, *Phytopath.* 77 : 882-886.

Liu, L., Kloepper, J. W. and Tuzun S. 1992, Induction of systemic resistance against cucumber mosaic virus by seed inoculation with select rhizobacteria strains (Abstr.), *Phytopath.* 82 : 1108.

Liu, L., Kloepper, J. W. and Tuzun, S. 1995, Induction of systemic resistance in cucumber by plant growth-promoting rhizobacteria: Duration of protection and effect of host resistance on protection and root colonization, *Phytopath.* 85 : 1064-1068.

Loper, J. E. 1988, Role of fluorescent siderophore production in biological control of *Pythium ultimum* by a *Pseudomonas fluorescence* strain, *Phytopath.* 78 : 166-172.

Maloney, A. 1995, Three sources for non-chemical management of plant disease: Towards an ecological framework, *Adav. Pl. Pathol.* 11 : 103-130.

Madi,L., Katan, T., Katan, J. and Henis, Y. 1997, Biological control of *Sclerotium rolfsii* and *Verticillium dahliae* by *Talaromyces flavus* is mediated by different mechanisms, *Phytopath.* 87 : 1054-1060.

M'piga, P., Belanger, R. R. , Paulitz, T. C. and Benhamou, N. 1997, Increased resistance to *Fusarium oxysporum* f. sp. *radicis-lycopersici* in tomato plants treated with the endophytic bacterium *Pseudomonas fluorescence* strain 63-28, *Physiol. Mole. Pl. Pathol.* 50 : 301-320.

Mukerji, K.G., Chamola, B.P. and Upadhyay, R.K. (eds.) 1999, *Biotechnological Approaches in Biocontrol of Plant Pathogens,* Kluwer Academic/Plenum Publishers, New York, London, pp. 255.

Mukerji, K.G. and Garg, K.L. (eds.) 1988a, *Biocontrol of Plant Diseases,* Vol. I, CRC Press Inc., Florida, USA.

Mukerji, K.G. and Garg, K.L. (eds.) 1988b, *Biocontrol of Plant Diseases,* Vol. II CRC Press Inc., Florida, USA.

Naik, M. K. and Sen, B. 1992, Biocontrol of plant diseases caused by *Fusarium* species, In: *Recent Developments in Biocontrol of Plant Diseases,* eds. K. G. Mukerji, J. P. Tewari, D. K. Arora and Geeta Saxena, Aditya Books Pvt. Ltd., New Delhi, India, pp. 37-52.

Parke, J. L., Rand, R. E., Joy, A. E. and King, E. B. 1991, Biological control of *Pythium* damping off and *Aphanomyces* root rot of peas by application of *Pseudomonas capacia* or *P. fluorescense* to seed, *Plant Dis.* 75 : 987-992.

Rajeswari, E. and Mariappan, V. 1992, Effect of plant extracts on *in vitro* growth of rice blast (Bl) pathogen, *P. oryzae, Internat. Rice Res. Newsl.* 17 : 24-26.

Safeeulla, K. M. 1976, Biology and control of the downy mildews of pearl millet, sorghum and finger millet, Mysore, Karnataka, India.

Satyaprasad, K., Kanwar, I. K. and Rama Rao, P. 1998, Biological control of Fusarial disease. In : *Integrated Pest and Disease Management,* eds. R. K. Upadhyay, K. G., Mukerji, B. P. Chamola and O. P. Dubey, A. P. H. Publishing Corporation, New Delhi, India, pp. 563-580.

Schippers, B., Lugtenberg, B. and Weisbeek, P. J. 1987, Plant growth control by fluorescent *Pseudomonas,* In: *Innovative Approaches to Plant Disease Control,* ed. I. Chet, John Wiley and Sons, New York, USA, pp. 19-39.

Schippers, B., Bakker, P.A.H.M., Van Peer, R., Niemann, G. J. and Hoffland, E. 1993, *Pseudomonas* induced resistance in carnation against *Fusarium* wilt, In: *Mechanisms of Plant Defense Responses,* eds. B. Fritig and N. Legrand, Kluwer Academic Publishers, Dordrecht, The Netherland, p. 453.

Schippers, B., Scheffer, R. J., Lugtenberg, B.J.J. and Weisbeek, P.J. 1995, Biocoating of seeds with plant growth-promoting bacteria to improve establishment, *Outlook Agricul.* 24 : 179-185.

Schroth, M. N. and Hancock, J. G. 1981, Selected topics in biological control, *Ann. Rev. Microbiol.* 35: 453-476.

Shetty, H. S. 1987, Biology and epidemiology of downy mildew of pearl millet, In : *Proc. Internat. Pearl Millet Workshop,* ICRISAT Center, India.

Shetty, H. S. 1990, Basic research on downy mildews of cereals and disease management, In: *Basic Research for Crop Disease Management,* ed. P. Vidyaskaran, Daya Publishing House, New Delhi, India, pp. 316-342.

Shetty, S. A., Shetty, H. S. and Mathur, S. B. 1995, Downy Midew of Pearl Millet, Technical Bulletin, University of Mysore, India.

Shishupala, S. 1988, *Management of Downy Mildew Disease in Pearl Millet,* M. Phil. dissertation, University of Mysore, Mysore, India, 154 pp.

264

Shishupala, S. 1994, *Variability in Pearl Millet Downy Mildew Pathogen and the Disease Management*, Ph. D. thesis, University of Mysore, Mysore, India, pp. 349.

Shishupala, S. and Shetty, H. S. 1989, Use of antagonists to control oospore infection in pearl millet downy mildew, *J. Biol. Cont.* 3 : 117-118.

Strausgaugh and Forster, R. L. 1992, The influence of fall soil solarization on soil-borne diseases of potatoes and beans, *Phytopath.* 82: 1172.

Subramanya, S., Safeeulla, K. M. and Shetty, H. S. 1982, Importance of sporangia in the epidemiology of downy mildew of pearl millet, In : *Proc. Ind. Nat. Scie. Acad.* 48: 824-833.

Sutton, J. C. and Peng, G. 1993, Manipulation and vectoring of biocontrol organisms to manage foliage and fruit diseases in cropping systems, *Ann. Rev. Phytopath.* 31 : 473-493.

Thomashow, L. S. and Weller, D. M. 1988, Role of phenazine antibiotic from *Pseudomonas fluorescens* in biological control *Gaeumannomyces graminis* var. *tritici, J. Bacteriol.* 170 : 3499-3508.

Trapero-Casas, A., Kaiser, W. J. and Ingram, D. M. 1990, Control of *Pythium* seed rot and preemergence damping-off of chickpea in the U. S. Pacific North West and Spain, *Plant Dis.* 74 : 563-569.

Tuzun, S. and Kuc, J. 1991, Plant immunization: An alternative to pesticides for control of plant diseases in the green house and field, In : *The Biological Control of Plant Diseases, Proc. Internat. Seminar* Biological Control of Plant Diseases and Virus Vectors, ed. J. Bay-Patersen, Food and Fertizer Technology Center, Taipei, Republic of China, pp. 30-40.

Tuzun, S. and Kloepper, J. K. 1995, Practical application and implementation of induced resisatnce, In: *Induced Resisatance to Disease in Plants,* eds. R. Hammerschmidt and J. Kuc, Kluwer Academic Publishers, Dordrecht, Boston, London, pp. 152-168.

Twombly, R. 1990, Genetic engineers at crop genetics and mycogen place hopes on two different strategies for pest control, *The Scientist* (July 9): 1,8,9,28.

Verma,H. N., Awasthi, L. P. and Mukerjee, K. 1981, The inhibitory activity of an interfering agent, extracted from the leaves of host plants treated with Datura leaf extract, on plant virus infection, *Zeitschrift fur Pflanzenschutz* 88 : 228-234.

Vidyasekaran, P. and Muthamilan, M. 1995, Development of formulations of *Pseudomonas fluorescens* for control of chickpea wilt, *Plant Dis.* 79 : 782-786.

Wei, G., Kloepper, J. W. and Tuzun, S. 1991, Induction of systemic resistance of cucumber to *Colletotrichum orbiculare* by select strains of plant growth-promoting rhizobacteria, *Phytopath.* 81 : 1508-1512.

Wei, G., Kloepper, J. W. and Tuzun, S. 1996, Induced systemic resistance to cucumber diseases and increased plant growth by plant-growth promoting rhizobacteria under field conditions, *Phytopath.* 86: 221-224.

Weller, D. M. 1988, Biological control of soilborne plant pathogens in the rhizosphere with bacteria, *Ann. Rev. Phytopath.* 26: 319-407.

Weller, D. M. and Cook, R. J. 1985, In : *Iron, Siderophores and Plant Disease*, ed. T. R. Swineburne, Plenum Publishing Corporation, New York, USA, pp. 71-82.

Williams, S. T. 1982, Are antibiotics produced in soil? *Pedobiologia* 23: 427-435.

BIOLOGICAL CONTROL OF MAJOR FUNGAL DISEASES OF RICE AND OTHER FOOD GRAINS WITH BACTERIAL ANTAGONISTS

S. S. Gnanamanickam, K. Krishnamurthy and A. Mahadevan

Centre for Advanced Studies in Botany
University of Madras, Guindy Campus
Chennai - 600 025, Tamil Nadu, INDIA

1. INTRODUCTION

Rice is the world's most important and staple food for more than a third of the world's population. Understandably, then there is a consistent demand for higher yields of rice grains through breeding and genetic manipulations, judicious use of fertilizers and water, as well as control of insect pests and diseases.

Several diseases, notably the blast disease caused by *Pyricularia grisea*, sheath blight caused by *Rhizoctonia solani*, sheath rot caused by *Sarocladium oryzae*, bacterial blight caused by *Xanthomonas oryzae* pv. *oryzae* and the tungro disease caused by the rice tungro bacilliform virus (RTBV) and rice tungro spherical virus (RTSV) can cause serious losses of grain yield worldwide. A high degree of control of these diseases is possible by planting disease resistant varieties. But, in only a few cases breeders have been successful in breeding cultivars with disease resistance that is durable. Eventhough chemical control is quite effective against most of the diseases, their consistent use on a regular basis is undesirable from economic and environmental considerations. Another important point to be considered is the possible emergence of fungicide resistant pathogens. These considerations have prompted a search for biocontrol agents effective against the rice pathogens.

The discovery of plant growth-promoting bacteria had renewed hopes of developing effective biological control systems which would be ecology-conscious, environment-friendly and cost-effective. Of the various antagonistic microbes so far reported *Bacillus* spp. and fluorescent pseudomonads appear most promising. Indeed, the results have been quite promising and successful for the control of various diseases including take-all of wheat (Weller and Cook, 1983), damping-off of cotton (Howell and Stipanovic, 1979: Howie and Suslow, 1991), and black root-rot of tobacco (Defago *et al.*, 1990). These studies have clearly demonstrated the possibilities of using bacterial antagonists to control plant diseases. However, compared to a number of reports regarding bacteria that are found effective in controlling diseases of several other crops, not much is known about the biocontrol agents for rice pathogens. Therefore, the purpose of this review is to focus on the recent work on rhizosphere bacteria that are active against the pathogens of rice, wheat and other cereal crops, their modes of action in biological disease suppression and their application for wider use.

2. BACTERIAL ANTAGONISTS INVOLVED IN DISEASE CONTROL

Bacteria thought or shown to have potential for biocontrol occur in many genera. Obviously biocontrol agents are not limited to a specific group; however, given the diversity of the rhizosphere microflora, it is probable that the full spectrum of potentially effective strains has barely been explored. Among the explored microbes, *Bacillus* sp. and *Pseudomonas* sp. appear most promising.

Bacillus species are appealing candidates for biological control because they produce endospores and are tolerant to heat and desiccation. Fluorescent pseudomonads make up a diverse group of bacteria which can be visually distinguished from other pseudomonads by their ability to produce a water-soluble yellow- green pigment that fluoresce under UV light. They are gram-negative, chemoheterotrophic, motile rods with polar flagella and they have simple nutritional requirements. They have been used as seed inoculants for biological control of several plant pathogens including *Fusarium, Gaeumannomyces graminis* var. *tritici, Pythium* spp., *Sclerotium rolfsii* and *Pseudomonas solanacearum* (Anuratha and Gnanamanickam, 1990; Ganesan and Gnanamanickam, 1987; Howell and Stipanovic, 1979, 1980; Keel *et al.*, 1992; Thomashow and Weller, 1988). Since the realization that fluorescent pseudomonads are effective colonizers of the rhizosphere of many crop plants and have the ability to inhibit the growth of a number of phytopathogens, this group has received increasing attention as potential biocontrol agents. The work on the biological control of diseases of rice, wheat and finger millet with bacterial agents is reviewed in this paper.

2.1. Suppression of Rice and Finger Millet Diseases

The survey made by Mew and Rosales (1986) showed the presence of a large number of antagonistic bacteria in rice fields in the tropics. They isolated bacteria that produced fluorescent and non-fluorescent pigments on King's B agar medium, from sclerotia, rice field flood water, rhizosphere soils of upland and low land rice fields and diseased and healthy rice plants. Among them, 91% of the fluorescent isolates and 33% of non-fluorescent isolates inhibited the mycelial growth of *R. solani* on PDA medium and the diameter of inhibition zone ranged from 4 to 30 mm for fluorescent isolates and 2 to 32 mm for non-fluorescent isolates. When used in seed bacterization, the antagonists (both fluorescent and non-fluorescent) suppressed rice sheath blight (ShB) severity and protected rice plants from ShB infection in greenhouse and field experiments.

In our laboratory a large number of Indian strains of fluorescent pseudomonads showing antagonism towards the rice pathogens such as *P. grisea, R. solani* and *S. oryzae*, have been isolated from rice rhizosphere samples collected from rice ecosystems in India and have been evaluated for their abilities to protect the rice plants from blast, sheath blight and sheath rot (Sakthivel and Gnanamanickam, 1987; Thara and Gnanamanickam, 1994; Valasubramanian, 1994).

Sakthivel and Gnanamanickam (1987) did isolate *Pseudomonas fluorescens* strains active against the sheath rot pathogen *S. oryzae* and evaluated them in greenhouse and field experiments for sheath rot suppression and grain yields of rice. In the greenhouse experiment, *P. fluorescens* reduced the sheath rot by 54% on IR20 rice cultivar. In field experiment the same strains reduced sheath rot severity by 20 to 42% in five rice cultivars tested. They also reported that bacterization of rice plants enhanced plant height, number of tillers and grain yields from 3 to 160%.

Native bacterial strains, both fluorescent and non-fluorescent, antagonistic to rice sheath blight pathogen *R. solani* were identified by Thara and Gnanamanickam (1994). Seed bacterization of IR50 plants with two efficient strains (*P. putida* and *P. fluorescens*) reduced the sheath blight severity by 68% and 52% in a hot-spot location in Kerala state of Southern India. These authors also reported the lack of correlation between chitinase production by fluorescent pseudomonads and sheath blight suppression.

We have also isolated a large number of fluorescent pseudomonad strains from rice rhizosphere samples and have screened them for their antagonism towards the rice blast pathogen, *P. grisea*. Out of 1,210 isolates screened on PDA medium 243 (20.08%) inhibited the mycelial growth of *P. grisea* and the diameter of inhibition zone varied from 1 to 40 mm (Valasubramanian, 1994). Ten of the 24 strains which induced more than 30 mm dia. inhibition, when screened in a seed-bed assay afforded 70 to 80% reduction of leaf blast in a IR50 rice crop.

The only other earlier study available on the biocontrol of rice blast is that of Gnanamanickam and Mew (1992) who reported the potential of *P. fluorescens* strains for suppression of rice blast and also suggested that such suppression was mediated by the production of antibiotic(s).

More recently, Viji and Gnanamanickam (1996) examined the potential of bacterial antagonists for the control of finger millet blast. When several strains of *P. fluorescens* and *P. putida* were screened in the laboratory for their ability to inhibit the ragi blast fungus. *P. grisea*, they showed fungal inhibition. In a field test, leaf and neck blasts were suppressed by 59 to 72% and 33 to 63% in finger millet cv. PR 202. *P. fluorescens* 7-14 was most effective among 6 strains of bacteria evaluated in the field (Viji and Ganamanickam, 1996).

3.   MECHANISMS OF PLANT DISEASE SUPPRESSION

Understanding the mechanisms of disease suppression by the bacterial antagonists is critical to the eventual improvement and wider use. A primary mechanism of pathogen inhibition is the production of antibiotics (Fravel, 1988; Keel *et al.*, 1992). However, other factors such as siderophore production, microbial cyanide, competition for nutrients (O'Sullivan and O'Gara, 1992) or physical displacement (Cook and Baker, 1983) may also play a role.

3.1. Antibiotic-mediated Disease Suppression in Rice and Wheat

Antibiotics are generally considered to be organic compounds of low molecular weight produced by microbes. At low concentration, antibiotics are deleterious to the growth or other metabolic activities of other microorganisms. Although purified antibiotics are applied as pesticides to control plant diseases, for many years it was questioned whether or not antibiotics were produced in the soil in sufficient quantities to mediate disease suppression. Now there are sufficient number of reports which substantiate this and these evidences are mostly based on genetic methods.

The antagonistic fluorescent pseudomonads produce one or more antibiotics such as phenazine-1-carboxylic acid (PCA) (Thomashow and Weller, 1988), pyoluteorin (Howell and Stipanovic, 1980), pyrrolnitrin (Howell and Stipanovic, 1979), 2,4-diacetyl phluoroglucinol (Keel *et al.*, 1992) and oomycin A (Howie and Suslow, 1991). At least part of the disease suppressive capacity of these pseudomonads has been attributed to one or more of these compounds.

Support for role of bacterial antibiotics in biological control has been from a correlation of antibiotic production and pathogen inhibition *in vitro* and disease suppression *in vivo*. In recent years, the role of antibiotics in disease suppression has been established using a classical method which involves the construction of transposon insertion mutants that are deficient in antibiotic production. These mutants are applied to seeds in a standard disease control assay and disease control is measured and compared with the wild type strain. Since antibiotics can have broad spectrum activity, it is possible that the loss of antibiotic biosynthesis could affect the ability of a strain to successfully proliferate in a natural soil environment. Thus rhizosphere populations are measured for mutant and parent strains. If the mutants have unaltered rhizosphere competence but have lost some or all of the effect on disease suppression, it is concluded that antibiotic biosynthesis has a role in disease control.

Thomashow and Weller (1988) indicated that PCA produced by *P. fluorescens* 2-79 was the primary factor in the control of take-all disease of wheat caused by *G. graminis* var. *tritici*. Six Tn5 mutants which were unable to produce PCA *in vitro* and showed reduced (36 to 80%) protection against the pathogen in the field. *P. aureofaciens* 30-84 was shown to produce three phenazines (PCA, 2-OH-PCA, and 2-OH-Pz) by these researchers. The Tn5 mutants deficient in the production of one or more of these antibiotics were significantly less efficient than the parent strain. Similarly, *P. fluorescens* CHAO suppressed *Thielaviopsis basicola*-induced black root rot of tobacco and *G. graminis* var. *tritici*-induced take-all of wheat by the production of 2,4-diacetyl phloroglucinol (Phl) (Keel *et al.*, 1992).

For the first time Gnanamanickam and Mew (1992) and Valasubramanian (1994) reported on the role of antibiotic(s) produced by antagonistic pseudomonads in rice disease (blast and sheath blight) control. Antibiotic isolation was made by following previously published protocols (Gurusiddaiah *et al.*, 1986) from six efficient strains of fluorescent pseudomonads (TNC82, TK8, U227, KE26, A55, TNVI115) which were efficient both in *in vitro* and seed-bed assays.

Of the 6 strains studied, 3 of the *P. fluorescens* strains, TNC82, TK8 and U227, produced a greenish-yellow coloured antibiotic of PCA nature. The purified compound was soluble in water and appeared as a brown non-fluorescent band under UV light (366 nm). The antibiotic(s) produced by strains, TNC82, TK8 and U227 were identical and matched with the properties of a reference sample of PCA. The purified antibiotic also inhibited the conidial germination and mycelial growth of *P. grisea in vitro* (Table 1).

Three other srains (KE26, A55, TNVI115) studied produced another kind of antibiotic(s) whose chemical nature remains to be determined. A preliminary analysis shows that the antibiotic principles of these strains are different from phenazines and were biologically active in inhibiting the conidial germination and mycelial growth of *P. grisea in vitro* (Valasubramanian, 1994).

Table 1. Strains of *Pseudomonas fluorescens* and their useful traits for the biological suppression of rice blast (Valasubramanian, 1994)

P. fluorescens strain	*In vitro* antibiosis mm^s	% Leaf blast suppression	Antibiotic
TNC82	26	68.6	Phenazine-1-carboxylic acid (PCA)
TK8	39	75.8	PCA
U227	30	62.3	PCA
KE26	32	74.5	Non-phenazine
A55	37	70.9	Non-phenazine
TNVI115	26	80.6	Non-phenazine

^{s} Mean of three replications, Test organism :*Pyricularia grisea*

270

In the dual plate assays done on PDA medium, the diameter of fungal inhibition zone induced by the bacterial antagonists ranged from 26 to 39 mm (Table 1). In a seed-bed experiment, these strains afforded 62 to 80% of leaf blast suppression in IR50 plants (Valasubramanian, 1994). These studies clearly demonstrate the existence of positive correlation between antibiotic production and pathogen inhibition *in vitro* and disease suppression *in vivo*. Hence, we have suggested that antibiotic production by these efficient pseudomonads mediate the biological disease suppression.

We have tried to confirm the involvement of antibiotics in the biological suppression of blast and sheath blight by a thorough genetic analysis (Chatterjee *et al.*, 1996). *P. fluorescens* strain 7-14, isolated from rice rhizosphere (Gnanamanickam and Mew, 1992), produces antibiotic(s) active against *P. grisea, R. solani, Pythium* spp., *G. graminis* var. *tritici* and an array of several fungal pathogens and controls blast and sheath blight in field. Transposon mutants of this strain which either totally or partially lost their *in vitro* antibiosis, were tested in a field experiment to compare the biocontrol efficiency. In this experiment the wild-type strain afforded 79 and 82% leaf blast and neck blast reductions while the mutants afforded 24 to 40% and 2.9 to 25% leaf and neck blast reductions (Fig.1) (Table 2). In the same experiment, the wild-type strain controlled sheath blight in IR50 rice crop by 82% while the mutants showed a 10% ShB reduction (Chatterjee *et al,* 1996; Valasubramanian, 1994). Although we have not attempted the isolation of antibiotics from plants treated with bacterial strains, differences in the efficiency of disease control *in vivo* by the wild-type strain and antibiotic-deficient mutants demonstrates that the antibiotic(s) produced by this particular *P. fluorescens* strain mediates the disease control. It is noteworthy that the extent of disease control with *P. fluorescens* 7-14 in the field was better than that achieved with a commercial fungicide, tricyclazole (beam) (Table 2) (Valasubramanian, 1994). Thus this strain and its derivatives are excellent candidates for use as biocontrol agents for the major rice diseases, blast and sheath blight.

Table 2 : Biological control of rice blast : evaluation of *Pseudomonas fluorescens* strain 7-14 and its mutants in the field for the suppression of leaf and neck blasts in IR50 rice (Chatterjee *et al.*, 1996)

Bacterial strain	Number of lesions per hill[*]	Leaf blast control (%)	Neck blast control (%)
Wild type 7-14	37	79	82
ant[-] mutants			
AC2000	118	32	3
AC2001	97	42	9
AC2002	114	35	9
AC2003	131	25	25
ant[leaky] mutants			
AC2007	92	47	35
AC2011	93	47	35
AC2013	110	37	52
Fungicide: tricyclazole	54	69	70
Check	174		
LSD	54.7 (0.05)		
	74.6 (0.01)		

[*]Mean of three replications.

Figure 1. Suppression of neck blast in IR50 rice by *Pseudomonas fluorescens* 7-14. Panicles from plants treated with Pf7-14 are free from neck blast (on the right) and panicles from an untreated check (on the left) show severe neck blast incidence. Field experiment, Madras, 1993.

3.2. Production of Chitinase and other Antifungal Proteins

Chitin is a linear polymer of β-1,4-linked N-acetylglucosamine and is a major component of the cell walls of fungi (except those in Class Oomycetes). It is also a chief constituent of exoskeleton of arthropods. Vascular plants and mammals do not contain chitin. All organisms that contain chitin and also the plants have been found to contain chitinase. In the plants they are produced in response to microbial infections or other injuries and thus chitinases are part of the "pathogenesis-related proteins". It has been postulated that plants produce chitinase to protect themselves from chitin-containing parasites such as fungi and insects (Boller, 1985). Bacteria also are producers of chitinases (exo and endo) and if the bacterial strain used as a biological control agent secretes chitinase it would hydrolyse the chitin in the cell wall of rice pathogens such as *R. solani, P. grisea* and others and thus reduce the aggressiveness of the pathogen.

The involvement of chitinase in the control of *R. solani* has been considered as a possible mechanism. *Sclerotium rolfsii* is another pathogen that contains chitin in its cell wall and may be controlled by chitinase secreted by a bacterial antagonist (Ordentlitch *et al.*, 1988). A large number of reports on the use of chitinase from various sources are available. These include the use of purified chitinase from a microbe, molecular cloning of a chitinase gene from a chitinase-producing *Serratia* strain into another efficient root-colonizing bacterium (Sundheim *et al.*, 1988) or the introduction of chitinase gene from bacteria or plants into crop plants such as tobacco (Broglie *et al.*, 1991), rice (Lin *et al.*, 1995) and others most susceptible to *R. solani* attack.

272

We have evaluated chitinase production by bacterial antagonists as a criterion to select an efficient biological control strain. When a total of 1,757 bacterial strains both of fluorescent and non-fluorescent groups were screened by a rapid assay for chitin production, 31% of the strains were positive for chitinase activity. However, there was no correlation between chitinase production *in vitro* and levels of rice ShB suppression *in vivo*. In particular, efficient biocontrol agents of the *P. fluorescens / P. putida* groups were very poor producers of chitinase (Thara and Gnanamanickam, 1994).

We have also encountered bacterial antagonists of the non-fluorescent group (eg. *Bacillus megaterium*) which produced chitinase and had reasonable ShB suppressing capability (Thara, 1994). This is not common. When chitinase produced by this strain of *Bacillus* was isolated and purified, the purified protein had antifungal activity towards the rice ShB pathogen in laboratory assays and in greenhouse experiments (Thara, 1994). Our results indicate that if a strain of bacterium would have the combined ability to secrete both antibiotic(s) and chitinase(s), the level of protection it affords against plant pathogens such as *R. solani* is greatly enhanced.

It has been difficult, however, to mobilize bacterial or plant chitinase genes into a *P. fluorescens* strain that produces antibiotic(s). It is known that plant chitinases are more antifungal than the bacterial chitinases. When two different rice chitinase genes (pBZ7-1 and RC-7) were mobilized separately into a strain of *P. fluorescens* that afforded 68% ShB suppression in the field, the transconjugants did not show chitinase activity (Table 3) (Thara, 1994).

Table 3. Mobilization of plant chitinase genes into *Pseudomonas fluorescens / P. putida* strains (Thara, 1994)

Donor	Recipient	No. of trans-conjugant	Frequency	No. screened for chitinase	No. positive
pBZ7-1/ *E. coli* HB101[a]	PfU113b	2,162	6.75×10^{-5}	2,162	0
	PpV14i	2,444	7.46×10^{-5}	2,444	0
pDSK519/ *E. coli* JM109[b]	PfU113b	2,350	7.31×10^{-5}	2,350	0

[a] Supplied by C. Lamb, Salk Institute
[b] Supplied by S. Muthukrishnan, Kansas State University.

3.3. Induction of Systemic Resistance (ISR) by Bacterial Biocontrol Agents

In a recent study, Krishnamurthy and Gnanamanickam (1997) demonstrated that pseudomonad strains expressing the lacZY genes survived endophytically when introduced into rice stem and induced a systemic resistance response. In rice tissues showing ISR response, salicylic acid levels doubled (Table 4). This led to 18% suppression of sheath blight and 21-29% suppression of rice blast in separate greenhouse experiments. The rice phytoalexin, momilactone-A was not associated with this ISR response or disease suppression (Krishnamurthy, 1997).

The first evidence for the direct involvement of antibiotic produced by a fluorescent pseudomonad in the biological control of "take all" disease of wheat was introduced by

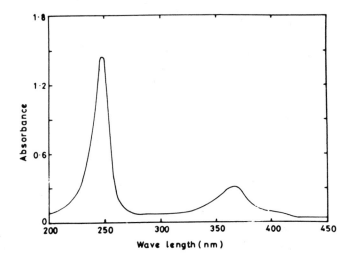

Figure 2. UV-absorption spectrum of phenazine-1-carboxylic acid (PCA) produced by *Pseudomonas fluorescens* Pfcp. This antifungal antibiotic produced by the biocontrol agent was recovered from soil samples of rice rhizosphere both in green house and field experiments after the bacterium was applied as a seed treatment on IR50 rice.

Table 4. Estimation of salicylic acid (SA) levels by HPLC in rice plants that received *Pseudomonas fluorescens* 7-14 and *P. putida* V14i treatments as root-dip and stem infiltration (Krishnamurthy, 1997)

Treatment	SA in rice tissue (ig/g tissue)									
	cv.IR50[a]					cv.C101LAC[b]				cv.C101PKT[b]
	S1	S5	S10	R5	L5	S5	S10	R5	L5	S5
Root-dip treatment										
Pf7-14	21	24	25	9	27	35	43	13	44	30
PpV14i	18	21	23	11	22	37	39	13	43	31
*E. coli*CE1	17	18	17	7	17	29	31	10	30	25
H₂O + cmc	14	14	13	6	13	25	26	9	25	24
Check (untreated)	12	13	13	6	12	24	24	10	25	22
Stem-infiltration treatment										
Pf7-14	23	22	18	9	21	37	39	17	38	36
PpV14i	24	22	19	8	20	38	43	13	41	34
*E. coli*CE1	18	17	14	7	14	30	31	11	30	29
H₂O	14	15	13	6	12	25	25	11	27	24
Check (untreated)	13	12	13	6	11	24	24	10	24	22

[a] Blast-susceptible cultivar; [b] Blast-resistant cultivar
S1,S5,S10 refer to stem samples collected, 1,5 and 10 days after the treatment
R and L refer to root and leaf tissue samples collected 5 days after treatment
cmc : carboxymethylcellulose
The experiment was repeated once with matching results. Each SA values is the mean of two replications.

274

Thomashow *et al.* (1990). By using a high performance liquid chromatography based assay, they detected phenazine antibiotic in the wheat rhizosphere colonized by the phenazine-producing fluorescent *Pseudomonas* sp.

We detected the antibiotic PCA in the rice rhizosphere colonized by a *P. fluorescens* strain Pfcp (Krishnamurthy, 1993). This strain was originally isolated from citrus leaves (Sakthivel and Gnanamanickam, 1987), produces PCA that inhibits several plant pathogens including *S. oryzae, R. solani, P. grisea, P. solanacearum* and *X. oryzae* pv. *oryzae* (Anuratha and Gnanamanickam, 1990; Sakthivel and Gnanamanickam, 1987; Sivamani and Gnanamanickam, 1988; Valasubramanian, 1994). A rifampicin and nalidixic acid-resistant mutant of this strain was applied to the IR50 rice seeds and they were planted in sterile and non-sterile soil. Seedlings were removed at regular intervals and antibiotic was isolated by following the method described previously (Thomashow *et al.*, 1990). The rice rhizosphere samples colonized by *P. fluorescens* strain Pfcp were washed in phosphate buffer and the antibiotic was isolated by organic solvent extraction with methylene chloride. Purification of the crude compound by column chromatography yielded the biologically active PCA. The spectral characters and Rf values were found to be similar to that of the authentic PCA (Fig. 2) (Krishnamurthy, 1993). When the IR50 plants treated with this bacterium were inoculated with the blast pathogen *P. grisea*, 60 to 70% reduction of leaf blast severity was observed (Krishnamurthy, 1993). Thus, the results of our studies suggest that the phenazine antibiotic produced by efficient *Pseudomonas* strains mediates blast disease suppression (Table 1).

4. BACTERIZATION TECHNIQUES

There are two main methods for the inoculation of bacteria to plants: (i) direct inoculation with bacterial culture (dipping seeds in a bacterial suspension), foliar spraying of bacteria or spreading the bacteria in the sowing furrow by various drip systems mounted on the sowing drill and (ii) use of various solid-phase bacterial inoculants based on organic materials including organic granular particles.

We generally apply the biocontrol agents to the seeds before sowing and or as foliar sprays to the plant. For seed-bacterization, we soak the surface sterilized seeds in the bacterial suspension (OD 0.1 at 600 nm) prepared in 1% sterile carboxymethylcellulose overnight and then air-dry the seeds before sowing. Besides seed bacterization we give 2 to 3 foliar sprays with the bacterial suspension prepared in 1% sterile carboxymethylcellulose solution. More recently, a methylcellulose : talc (mc:talc, 1:4) formulation of efficient strains of *P. fluorescens* and *P. putida* was evaluated in the field (Krishnamurthy, 1997). It afforded 68.5% leaf blast control and 60% ShB control when applied as seed treatment and three foliar sprays.

4.1. Possible Reasons for the Inconsistent Performance of Biocontrol Agents

It is encouraging that there are now so many examples of biocontrol with bacteria in field. Unfortunately, one characteristic that is common to most biocontrol systems with introduced bacteria is the inconsistency of disease control. A multitude of factors could account for inconsistent results, given the complex interactions among host, pathogen, antagonists and the environment. Three possible reasons are

4.1.1. Loss of ecological competence

Ecological competence is the ability of a bacterium to compete and survive in nature. Many bacterial traits contribute to the ecological competence in the rhizosphere and loss of any one can reduce the ability of the bacteria to become established or function on or near the

275

root. Important traits can be lost when a bacterium is grown *in vitro*. Repeated culturing of fluorescent pseudomonads *in vitro* can result in a loss of efficacy, possibly related to changes in cell and colony morphology, loss of cell surface structures or reduction in antibiotic production.

4.1.2. Absence of target pathogen or interference by non-target pathogens

Bacterial biocontrol agents improve plant growth by reducing damage from pathogens. A positive response to their introduction does not occur when a target pathogen is absent or when environmental conditions are unsuitable for disease development. The effect of pathogens other than the target pathogen is another concern. If a bacterium suppresses only one pathogen and another pathogen becomes predominant, then the treatment will appear ineffective. Hence understanding of the pathogens in the agroecosystems and the conditions that favor each is essential.

4.1.3. Variable root colonization by introduced bacterium

Variability in the colonization from plant to plant and root to root on a given plant is probably an important reason for inconsistent disease control by biocontrol agents.

5. CONCLUSIONS

A great deal of progress has been made in the selection and use of bacterial antagonists for the control of major rice and wheat diseases. In order to develop bacterial biocontrol agents for commercial use, the consistency of their performance must be improved. Accomplishing this will require research in many diverse areas, because biological control is the culmination of complex interactions among the host, pathogens, antagonists and environment. More research on formulation and delivery of the biocontrol agent is required.

Engineering bacterial strains with chitinases or other antifungal proteins (such as thaumatin-like proteins) (Vigers *et al.,* 1992) or a combination of these genes is a possible direction for future line of research. Such engineered strains which naturally produce antibiotics and express antifungal protein(s) can be hoped to be highly efficient in enhancing the level of biological disease suppression.

Transgenic rice generated to express insect or plant chitinase genes as has been done in recent years may ultimately prove to be the best means of insect and disease control. Most recently Lin *et al.* (1995) have cloned a 1.1 kb rice genomic DNA fragment containing the chitinase gene under the control of CaMV 35S promoter into a rice transformation vector (pGL2) to transform the protoplasts of a indica rice variety. The transgenic plants displayed a degree of resistance to *R. solani* which correlated with the level of expression of chitinase.

Therefore, there are exciting possibilities for the future. Strategies described in this paper are the future tools for disease and insect management in rice and other economic crops.

REFERENCES

Anuratha, C.S. and Gnanamanickam, S.S. 1990, Biological control of bacterial wilt caused by *Pseudomonas solanacearum* in India with antagonistic bacteria, *Plant Soil* 124:109-116.
Boller 1985, Induction of hydrolases as defense reaction against pathogens, In: *Cellular and Molecular Biology of Plant Stress*, eds. J. L. Key and T. Kosuge, Alan Liss Inc, New York, USA, pp. 247-262.

Broglie, K., Chet, I., Holliday, M., Cressman, R., Biddle, P., Knowlton, S., Mauvais, C.J. and Broglie, R. 1991, Transgenic plants with enhanced resistance to the fungal pathogen *Rhizoctonia solani*, *Science* 254: 1194-1197.

Chatterjee, A., Valasubramanian, R., Ma, W.L, Vachhani, A.K., Gnanamanickam, S.S. and Chatterjee, A.K. 1996, Biocontrol of rice diseases with *Pseudomonas fluorescens* strain Pf7-14: isolation of ant mutants altered in antifungal antibiotic production, cloning of ant[+] DNA and evaluation of a role of antibiotic production in the control of blast and sheath blight, *Biol. Con.* 7:185-195.

Cook, R.J. and Baker, K.F. 1983, *The Nature and Practice of Biological Control of Plant Pathogens*, American Phytopathological Society, St.Paul, Minnesota, USA, 539 p.

Defago, G., Berling, C.H., Burger, U., Hass, D., Kahr, G., Keel, C., Voisard, C., Wirthner, P. and Wuthrich, B. 1990, Suppression of black root rot of tobacco and other root diseases by strains of *Pseudomonas fluorescens*: potential and applications and mechanisms, In: *Biological Control of Soil-borne Plant Pathogens*, eds. R.J. Cook, Y. Henis and D. Hornby, CAB International, USA, pp. 93-108.

Fravel, D.R. 1988, Role of antibiosis in the biocontrol of plant diseases, *Ann. Rev. Phytopath.* 26:75-91.

Ganesan, P. and Gnanamanickam, S.S. 1987, Biological control of *Sclerotium rolfsii* Sacc. in peanut by inoculation with *Pseudomonas fluorescens*, *Soil Biol. Biochem.* 19:35-38.

Gnanamanickam, S.S. and Mew, T.W. 1992, Biological control of blast disease of rice (*Oryza sativa*, L.) with antagonistic bacteria and its mediation by a *Pseudomonas* antibiotic, *Ann. Phytopath. Soc. Japan* 58:380-385.

Gurusiddaiah, S., Weller, D.M., Sarkar, A. and Cook, R J. 1986, Characterization of an antibiotic produced by a strain of *Pseudomonas fluorescens* inhibitory to *Gaeumanomyces graminis* var. *tritici* and *Phytium* spp., *Antimicro. Ag. Chemoth.* 29:488-495.

Houghland, G.V.C. and Cash, L.C. 1954, The use of vermiculite in providing scab inoculation for potatoes, *Pl. Dis. Rep.* 38:460-461.

Howell, C.R. and Stipanovic R.D. 1979, Control of *Rhizoctonia solani* on cotton seedlings with *Pseudomonas fluorescens* and with an antibiotic produced by the bacterium, *Phytopath.* 69:480-482.

Howell, C.R. and Stipanovic, R.D. 1980, Suppression of *Pythium ultimum* inudced damping-off of cotton seedlings by *Pseudomonas fluorescens* and its antibiotic, pyoluteorin, *Phytopath.* 70:712-715.

Howie, W.J. and Suslow T.V. 1991, Role of antibiotic biosynthesis in the inhibition of *Pythium ultimum* in the cotton spermosphere and rhizosphere by *Pseudomonas fluorescens*, *Mol. Plant-Microbe Interac.* 4:393-399.

Keel, C., Schnider, U., Maurhofer, M., Voisard, C., Laville, J., Burger, U., Wirthner, P., Hass and Defago, G. 1992, Suppression of root disease by *Pseudomonas fluorescens* CHAO: importance of the bacterial secondary metabolite 2, 4-diacetylphloroglucinol, *Mol. Plant-Microbe Interac.* 5:4-13.

Krishnamurthy, K. 1997, Biological control of rice blast and sheath blight with *Pseudomonas* spp. : Survival and migration of the biocontrol agents and the induction of systemic resistance in biological disease suppression, Ph.D. dissertation, Univ.of Madras, India, 109 p.

Krishnamurthy, K. 1993, Monitoring the production of antifungal antiobiotic (Afa) by *Pseudomonas fluorescens* in rice rhizosphere, M.Phil dissertation, Univ. of Madras, India, 36 p.

Krishnamurthy, K. and Gnanamanickam, S.S. 1997, Biological control of sheath blight of rice : induction of systemic resistance in rice by plant-associated *Pseudomonas* spp., *Curr. Scie.* 72:331-334.

Lin, W., Anuratha, C.S., Datta, K., Potrykus, I., Muthukrishnan, S. and Datta, S. 1995, Genetic engineering of rice for resistance to sheath blight, *Biotech.* 13:686-691.

Mew, T.W. and Rosales, A.M. 1986, Bacterization of rice plants for control of sheath blight caused by *Rhizoctonia solani*, *Phytopath.* 76:1260-1264.

Ordentlich, A., Elad, Y. and Chet, I. 1988, The role of chitinase of *Serratia marcescens* for the control of *Sclerotium rolfsii*, *Phytpath.* 78:84-88.

O'Sullivan, D.J. and O'Gara, F. 1992, Traits of fluorescent *Pseudomonas* spp. involved in suppression of plant root pathogens, *Microbiol. Rev.* 56:662-676.

Sakthivel, N. and Gnanamanickam, S.S. 1987, Evaluation of *Pseudomonas fluorescens* for suppression of sheath rot disease and for enhancement of grain yields in rice, *Oryza sativa*, L., *App. Environ. Microbiol.* 53:2056-2059.

Sivamani, E. and Gnanamanickam, S.S. 1988, Biological control of *Fusarium oxysporum* f.sp. *cubense* in banana by inoculation with *Pseudomonas fluorenscens*, *Plant Soil* 107:3-9.

Sundheim, L., Poplawsky, A.R. and Ellingboe, A.H. 1988, Molecular cloning of two chitinase genes from *Serratia marcescens* and their expression in *Pseudomonas* species, *Physiol. Mol. Pl. Path.* 33:483-491.

Thara, K.V. 1994, Biological control of rice sheath blight: mechanisms of disease suppression, Ph.D dissertation, University of Madras, Madras, India, 108 p.

Thara, K.V. and Gnanamanickam, S.S. 1994, Biological control of sheath blight in India: lack of correlation between chitinase production by bacterial antagonists and sheath blight suppression, *Plant Soil* 160:277-280.

Thomashow, L.S. and Weller, D.M. 1988, Role of a phenazine antibiotic from *Pseudomonas fluorescens* in biological control of *Gaeumannomyces graminis* var. *tritici, J. Bacteriol.* 170:3499-3508.

Thomashow, L.S., Weller, D.M., Bonsall, R.F. and Pierson, L.S.III. 1990, Production of the antibiotic phenazine-1-carboxylic acid by fluorescent *Pseudomonas* species in the rhizosphere of wheat, *App. Environ. Micribiol.* 56:908-912.

Valasubramanian, R. 1994, Biological control of rice blast: role of antifungal antibiotic in disease suppression, Ph.D dissertation, University of Madras, Madras, India.

Vigers, A. J., Wiedemann, S., Roberts, W.K., Legrand, M., Selitrennikoff, C.P. and Fritig, B. 1992, Thaumatin-like pathogenesis-related proteins are antifungal, *Pl. Scie.* 83:155-161.

Viji Sitther and Gananamanickam, S.S. 1996, Biological control of blast disease of finger millet (*Eleusine coracona* L.) and an analysis of the fertility status of *Magnaporthe grisea, Curr. Scie.* 71:144-147.

Weller, D.M. and Cook, R.J. 1983, Suppression of take-all of wheat by seed treatments with fluorescent pseudomonads, *Phytopath.* 73:463-469.

INNOVATIVE APPROACHES IN RICE SHEATH BLIGHT MANAGEMENT

K. Manibhushanrao and U.I. Baby

Centre for Advanced Studies in Botany
University of Madras, Guindy Campus
Chennai - 600 025, Tamil Nadu, INDIA

1. INTRODUCTION

Sheath blight (ShB) disease of rice caused by *Rhizoctonia solani* Kuhn is one of the major diseases of rice. The disease became a serious problem after the introduction of high yielding semi- dwarf cultivars. Further, intensive cultural practices and heavy application of nitrogenous fertilizers intensified the disease situation. The yield loss can be even upto 50% due to the disease (Tsai, 1976). The disease appears at seedling, tillering and booting stage of the plant and the damage to crop is most severe when the disease appears at later stages (Cu *et al.*, 1996).

The pathogen *R. solani* is a versatile soilborne saprophyte, survives in soil as sclerotia or thick walled mycelia (Endo, 1931). Sclerotia represent the primary source of inoculum (Lee and Rush, 1983; Leu and Yang, 1985) and they remain viable in soil for several months over a wide range of temperature and moisture (Manian and Manibhushanrao, 1990). Crop residues colonized by the pathogen also play an important role in ShB epidemics (Cu *et al* ., 1996). A comprehensive monograph on rice ShB disease has appeared covering all aspects of the disease including the strategies of biological control (Manibhushanrao 1995). The management of ShB disease is very difficult due to the versatile nature of the pathogen with wide host range, high competitive saprophytic ability. Further, none of the commercial rice cvs are resistant to the disease. Although some effective chemical fungicides are available, their use is limited due to pollution problem, cost factor and nontarget effect on beneficial organisms. In this scenario, it is necessary to explore other methods to manage the disease and a comprehensive account on various viable alternative strategies are discussed from time to time. (Mukerji and Garg, 1988 a,b; Mukerji *et al.*, 1992)

2. ORGANIC SOIL AMENDMENTS

The use of organic manures and crop residues as an approach of biological control is gaining importance in agriculture. The impact of organic amendments on soilborne plant pathogens has been reviewed (Linderman, 1989; Lumsden et al., 1983). Organic soil amendments effectively controlled rice ShB in greenhouse and field conditions (Kannaiyan and Prasad, 1981; Rajan, 1980; Rajan and Menon, 1975). The effect of various organic amendments such as oilseed cakes, green leaf manures and agroindustrial wastes on ShB incidence was critically studied (Baby, 1991). Incorporation of organic amendments improved the physico-chemical and biological properties of the soil. The amendments stimulated soil microflora including the antagonistic fungi such as *Trichoderma*, with a consequent reduction in the survivability of the pathogen *R. solani* and ShB incidence (Baby and Manibhushanrao, 1993a; Manibhushanrao et al., 1989a). In addition, there was a significant increase in the growth and yield of the plant (Rajan and Menon, 1975).

It is opined that the control achieved with organic soil amendments is through a complex process and interactions. The organic amendments can have a range of effects on the pathogen which depend on the type and stage of decomposition of organic materials. Enhancement of microbial population including the antagonists results in competition and high level of fungistasis and propagule lysis (Lumsden et al., 1983; Singh and Singh, 1984). Production of antifungal substances in amended soil has also been reported (Chun and Lockwood, 1985).

3. VESICULAR ARBUSCULAR MYCORRHIZAE (VAM)

It has already been established that VAM fungi, enhance nutrient uptake (Clarkson, 1985) and plant growth (Manjunath and Bagyaraj, 1986), impart stress tolerance (Marx and Schenck. 1983) and resistance to soil borne plant pathogens (Caron, 1989; Schenck, 1981).

Association of VAM fungi with upland (Ammani et al., 1985; Gangopadhyay and Das, 1982) and low land (Sivaprasad et al., 1990) rice has been reported. Further, VAM fungi as a tool in the biocontrol of rice ShB has been studied by Baby and Manibhushanrao (1996a). The results indicated that pre-inoculation with VAM fungi could effectively suppress ShB disease and promote the growth and biomass of rice plants and moreover, the efficacy was significantly superior compared to simultaneous and post application. The impact of organic amendments on VAM fungal development has also been reported (Baby and Manibhushanrao, 1996b).

Physiological alterations in the host carrying VAM lead to the formation of physical and/or chemical barriers rendering the plant more resistance to a given pathogen. Increased amino acids (Baltruschat and Schonbeck, 1975), accumulation of phenolic compounds (Krishna and Bagyaraj, 1983), greater activity of biosynthetic and oxidative enzymes(Smith and Ginianazzi-Pearson, 1988) have been ascribed in creating condition detrimental to the pathogen. Further, increase in wound barrier formation (Wick and Moore, 1984) and enhanced lignification of the cell wall (Dehne and Schonbeck, 1979) are considered as the physical barriers countering the pathogen invasion.

4. ANTAGONISTIC MICROBES

There is a lot of scope in employing residential and introduced antagonists for the control of soil borne plant pathogens and is proved successful with many crops. Members

280

of the genus *Trichoderma, Gliocladium, Bacillus* and *Pseudomonas*, which are very common soil inhabitants, have gained considerable importance due to their antagonistic potential to an array of phytopathogens (Cook and Baker, 1983; Papavizas, 1985; Schippers *et al.,* 1987). IRRI (1978) recorded the susceptibility of rice ShB pathogen *R. solani* to many fungi, bacteria and nematodes .

Seed bacterization and spray application of bacterial antagonists suppressed rice ShB and increased grain yield (Mew and Rosales, 1992; Vasanthadevi *et al.,* 1989). The main mechanism of the antagonism seems to be competition with the harmful microorganisms for iron by the release of siderophores, secondary metabolites with a strong affinity to Fe^{3+}. In addition to the production of siderophores, effective strains of *P. fluorescence* are capable of producing antibiotics also (Schippers *et al.,* 1987).

Mycoparasitic potential of fungal antagonists such as *Trichoderma* (Manibhushanrao *et al.,* 1989b; Roy and Sayre, 1984) and *Gliocladium* (Manibhushanrao *et al.,* 1989b; Gokulapalan and Nair, 1986) on rice ShB pathogen *R. solani* is known. Apart from these, other fungal species like *Aspergillus terreus* (Roy, 1991a); *A. niger* (Gokulapalan and Nair, 1984); *Laetisaria arvalis* (Burdsall *et al.,* 1980) and *Penicillium ehrlichii, P. vermiculatum* (Roy, 1991b) also found to be antagonistic to ShB pathogen. The role of wall lytic enzymes like glucanase, chitinase and cellulase in mycoparasitism (Howell, 1987; Manocha, 1987; Sreenivasaprasad and Manibhushanrao, 1990), production of antibiotics such as trichodermin, suzukacillin and alamethicin by *T. harzianum* (Dennis and Webster, 1971a,b) and gliovirin by *G. virens* (Howell and Stipanovic, 1983) have been reported. The biocontrol potential as well as growth promotive effect of *G. virens* and *Trichoderma* through soil application have been established under greenhouse and field conditions (Baby and Manibhushanrao, 1993b; Elavarasan, 1989; Kumaresan and Manibhushanrao, 1991). Further, Das *et al.* (1995) observed effective ShB control through both soil application and spraying of spore suspension of *T. viride* on to the aerial parts of the plant.

Genetic manipulation of biocontrol agents to improve their efficacy is emphasised (Baker, 1989; Baker and Scher, 1987). Improved strains of *Trichoderma* spp. were raised by UV irradiation (Viji *et al.,* 1993) and by protoplasmic fusion (Mrinalini and Lalithakumari, 1996). These strains possessed enhanced biocontrol potential against ShB disease (Mrinalini and Lalithakumari, 1996).

5. CROSS PROTECTION AND INDUCED RESISTANCE

Resistance to pathogen can sometimes be induced by pre- inoculating plants with avirulent or mildly virulent strains of the targeted pathogen (immunizing microorganism). It is also possible through the metabolic products of the pathogen or non-pathogen, inducer chemicals or physical stimuli. The former being termed as cross protection and the latter as induced resistance. Induced resistance may result from the formation of, or increase in, certain resistance inducing compounds (phytoalexins) (Kuc, 1977). Additionally, depletion of host nutrients from tissues pre-infected with an immunizing microorganism may cause such tissues to be less suitable for infection, thus protecting the tissue from subsequent colonization by pathogens (Matta, 1971). Manibhushanrao *et al* (1988) found that when excised rice leaves were pre- inoculated with an avirulent strain of *R. solani* (R7) and challenge inoculated with a virulent isolate (R5), the incompatible interaction imparted 81% protection.

The phytotoxins Phenyl acetic acid (PAA) and its hydroxy derivatives (m-HPAA, o-HPAA and p-HPAA) produced by the pathogen *R. solani* at higher concentrations,

(>100 ppm) produce typical ShB lesions comparable to those caused by the pathogen *R. solani*; but at lower concentrations (0.1 to 5 ppm) resulted in rapid accumulation of phenolic compounds and provided 25- 30% protection from the disease on excised rice leaves (Waheeta *et al.*, 1987). These metabolites were neither phytotoxic nor fungitoxic. The acquired resistance may presumably due to the physical and/or chemical barriers elicited in the host.

Plant immunization using inducer chemicals is also well established. Manibhushanrao *et al.* (1990) tested an array of chemical elicitors *viz.*, lithium sulphate, sodium azide, sodium selenite, thioglycollic acid, cycloheximide and penicillin on rice ShB. The seed and root dip treatment of the chemicals rendered protection even upto 60 %. Further, the leaf diffusate bioassay against *R. solani* revealed a positive correlation between enhanced fungitoxicity and disease reduction implicating the involvement of phytoalexin type substance(s).

6. BOTANICAL PESTICIDES

Pesticides of plant origin are less toxic to plants , systemic in nature and are easily biodegradable. The screening of several plant species has led to the identification of new plants possessing compounds toxic to rice pathogens. Leaf extract of *Lawsonia inermis* contain an alkaloid 'lawsone' (2- hydroxy- 1,4- naphthoquinone) found to be toxic to many pathogens including *R. solani* (Sridhar *et al.*, 1989). Leaf and seed extracts of many plant species inhibited sclerotial germination, mycelial growth and sclerotial production of *R. solani* (Kotasthane and Lakpale, 1994) and thermolabile nature of the fungitoxic substances has also been reported (Kurucheve *et al.*, 1997).

7. SOIL SOLARIZATION

Soil solarization is a technique to control soil borne pathogens by disinfesting the soil through solar energy. This is possible by covering (tarping or mulching) the soil with transparent polyethylene sheets under appropriate climatic conditions (Katan, 1987). It is a hydrothermal process simultaneously causing many changes in the biotic and abiotic components of the soil, during and after the process leading to changes in disease incidence, plant growth and yield (Stapleton *et al.*, 1985). The basic process involved in solarization is the heating of the soil to relatively mild levels, usually 36- 56°C in the upper 30cm level. Though heat is the major factor, evidences show the involvement of biological processes in achieving surprisingly good result with solarization. Higher temperature, accumulation of volatile gases such as CO_2, ethylene and others, under the polythene cover, become unfavourable to many soil borne organisms to withstand the condition. Soil solarization during summer months has found to control *R. solani* in many crops. The combination of solarization with antagonists is promising (Katan, 1987). When soil solarization was followed by the application of *T. harzianum*, in *R. solani* infested soil, there was an increase in the population of *Trichoderma* and a reduction in *R. solani* propagules followed by improved disease control (Chet *et al.*, 1982; Elad *et al.*, 1980). The possibilities of exploiting this method in rice ShB system is stressed (Gangopadhyay and Chakrabarthi, 1982; Roy, 1996). Soil solarization might be combined with many biological, chemical and cultural methods of control (Katan, 1987; Mukerji and Garg, 1988a,b).

8. INTEGRATED CONTROL

Integrated control is the combined use of chemicals and biocontrol agents and traditional cultural practices to control the disease. It is proved that a combination of two or more methods provide high level of control than that of a single system (Upadhyay *et al.*, 1997)

Rajan and Alexander (1988) studied the effect of combination of soil amendments (rice husk, punna cake, glyricidia leaf, lime and gypsum), antagonistic fungus (*T. viride*) and fungicide (carbendazim) on rice ShB and found that the disease incidence was least in the combination treatments with a subsequent reduction in *R. solani* population and an increase in general microflora (fungi and bacteria). The efficacy of the combination of organic amendments and fungal antagonists in controlling rice ShB was studied under field condition (Baby and Manibhushanrao, 1993b; Manibhushanrao and Baby, 1991). Two fungal antagonists (*G. virens* and *T. longibrachiatum*) and two organic amendments (Glyricidia and neem cake) were tested. Integration of these systems significantly enhanced the efficacy. In addition, the growth as well as grain yield were also increased in varying degrees in different treatments. When the antagonists were supplemented with organic substrates, an increase in their colony forming units and subsequent reduction in the pathogen population were noticed. Further, when the antagonist (*T. harzianum*) was used with sub-lethal dose of the fungicide thiram, the efficacy of control was better than thiram alone at recommended dose (Borah, 1992). Fungicides active against relatively narrow spectrum of plant parasites, but not against biocontrol agents afford an opportunity to integrate these systems. During the screening of fungicides against ShB pathogen *R. solani* and its fungal antagonists *Trichoderma* and *Gliocladium*, a differential sensitivity was noticed between these two groups of fungi (Viji *et al.*, 1997). Even if the biocontrol agent is sensitive to a fungicide, it is possible to produce mutants that are tolerant to the potent chemicals. Thus benomyl tolerant biotypes of *T. harzianum* (Papavizas *et al.*, 1982) and bavistin tolerant biotypes of *T. harzianum* and *T. longibrachiatum* (Viji *et al.*, 1993) were raised by mutation. Further, fungicide resistant biotypes were found to be superior to their parents in biocontrol potential (Papavizas *et al.*, 1982).

9. CONCLUSION

Indiscriminate use of toxic pesticides led to serious environmental and ground water pollution. Due to the public awareness on the pollution problems caused by toxic pesticides and their residues, there is an increasing interest in the field of biological control. Biocontrol has been proved successful in many crops including rice diseases in recent years. Due to the peculiar nature and versatility of the pathogen *R. solani*, suppression of the propagules through biocontrol methods seems to be more promising than chemical control. Various biocontrol strategies discussed are ecologically, environmentally and economically viable. The management of resident antagonists by cultural practices is of prime importance. But alien antagonists have to be introduced in many cases, as the potent antagonists effective against a specific pathogen need not be present in all soils. Maintenance of soil organic matter/or the use of organic amendments deserves attention in this regard, as they trigger beneficial soil saprophytes and also favour root health by improving physical and chemical properties of the soil. Incidently the VAM population is also highly depended on the soil organic matter. Genetic manipulation of the antagonists is an important tool to introduce desirable traits to improve their biocontrol potentialities. Induction of host resistance through avirulant strains or foliar sprays of inducer chemicals, at very low concentration; and the use of botanical pesticides are also

283

effective. Soil solarization is also an effective tool to disinfest the soil and thereby suppress soil borne pathogens. soil solarization is compatible with many other chemical and nonchemical methods. An integration of one or more of these methods along with need based use of chemicals can be adopted as a part of integrated disease management (IDM) programme. Thus pesticides should be considered as the last resort and not the first step in plant protection to minimise environmental pollution and residue problems.

REFERENCES

Ammani, K., Venkateswarlu, K. and Rao, A.S. 1985, Development of vesicular-arbuscular mycorrhizal fungi of upland rice, *Curr. Sci.* 59: 1120- 1122.

Baby, U.I. 1991, Studies on the control of rice sheath blight disease with organic soil amendments, Ph. D.thesis, University of Madras, Madras, India.

Baby, U.I. and Manibhushanrao, K. 1993a, Control of rice sheath blight disease by organic manures, In: *Proc. Fifth Kerala Sci. Cong.*, ed. R. Ravikumar, State Comm. Sci. Tech. Environ., Govt. Kerala, Thiruvananthapuram, pp. 158- 161.

Baby, U.I. and Manibhushanrao, K. 1993b, Control of rice sheath blight through the integration of fungal antagonists and organic amendments, *Trop. Agric.* (Trinidad) 70: 240- 244.

Baby, U.I. and Manibhushanrao, K. 1996a, Fungal antagonists and VA mycorrhizal fungi for biocontrol of *Rhizoctonia solani*, the rice sheath blight pathogen, In: *Recent Developments in Biocontrol of Plant Pathogens, Current Trends in Life Sciences* Vol. XXI, eds K.Manibhushanrao and A. Mahadevan, Today and Tomorrow's Printers and Publishers, New Delhi, India.

Baby, U.I. and Manibhushanrao, K. 1996b, Influence of organic amendments on arbuscular mycorrhizal fungi in relation to rice sheath blight disease, *Mycorrhiza* 6: 201- 206.

Baker, R. 1989, Some perspectives on the application of molecular approaches to biocontrol problems, In : *Biotechnology of Fungi for Improving Plant Growth.*, eds J. M. Whipps and R.D. Lumsden, Cambridge Univ. Press, Cambridge, pp. 220- 223.

Baker, R. and Scher, F.M. 1987, Enhancing the activity of biological control agents, In : *Innovative Approaches to Plant Disease Control*, ed. I. Chet, John Wiley & Sons, New York, USA, pp. 1-19.

Baltruschat, H. and Schonbeck, F. 1975, The influence of endotrophic mycorrhiza on the infestation of tobacco by *Thielaviopsis basicola*, *Phytopath.* 84: 172- 188.

Borah, K. 1992, Integrated management of sheath blight of rice with mycoparasite and chemical, M. Sc (Ag) thesis, Assam Agricultural University, Jorhat, Assam, India.

Burdsall, H.H. Jr., Hock, H.C., Boosalis, M.C. and Setliff, E.C. 1980, *Laetisaria arvalis* (Aphyllophorales, Corticiaceae): a possible biocontrol agent for *Rhizoctonia solani* and *Pythium* species, *Mycologia* 72: 728- 736.

Caron, M. 1989, Potential use of mycorrhizae in control of soil-borne diseases, *Can. J. Pl. Pathol.* 14: 177- 179.

Chet, I., Elad, Y., Kalfon, A., Hadar, Y. and Katan, J. 1982, Integrated control of soil borne and bulb borne pathogens in iris, *Phytoparasitica* 10: 229- 236.

Chun, D. and Lockwood, J.L. 1985, Reduction of *Pythium ultimum, Thielaviopsis basicola* and *Macrophomina phaseolina* populations in soil associated with ammonia generated from urea, *Plant Dis.* 69: 154- 158.

Clarkson, D.T. 1985, Factors affecting mineral nutrient acquisition by plants, *Annu. Rev. Pl. Physiol.* 36: 77- 115.

Cook, R.J. and Baker, K.F. 1983, *The Nature and Practice of Biological Control of Plant Pathogens*, Amer. Phytopath. Soc., St. Paul, Minnesota, USA.

Cu, R.M., Mew, T.W., Cassman, K.G. and Teng, P.S. 1996, Effect of sheath blight on yield in tropical intensive rice production system, *Plant Dis.* 80: 1103- 1108.

Das, B.C., Bhuyan, S.A. and Bora, L.C. 1995, Comparative efficacy of *T. viride* in suppressing sheath blight of rice by different method of application, *Plant Health* 1 : 7-11.

Dehne, H.W. and Schonbeck, F. 1979, The influence of endotrophic mycorrhizae on plant disease, II, Phenol metabolism and lignification, *Phytopath. Z.* 95: 210-216.

Dennis, C. and Webster, J. 1971a, Antagonistic properties of species- groups of *Trichoderma*, 1, Production of non-volatile antibiotics, *Trans. Brit. Mycol. Soc.* 57: 25-39.

Dennis, C. and Webster, J. 1971b, Antagonistic properties of species- groups of *Trichoderma*, 2, Production of volatile antibiotics, *Trans. Brit. Mycol. Soc.* 57: 41- 48.

Elad, Y., Katan, J. and Chet, I. 1980, Physical, biological and chemical control integrated for soil-borne diseases of potatoes, *Phytopath.* 70: 418- 422.

Elavarasan, A. 1989, Studies on the biocontrol potential of *Gliocladium virens* against rice sheath blight disease, M. Phil. dissertation, University of Madras, Madras, India.

Endo, S. 1931, Studies on sclerotium diseases of the rice plant, V, Ability of over wintering certain important fungi causing sclerotium diseases of rice plant and their resistance to dry conditions, *Forschn Geb. Pflkrankh*, Tokyo 1: 149-167.

Gangopadhyay, S. and Chakrabarthi, N.K. 1982, Sheath blight of rice, *Rev. Plant Pathol.* 61: 451- 460.

Gangopadhyay, S. and Das, K.M. 1982, Occurrence of vesicular-arbuscular mycorrhizae in rice in India, *Indian Phytopath.* 35: 83-85.

Gokulapalan, C. and Nair, M.C. 1984, Antagonism of few fungi and bacteria against *Rhizoctonia solani* Kuhn., *Indian J. Microbiol.* 24: 57- 58.

Gokulapalan, C. and Nair, M.C. 1986, Mycoparasites of *Rhizoctonia solani* and control of sheath blight of rice, (Abstr.), Fourteenth Annual meeting of the Mycological Society of India and Seminar on Applied Mycology, Thanjavur,India, p.18.

Howell, C.R. 1987, Relevance of mycoparasitism in the biological control of *Rhizoctonia solani* by *Gliocladium virens*, *Phytopath.* 77: 992- 994.

Howell, C.R. and Stipanovic, R.D. 1983, Gliovirin, a new antibiotic from *Gliocladium virens*, and its role in the biological control of *Pythium ultimum*, *Can. J. Microbiol.* 29: 321- 324.

IRRI, 1978, *Annual Report 1977*, Los Banos, Philippines, pp. 118-121.

Katan, J. 1987, Soil solarization, In: *Innovative Approaches of Plant Disease Control*, ed. I. Chet, John Wiley and Sons, New York, pp. 77- 105.

Kannaiyan, S. and Prasad, N.N. 1981, Effect of organic amendments on seedling infection of rice caused by *Rhizoctonia solani*, *Plant Soil* 62: 131- 133.

Kotasthane, A.S. and Lakpale, N. 1994, Effect of plant and seed extract on growth and sclerotial production of *R. solani* (*Corticium sasakii*) causing sheath blight of rice, *Adv. Pl. Sci.* 7: 407- 410.

Krishna, K.R. and Bagyaraj, D.J. 1983. Interaction between *Glomus fasciculatum* and *Sclerotium rolfsii* in pea nut, *Can. J. Bot.* 61: 2349- 2351.

Kuc, J. 1977, Plant protection by the activation of latent mechanism for resistance, *Neth. J. Plant Pathol.* 83(Suppl. 1): 463- 471.

Kumaresan, S. and Manibhushanrao, K. 1991, Studies on the biological control of rice sheath blight disease, *Indian J. Pl. Pathol.* 9: 64- 70.

Kurucheve, V., Ezhilan, J.G. and Jayaraj, J. 1997, Screening of higher plants for fungitoxicity against *Rhizoctonia solani in vitro*, *Indian Phytopath.* 50: 235- 241.

Lee, F.N. and Rush, M.C. 1983, Rice sheath blight: a major rice disease, *Plant Dis.* 67: 829- 832.

Leu, L.S. and Yang, H.C. 1985, Distribution and survival of sclerotia of rice sheath blight fungus, *Thanatephorus cucumeris*, in Taiwan, *Ann. Phytopath. Soc. Japan* 51: 1-7.

Linderman, R.G. 1989, Organic amendments and soil-borne diseases, *Can. J. Pl. Pathol.* 11: 180- 183.

Lumsden, D.R., Lewis, J.A. and Papavizas, G.C. 1983, Effect of organic amendments on soilborne plant diseases and pathogen antagonists, In: *The Environmentally Sound Agriculture*, ed. W. Lockeretz, Praeger Press, New York, USA pp. 51-70.

Manian, S. and Manibhushanrao, K. 1990, Influence of some factors on the survival of *Rhizoctonia solani* in soil, *Trop. Agric.* 67: 207- 208.

Manibhushanrao, K. 1995, *Sheath Blight Disease of Rice*, Daya Publishing House, New Delhi, India.

Manibhushanrao, K. and Baby, U.I. 1991, Management of rice sheath blight using fungal antagonists and organic amendments, *Int. Rice Res. Newsl.* 16: 19- 20.

Manibhushanrao, K., Baby, U.I. and Joe, Y. 1989a, Effect of organic amendments on the saprophytic survival of rice sheath blight pathogen and the soil microflora, *Oryza* 26: 71- 89.

Manibhushanrao, K., Joe, Y. and Madathiammal, P. 1990, Elicitation of resistance in rice to sheath blight disease, *Int. J. Trop. Pl. Dis.* 8: 193- 197.

Manibhushanrao, K., Sreenivasaprasad, S., Baby, U.I. and Joe, Y. 1989b, Susceptibility of rice sheath blight pathogen to mycoparasites, *Curr. Sci.* 58: 515- 518.

Manibhushanrao, K., Sreenivasaprasad, S., Chitralekha, R.S. and Kalaiselvi, K. 1988, Cross protection in rice to sheath blight, *J. Ind. Bot. Soc.* 67: 97- 100.

Manjunath, A. and Bagyaraj, D.J. 1986, Response of blackgram, chickpea and mungbean to vesicular arbuscular mycorrhizal inoculation in unsterile soil, *Trop. Agric.* 63: 33- 35.

Manocha, M.S. 1987, Cellular and molecular aspects of fungal host- mycoparasite interaction, *Z. Pflkr. Pflschutz.* 94: 431- 444.

Marx, D.H. and Schenck, N.C. 1983, Potential of mycorrhizal symbiosis in agricultural and fungal productivity, In: *Challenging Problems in Plant Health*, eds. T. Kommendahl and P.H. Williams, 75th Annual publication of the Ammerican Phytopathological Society, USA.

Matta, A. 1971, Microbial penetration and immunization of uncongenial host plants, *Annu. Rev. Phytopath.* 9: 387-410.

Mew, T.W. and Rosales, A.M. 1992, Control of *Rhizoctonia* sheath blight and other diseases of rice by seed bacterization, In: *Biological Control of Plant Diseases*, ed. E.J. Tjamos, Plenum Press, New York, USA, pp. 113-123.

Mrinalini, C. and Lalithakumari, D. 1996, Protoplast fusion: A biotechnological tool for strain improvement of *Trichderma* spp., In: *Recent Developments in Biocontrol of Plant Pathogens : Current Trends in Life Sciences* Vol. XXI, eds. K. Manibhushanrao and A. Mahadevan, Today and Tomorrow's Printers and Publishers, New Delhi, India, pp. 133-146.

Mukerji, K.G. and Garg, K.L. (eds.) 1988a, *Biocontrol of Plant Diseases,* Vol. I, CRC Press Inc. Florida, U.S.A.

Mukerji, K.G. and Garg, K.L. (eds.) 1988b, *Biocontrol of Plant Diseases,* Vol. II, CRC Press Inc. Florida, U.S.A.

Mukerji, K.G., Tewari, J.P., Arora, D.K. and Saxena, G. (eds.) 1992, *Recent Developments in Biocontrol of Plant Diseases*, Aditya Books Pvt. Ltd., New Delhi, India.

Papavizas, G.C. 1985, *Trichoderma* and *Gliocladium*: biology, ecology and potential for biocontrol, *Annu. Rev. Phytopath.* 23: 23- 54.

Papavizas, G.C., Lewis, J.A. and Abd-ElMoity, T. 1982, Evaluation of new biotypes of *Trichoderma harzianum* for tolerance of benomyl and enhanced biocontrol capabilities, *Phytopath.* 72: 126- 132.

Rajan, K.M. 1980, Soil amendments in plant disease control, *Int. Rice Res. Newsl.* 5: 515.

Rajan, K.M. and Alexander, S. 1988, Management of sheath blight disease of rice with *Trichoderma viride* and some soil amendments in relation to the population of pathogen in soil, *J. Biol. Cont.* 2: 36- 41.

Rajan, K.M. and Menon, R.M. 1975, Effect of organic amendments on plant growth and intensity of sheath blight of rice, *Agric. Res. J. Kerala* 13: 179- 181.

Roy, A.K. 1991a, Inhibitory effect of *Aspergillus terreus* on *Rhizoctonia solani* causing sheath blight of rice, *J. Ind. Bot. Soc.* 70: 95- 98.

Roy, A.K. 1991b, Antagonistic effect of some fungi on *Rhizoctonia solani* causing sheath blight of rice, In: *Recent Trends in Plant Disease Control*, ed. H.B. Singh, Today and Tomorrows Printers and Publishers, New Delhi, India, pp.263-266.

Roy, A.K. 1996, Innovative methods to manage sheath blight of rice, *J. Mycopathol. Res.* 34: 13- 19.

Roy, A.K. and Sayre, R.M. 1984, Electron microscopical studies of *Trichoderma harzianum* and *T. viride* and mycoparasitic activity of the former on *Rhizoctonia solani* f. sp. *sasakii, Indian Phytopath.* 37: 710-712.

Schenck, N.C. 1981, Can mycorrhizae control root disease? *Plant Dis.* 65: 230- 231.

Schippers, B., Bakker, A.W. and Bakker, P.H.A.M. 1987, Interaction of deleterious and beneficial rhizosphere microorganisms and their effect on the cropping of potatoes, *Annu. Rev. Phytopath.* 25: 339-358.

Singh, N. and Singh, R.S. 1984, Significance of organic amendment of soil in biological control of soilborne plant pathogens, In: *Progress in Microbial Ecology*, eds. K.G. Mukerji., V.P. Agnihotri· and R.P. Singh, Print House, Lucknow, India, pp. 303- 323.

Sivaprasad, P., Sulochana, K.K. and Salam, M.A. 1990, Vesicular arbuscular mycorrhizae (VAM) colonization in low land rice roots and its effect on growth and yield, *Int. Rice Res. Newsl.* 15: 14-15.

Smith, S.E. and Ginianazzi-Pearson, V. 1988, Physiological interactions between symbionts in vesicular- arbuscular mycorrhizal plants, *Ann. Rev. Plant Physiol. Mol. Biol.* 39: 221- 244.

Sreenivasaprasad, S. and Manibhushanrao, K. 1990, Antagonistic potential of *Gliocladium virens* and *Trichoderma longibrachiatum* to phytopathogenic fungi, *Mycopathologia* 109: 19- 26.

Sridhar, R., Manibhushanrao, K. and Sinha, A.K. 1989, Innovative approaches to management of some major diseases of rice, *J. Sci. Ind. Res.* 48: 181- 185.

Stapleton, J.J., Quick, J. and Devay, J.E. 1985, Soil solarization: effect on soil properties, crop fertilization and plant growth, *Soil Biol. Biochem.* 17: 369- 373.

Tsai, W.H. 1976, The influence of rice sheath blight on yield at different inoculation stages, *Ann. Rep. Taiwan Agric. Res. Inst.* pp. 115- 116.

Upadhyay, R.K., Mukerji, K.G. and Rajak, R.L. (eds.) 1997, *IPM System in Agriculture,* Vol.2. *Biocontrol in Emerging Biotechnology*, Aditya Books Pvt. Ltd., New Delhi, India.

Vasanthadevi, T., Malarvizhi, R., Sakthivel, N. and Gnanamanickam, S.S. 1989, Biological control of sheath blight of rice in India with antagonistic bacteria, *Plant Soil* 119: 325- 330.

Viji, G., Baby, U.I and Manibhushanrao, K. 1993, Induction of fungicidal resistance in *Trichoderma* spp. through UV irradiation, *Indian J. Microbiol.* 33: 125- 129.

Viji, G., Manibhushanrao, K. and Baby, U.I. 1997, Non- target effect of systemic fungicides on antagonistic microflora of *Rhizoctonia solani, Indian Phytopath.* 50: 324- 328.

Waheeta, A., Sreenivasaprasad, S. and Manibhushanrao, K. 1987, Induced resistance in rice to sheath blight disease, *Curr. Sci.* 56: 486-489.

Wick, R.L. and Moore, L.D. 1984, Histology of mycorrhizal and non- mycorrhizal *Ilex crenata* "Helleri" challenged by *Thielaviopsis basicola, Can, J. Plant Pathol.* 6: 146- 150.

Index

A

Abutilon striatum, 74, 78
Abutilon theophrasti, 193, 202
Acacia saligna, 190
Acaulospora laevis, 130
Acaulospora spinosa, 130
Acinetobacter baumanii, 70
Acremonium diospyri, 202
Acremonium kiliense, 144
Acremonium sp. 144
Acrobeloides spp., 162
Aeromonas caviae, 31, 35
Aeromonas hydrophilla, 70
Aeroponic, 129, 135, 136
Aerva sanguinolenta, 75
Aeschynomene virginica, 193, 202
Aeschynomene sp., 202
Agaricus bisporus, 164, 165
Agaricus hygrophila, 209
Agaricus sp. 86
Ageratina adenophora, 213, 214
Ageratina riparium, 190
Agrobacterium radiobacter, 13, 22, 25, 26, 51, 65, 98, 174, 181, 252
Agrobacterium rhizogenes, 19, 179
Agrobacterium tumefaciens, 19, 26, 54, 56, 64, 65
Agrobacterium sp. 25, 26, 51, 54, 56, 59, 64, 65
Albizia procera, 141, 150
Alcaligenes eutrophus, 68
Alcaligenes sp., 26, 35, 59
Alternanthera philoxeroides, 209, 209
Alternaria alternata, 242
Alternaria brassicola, 30
Alternaria cassiae, 193, 203
Alternaria sp., 193, 203
Ampelomyces quisqualis, 106
Anastomosis, 86
Antagonism, 13, 17, 59, 62, 97, 113, 155, 201, 223, 227, 229, 244, 245, 252, 269
Antagonist, 10, 11, 12, 13, 14, 19, 22, 26, 27, 29, 33, 34, 37, 39, 40, 41, 42, 45, 46, 47, 50, 51, 95, 97, 98, 100, 101, 104, 105, 106, 107, 109, 112, 114, 115, 125, 133, 148, 151, 152, 155, 165, 180, 228, 230, 231, 232, 240, 244, 245, 254, 261, 268, 271, 272, 273, 275

Antagonistic, 39, 40, 41, 45, 46, 56, 59, 62, 69, 146, 147, 164, 173, 242
Antibiosis, 41, 46, 56, 57, 61, 84, 243, 252
Antibiotics, 41, 42, 43, 46, 48, 81, 84, 233
Antibody, 176
Aphanomyces euteichus, 143
Aphelenchus sp., 181
Apion brunneonigrum, 214, 215
Apis mellifera, 107
Aporcellaimellus sp., 175
Appresoria, 162
Appresorium, 157
Arachis hypogea, 203
Arbuscules, 119, 124, 140, 145, 147
Armillaria mellea, 101
Arthrobacter sp., 26, 59
Arthrobotrys cladoides, 164, 165
Arthrobotrys dactiloides, 165
Arthrobotrys irregularis, 164, 165
Arthrobotrys musiformis, 165
Arthrobotrys oligospora, 164
Arthrobotrys robusta, 165
Arthrobotrys thaumasia, 165
Arthrobotrys spp., 164
Ascochyta fabae f. sp. *lentis,* 239
Ascochyta pisi, 245
Ascochyta rabiei, 239
Asparagus officinalis, 28, 30
Asparagus sp., 28, 29, 30, 33
Aspergillus candidus, 242, 246
Aspergillus flavus, 66, 241
Aspergillus foetidus, 85, 91
Aspergillus fumigatus, 148, 240
Aspergillus luchuensis, 243
Aspergillus niger, 241, 242, 243, 254, 256, 257
Aspergillus sp., 41, 82, 84, 85, 86, 91, 226
Azospirillum sp., 105, 144, 181, 240
Azotobacter chroococcum, 60, 67, 174
Azotobacter vinelandii, 67
Azotobacter sp., 26, 36, 59, 60, 67, 69, 181

B

Bacillus cereus, 27, 30, 37, 174
Bacillus megaterium, 69, 273
Bacillus pabuli, 31, 34

287

Intercellular, 124
Intracellular, 2

L

Laetisaria arvalis, 97, 226, 227
Lantana camara, 209, 213
Lantana sp., 209
Lathyrus sativus, 239
Lens culinaris, 203, 239
Lentinus edodes, 72
Leptobyrsa decora, 213
Linum utitatissimum, 90
Lipophilic, 56
Luffa aegyptica, 157
Luffa sp., 159
Lycopersicon esculentum, 77, 159

M

Macrophomina phaseolina, 28, 33, 95, 97, 100,
    101, 103, 105, 112, 113, 115, 143, 144, 159,
    240, 242, 243, 244, 245
Macrophomina spp., 28
Macrotyloma biflorus, 239
Magnaporthe grisea, 83
Malva pusilla, 193, 203
Maravalia cryptostegiae, 190
Medicago sp., 147
Melampsora lini, 90
Meloidogyne hapla, 160
Meloidogyne incognita, 120, 121, 124, 125, 134,
    135, 136, 137, 138, 156, 158, 159, 160, 162,
    164, 165, 178, 179, 180, 181, 225, 229, 240,
    243, 245
Meloidogyne incognita acrita, 157
Meloidogyne javanica, 157, 159, 160, 161, 176, 178,
    179, 181
Meloidogyne sp., 159, 162, 174, 175, 179, 181, 229,
    244
Mescinia parvula, 214 , 215
Mesodorylaimus japonicus, 175
Microbial pesticide, 110
Micrococcus luteus, 244
Microsporum punctatum, 241
Mirabilis jalapa, 72, 75, 78
Mirabilis sp., 72, 75, 76, 78
Monacrosporium cionopagum, 165
Monacrosporium ellipsosporum, 165
Monacrosporium spp., 165
Morrenia odorata, 193, 200, 202
Mucor mucedo, 108
Mycobacterium smegmatis, 69
mycoherbicide, 194, 195, 200, 201, 202, 203
Mycoparasite, 61, 105, 230
Mycoparasitism, 41, 42, 46, 61, 66
Mycophage, 85, 93
Mycorrhizosphere, 117, 118, 125, 133, 136, 140,
    146, 147
Mycovirus, 81, 83, 84, 85, 86, 88, 91, 92, 93

N

Nacobbus aberrans, 157
Nacobbus sp., 156
Nematophagous, 164, 165
Nematophthora gynophila, 162
Nematophthora sp., 163
Neochetina bruchi, 217
Neochetina eichhorniae, 217
Neurospora crassa, 43, 68, 99
Niche, 11, 16, 17, 18, 29, 43
Nicotiana benthamiana, 74
Nicotiana tabaccum, 74, 120

O

Octotoma scabripennis, 213
oligophagous, 210
Olpidium brassicae, 143, 152
Oncogenic, 107
Ophiomyia lantanae, 213
Ophiostoma ulmi, 83, 88, 93
Ophiostoma novo-ulmi, 88
Opuntia inermis, 209
Opuntia stricta, 209
Opuntia triacantha, 210
Opuntia vulgaris, 208, 209
Opuntia spp., 213
Orthogalumna terebrantis, 217
Osmotolerance, 31

P

Paecilomyces lilacinus, 155, 156, 157, 158, 159,
    166, 180, 229, 240, 241, 242, 243, 244, 245,
    246
Paecilomyces sp., 41, 140
Panagrellus sp., 181
Paracoccus denitrificans, 54
Paralongidorus sali, 175
parasexual, 44
Pareuchaetes pseudoinsulata, 214, 215
Parthenium hysterophorus, 190, 215, 216
Pasteuria nishizawae, 174, 180
Pasteuria penetrans, 159, 245
Pasteuria thomei, 174
Pasteuria spp., 173, 174, 175, 176, 177, 178, 179,
    180, 181, 182
Penicillium expansum, 27
Penicillium funiculosum, 240
Penicillium oxalicum, 4, 6, 241, 243, 244
Penicillium pinophilum, 240
Penicillium stolonifer, 85, 91
Penicillium sp., 41, 82, 84, 85, 86, 91, 93, 140,
    242, 245
Peniophora gigantea, 65
Pennisetum glaucum, 253
permeability, 117, 123
Phaseolus vulgaris, 239
Phoma betae, 106

U

Ulocladium sp., 193
Uncinula necator, 106
Urocystis tritici, 140, 151
Uromyces fabae, 239
Uromyces phaseoli, 106
Uromyces pisi, 239
Uromyces vignae, 90, 94
Uromycladium tapperianum, 190
Uroplata girardi, 213
Ustilago maydis, 90, 92, 93
Ustilago sphaerogena, 67
Ustilago sp., 90, 92, 93
Ustulina zonata, 57, 60

V

Variability, 174, 276
Vector, 86, 190, 201
Venturia inaequalis, 45
Verticillium albo-atrum, 143
Verticillium chlamydosporium, 159, 160, 161, 166, 180, 242
Verticillium sp., 119, 134, 143, 144, 149, 150, 151
Verutus mesoangustus, 175
Vesicular Arbuscular Mycorrhizae, 229
Viability, 100, 104, 106, 181, 201
Vibrio anguillarum, 66

Vibrio vulnificus, 69
Vigna aconitifolia, 239
Vigna mungo, 239
Vigna radiata, 159, 239, 243, 244
Vigna unguiculata, 239
Virulence, 82, 83, 84, 85, 87, 88, 89, 91, 92, 93, 94, 126, 166, 200, 202
Virulent, 56, 83, 85, 87, 88, 89, 91, 92, 93, 94, 201, 235

X

Xanthium pennsylvanicum, 203
Xanthium spinosum, 203
Xanthium strumarium, 218
Xanthium sp., 203
Xanthomonas compestris pv. *vignaeradiatae*, 244
Xanthomonas oryzae, 3, 8
Xanthomonas oryzae pv. *oryzae*, 4, 275
Xiphinema bakeri, 175
Xiphinema brasilensis, 175
Xiphinema brevicolle, 175

Y

Yersinia enterocolitica, 67, 68

Z

Zygogramma bicolorata, 215, 216, 218